Forensic Botany

Forensic Botany: From Crime Scene to Court outlines a history of the discipline and field to provide current forensic operational context, real-world applications to cases and how we arrived at this point.

Written by a leading expert, the book presents forensic botany in a realistic and practical framework, offering a schema – a four-stage classification model – for how to manage botanical evidence in a contemporary case management context. Across 12 chapters, Robertson discusses aspects vital to forensic botany, from the central role of crime scene investigators and specialists (in ensuring that data are not mishandled or ignored) to how to handle a range of botanical materials in a triage environment so that they are properly preserved for detailed examination by suitable experts. This book shines a spotlight on the many approaches and forensic applications to plant science that are not traditionally considered, such as conservation, art fraud and archaeology. In doing so, the coverage presents a novel and useful blueprint to employ a forensic approach in non-traditional areas.

Forensic Botany will assist investigators and crime scene examiners in recognising the potential for forensic botanical evidence. It outlines a framework for forensic professionals of varying levels of expertise to effectively manage botanical evidence and serves as an invaluable guide for those botany professionals undertaking forensic roles as subject matter experts.

Forensic Botany
From Crime Scene to Court

James Robertson

CRC Press
Taylor & Francis Group
Boca Raton London New York

CRC Press is an imprint of the
Taylor & Francis Group, an **informa** business

Cover images from Getty: Monty Rakusen

First edition published 2027
by CRC Press
2385 NW Executive Center Drive, Suite 320, Boca Raton FL 33431

and by CRC Press
4 Park Square, Milton Park, Abingdon, Oxon, OX14 4RN

CRC Press is an imprint of Taylor & Francis Group, LLC

ISBN: 978-1-032-31185-2 (hbk)
ISBN: 978-1-032-31184-5 (pbk)
ISBN: 978-1-003-30850-8 (ebk)

DOI: 10.4324/9781003308508

Typeset in Minion
by SPi Technologies India Pvt Ltd (Straive)

Contents

Preface and Dedication

The completion of this book marks a personal journey in the world of forensic science of close to 50 years since joining the University of Strathclyde in 1976. At the time, I was a postdoctoral researcher in London. I still have the advert inviting applications for a temporary lectureship for a period of five years in forensic science within the Department of Pharmaceutical Chemistry. Applicants were expected to have experience in plant microscopy and research interests relevant to phytochemistry or forensic science. With an undergraduate degree in agricultural botany and a PhD in seed dormancy, it did not look like a great match, but nothing ventured, nothing gained, so I duly applied, and much to my surprise and delight, I was asked to interview. It was an unusual process as the selection panel interviewed three applicants in the same morning. All three of us sat together nervously awaiting the outcome – to my even greater delight, I was offered the position.

During the interview, I was asked how I would cope with the examination of clothing involving an elderly lady who had defaecated during an assault. As an agriculture student, I guess I had a bit more experience with faeces than some, albeit animal and not human! In a post interview tour of the laboratories, and to the obvious delight of the interviewer, I was invited to look into some black plastic bags (or perhaps smell would be more accurate), and, yes, this was a real case.

Later that year, on joining Strathclyde, my duties included teaching the examination of a wide range of materials, including crude drugs, foodstuffs, vegetation, woods, hairs and fibres as part of the master's degree in forensic science. I was also expected to contribute to teaching pharmacognosy (the identification of powdered vegetable drugs and adulterants) and units on food microscopy to an undergraduate degree in food science. Coming from an undergraduate degree in agricultural botany, my early teaching experience opened up a very different perspective on botany with a focus on microscopy and trace materials.

What does it take to be a forensic botanist? As will become evident throughout this book, there is no single pathway to finding oneself applying botanical knowledge in a forensic context. There will be autobiographical elements that explain in more detail my own journey in forensic science. Note, I make no claim to be a forensic botanist. My core interest in forensic science has been in trace evidence with a focus on a criminalistics approach to case management. With the exception of Cannabis, my botanical expertise, such as it is, has been on trace botanical evidence. However, what I will attempt to do in this book is to provide the reader with a broad insight into how botanical expertise can be applied in a forensic context and, importantly, how individuals wishing to apply their botanical skills in a forensic context should operate to meet the expectations of the criminal justice system (CJS).

Equally, the aim of this book is to inform participants in the CJS, including crime scene examiners, forensic trace examiners, specialist botanists, investigators and legal players about the potential value of forensic botany to assist in investigations and, ultimately, contribute to CJS outcomes.

I wish to thank Professors Hilton Kobus and Dennis McNevin, who have read various chapters and offered useful feedback. Any errors remain my own responsibility. In preparing this book, I should have paid earlier attention to advice from my publisher that 'gaining permissions to use third party material can be complicated, time-consuming and expensive', advising authors to avoid third party material where possible. I can attest to the truth of that advice. Hence, there are images that I have not been able to use due to not being able to obtain permissions or where the cost was simply prohibitive. All I can add is thank goodness for Wikimedia Commons!

This book is dedicated to my fellow students in the class of 1972 in agricultural botany, Helena Butler (nee West) and James (Jim) Paterson, sadly now deceased, and recently deceased Alexander M.M. Berrie, formerly a Reader in Agricultural Botany at the University of Glasgow. Alex was responsible for kindling my interest in botany and was the supervisor of my PhD research. Without his influence, in my early career, I would not have had the interesting life I have had in science.

Helena, Jim and Alex Berrie at Glasgow University.

About the Author

James Robertson graduated from the University of Glasgow in 1972 with a BSc (Hons) in Agricultural Botany and in 1976 with a PhD in plant physiology. Following a short period as a postdoctoral researcher in London, he entered the world of forensic science as a lecturer at the University of Strathclyde, where he taught Master of Forensic Science from 1976 to 1985. James worked on a Royal Commission in South Australia in 1983 and in 1985, migrated to Australia to work as a senior forensic scientist in Adelaide, where one of his major roles was the examination of cannabis seizures and crops. He joined the Australian Federal Police (AFP) in 1989 as its first Director of Forensic Science. During the next 20 years, he established the AFP forensic group as a world-respected forensic organisation.

Despite occupying a senior managerial role, James always maintained his case work competency and his interest in all things academic! Hence, it was no surprise that on his retirement from the AFP, he returned to an academic role as a Professorial Fellow at the University of Canberra (UC) and Director of the National Centre for Forensic Studies. James 'retired' (again) in 2019 but remains as a Professor Emeritus at UC.

James has authored or co-authored over 200 academic papers and edited or co-edited several books on forensic science, including books on fibres and hairs. He is Editor Emeritus of the *Australian Journal of Forensic Sciences*, Life Member of the Australian Academy of Forensic Sciences and the Australian and New Zealand Forensic Science Society, having served both organisations as either President or Vice President. James has chaired all of the major forensic advisory groups in Australia during his career. His contributions to forensic science have been formally recognised with the Public Service Medal, Member of the Order of Australia and Doctor of the UC.

James is Fellow of the Royal Society of New South Wales.

James continues his active interest in, and passion for, the forensic aspects of trace evidence and criminalistics.

Scope and Application of Botany in Contemporary Forensic Practice

1

1.1 Introduction

The term 'forensic' is widely used and applied to many disciplines beyond those that may be considered 'science'. Forensic comes from the Latin *forensis*, meaning 'of or before the forum'. Science comes from the Latin *scientia*, meaning 'knowledge'. Taken together, the goal of forensic science is to produce (useful) knowledge before a forum, typically a court or tribunal, with the aim to assist in resolving legal disputes. Forensic botany is simply a subset of forensic science.

There are a number of textbooks that deal with forensic botany or forensic plant science (Coyle, 2005; Coyle, 2024; Hall and Byrd, 2012; Bock and Norris, 2016). Is there a difference between botany and plant science? This may seem a question of semantics, but it goes to the way organisms are classified. When I first studied botany in the late 1960s, plants included those that live on the land, all the algae, fungi and many microscopic organisms. A common feature that connected these organisms was that they photosynthesised. Today, arranging groups based on evolutionary history has gained almost universal acceptance (Simpson, 2019). It is also now understood that fungi are closer to the animal kingdom than plants and fungi are now recognised as a distinct kingdom.

Botany is the study of most organisms traditionally treated as plants. Plant science is the study of only land plants. In this sense, botany includes plant science but is broader than plant sciences (Simpson, 2019). Taking this approach, this book adopts the broader definition of botany. In Chapter 2, plant systematics, classification and taxonomy are considered in greater detail, as this is key to the naming and identification of plants.

The last 40 years or so have seen an emphasis on the study of biology at the molecular level through the analysis of DNA. As will be explored throughout this book, plant DNA is in many ways more complex than human DNA, and it is important to understand that the interpretation of plant DNA differs from that of human DNA. For a start, humans are a single species compared to 420,000+ angiosperm species (seed-bearing plants) and more than 1.5 million fungal species. Given the enormous variety of species, it is unrealistic

DOI: 10.4324/9781003308508-1

to expect any individual to have expert knowledge across this vast expanse of different plants and fungi. Botany is, hence, a world of specialisations and specialists.

It is also important to recognise that a plant species does not exist in isolation from other plants and other organisms. They are part of a broader ecosystem, subject to natural and imposed influences, with human impact often being the dominant factor in a forensic context. Forensic ecology attempts to consider all factors that create a complex, dynamic (living and changing) and interdependent chemical, physical and biological environment.

1.2 How Forensic Botany Can Contribute Useful Information?

Forensic scientists have an obvious and close working relationship with those who investigate crime and with the parties who are players in the regulatory and justice system. The latter may be civil or criminal in jurisdiction. The major focus of forensic science is often the criminal justice system (CJS).

Forensic scientists often have a dual role. One is to assist investigators, often police, but for forensic botany, investigators may come from a variety of agencies, such as environmental or wildlife crime agencies. Assisting an agency does not equate to telling an agency what it may wish to hear, and experts must remain impartial. The second role is to provide information that may be used as evidence in a tribunal or court of law. Once again, an expert is not an advocate for the instructing agency or party. Across multiple jurisdictions, there are codified rules for experts. These will be considered in Chapter 11 of this book.

There are four broad areas that forensic science can assist in, one or both of these dual roles:

- It can establish the elements of an alleged crime. For example, is it a sample of wood or timber from a protected species, or is it a seized plant material, cannabis.
- It can help to associate a suspect with an alleged crime.
- It can help to eliminate a suspect from an alleged crime.
- It can help to reconstruct what may have occurred in an alleged crime.

As was commented on earlier, there has been a major shift to using DNA to identify organisms, which has also seen a shift in how forensic resources are allocated towards personal identifications. I have argued that this has led to a reduction in the resourcing of other areas of forensic science. DNA, however, has an undeniable and increasing role to play in forensic botany (Robertson et al., 2021). This needs to be balanced with the fact that, insofar as forensic botany may contribute answers to the above four broad questions, the answers may be more of a *criminalistics* nature rather than identification alone.

The term criminalistics is credited to Hans Gross (1893). The late Paul Kirk defined criminalistics as '*the subject that is concerned with physical evidence, that is the examination and identification, individualization and interpretation of physical evidence of all types*'. As Kirk observed when discussing physical evidence, '*Only human failure to find it, study it and understand it, can diminish its value*' (Kirk, 1953).

Physical materials may be present in very small amounts, often too small to be measured. These are referred to as *trace* or *trace evidence*. Trace can be defined as '*the surviving evidence of a former occurrence of action of some event or agent*'. Botanical materials are physical evidence and may exist at trace levels and, as surviving evidence of a former occurrence, assist in event reconstruction.

Fundamental to understanding trace evidence is the *Locard Exchange Principle*, which is often quoted as 'every contact leaves a trace'. What Locard actually stated was '*Toute action de l'homme, et a fortiori violente qu'est un crime, ne puet pas se derouler sans laisser quelque marque*', which translates to '*any action of an individual, and obviously, the violent action constituting a crime, cannot occur without a trace*' (Locard, 1931).

Coyle and Robertson (2023) list examples of the types of plant material that may be encountered in forensic botany. These include:

- partially digested food such as stomach contents, vomit and faeces
- wood and wood fragments
- pollen and other spores
- diatoms and other algae
- illicit and licit drugs
- poisonous plants
- plant fibres found as textiles, ropes and in paper
- plants or parts of plants present as trace physical evidence

Coyle and Robertson (2023) list typical sample types encountered in forensic botany, giving case examples. Consideration of these sample types and the forensic investigation that may ensue will be left to individual chapters in this book.

Another way of looking at forensic botany is to think about the types of cases where botanical materials may help answer investigative questions. A useful way of thinking is to apply what is known as the 5 W's approach to the investigative process. These are:

- WHO? Who was the victim of the alleged crime? Who was the perpetrator, and who might have witnessed the alleged crime?

- WHEN did the crime occur?
- WHERE did key events take place at the scene of the alleged crime? For example, in the case of a body found in suspicious circumstances, there may be clues that indicate whether the person died at the scene or somewhere else and was transported to the scene. Where at the scene could individuals have entered or left the scene? Where in the scene could individuals have stood, sat or knelt, and where at the scene are any physical materials that may relate to the alleged crime that may have been introduced or taken away from the scene by a perpetrator?
- WHAT was the sequence of events that may have occurred during the commission of the alleged crime? What was the motive for the alleged crime? What was the modus operandi of the perpetrator? What inanimate objects, such as vehicles, tools or weapons, that may have been used, left at the scene or removed from the scene?
- WHY did the crime take place, where it did and when it did?

Forensic botany may contribute to providing answers to many of these questions, and it is useful if the forensic botanist thinks about how they can apply their expertise to assisting investigators without compromising impartiality.

1.3 Forensic Botany, a New Forensic Discipline?

Given the enormous potential case applications where forensic botany could contribute to the CJS, why is it that an almost universal comment in the literature is the underutilisation and unrealised potential of forensic botany?

Whilst often quoting the famous Lindbergh kidnapping case from the 1930s as the ‘first’ case in which forensic botany was recognised (see Chapter 7 for a discussion about this case), most recent authors assert that forensic botany is a new discipline or application. This is just not an accurate observation. For example, when I started my journey in forensic science in the mid-1970s, forensic botany was a significant discipline with most laboratories having in-house expertise. Interestingly, the first reported case involving forensic botany in Australia also involved a kidnapping, tragically leading to the death of an eight-year-old boy, Graeme Thorne (see Appendix A.1).

The Metropolitan Police Forensic Science Laboratory (MPFSL) Biology Methods Manual of 1982 contained chapters on urine and faeces; foodstuffs; fibres including plant fibres and ropes and cordage; paper and a standalone chapter simply called ‘botanical applications’ that included consideration of algae, fungi, lichens, mosses and liverworts; and ferns and higher plants including Gymnosperms (conifers) and Angiosperms (flowering plants). This chapter alone ran to 41 pages of methods and information.

At that time, the MPFSL was still dealing with about a case a week of identifying trace levels of wood resulting from safe blowing. Hard as it is to imagine today, workers received their pay in cash, usually on a Friday. Businesses would receive cash on a Thursday and keep it in a safe on their premises. These safes were in the main manufactured in the late 1800s or the early part of the 1900s. The purpose of a safe was to protect the contents not only from theft but, in the event of a fire, to protect the contents from destruction. Safes were constructed of essentially two boxes with the space between them filled with ballast intended to provide an insulating layer in the event of a fire. When a safe was blown open with a small explosive charge, the ballast material would be released and often transferred to the perpetrator's clothing. The ballast material in these safes included wood shavings, which would often include several species. Hence, the identification of small fragments of wood was a core expertise found in many forensic laboratories as late as the 1980s.

With the exception of the examination of seized Cannabis material, there are few laboratories worldwide that would now employ a specialist forensic botanist. One exception known to the author is the Victorian Police Forensic Laboratory in Australia, although, as far as I am aware, their main work is Cannabis identification. Hence, today practitioners of forensic botany are to be found in academic institutions, specialist herbaria and botanic research institutions. A small number of individuals from these backgrounds have established themselves as independent providers of forensic expertise.

To the best of my knowledge, there are no academic programmes specialising in forensic botany, and as I shall discuss further in Chapter 11, this is part of a larger problem faced by botany as an academic subject. As Bock and Norris (2016) observe, '*currently, there are no training programs in forensic botany*', and the NAS report (2009) does not consider forensic botany as a forensic discipline. Hall and Byrd (2012) make similar observations, drawing attention to the need for training courses for the advancement of forensic botany and the lack of standards for training or expertise to establish professional certification. The reality is that botanists '*scattered around the world have been involved in a case or two but only a few have worked on several to many cases*' (Hall & Byrd, 2012).

One of the individuals who has worked on 'several cases' is Dr Mark Spencer from the United Kingdom. Coming from a background as a curator of the British and Irish Herbarium of the Natural History Museum, he, like many others, could be considered the 'accidental forensic botanist'. In his book Murder Most Florid (Spencer, 2019), he describes his journey to becoming a forensic botanist, pointing out that there is nowhere in the world that offers a qualification in forensic botany and that some formal training would have eased his transition. Pat Wiltshire tells a similar story as she first came into contact with the world of forensic science in 1994, from being an environmental archaeologist at University College London (Wiltshire, 2019).

In conclusion, forensic botany, if not always labelled forensic, is not a new discipline and has been around in one form or another for centuries, if not millennia. The more recent significant demise of forensic botany is not unique, as non-DNA disciplines in forensic laboratories have been diminished or entirely disappeared from mainstream forensic laboratories under the pressure of resources being targeted at human DNA. Paradoxically, the emergence of enhanced DNA approaches in botany is the reason why there has been a resurgence of interest in forensic botany, along with a greater realisation of the scope of where a forensic approach to botany can assist in a wide range of applications.

Forensic botany, however, is highly unlikely to re-emerge as a strong discipline in mainstream forensic laboratories, and, hence, investigators will have to call on the services of individuals who have carved out careers in forensic botany, such as Mark Spencer and Patricia Wiltshire, or turn to major botany institutions where individuals may agree to undertake forensic work. The Baden-Clay case discussed in Appendix A.1 is a cautionary tale for those outside experts who may agree to become involved in investigations.

If the future is not to be one of '*failure to find it, study it and understand it*', the challenge will be to develop a framework that ensures that forensic botany is not ignored or forgotten in mainstream forensic practice and that individual experts or non-forensic organisations with expertise are assisted to understand the very specific requirements applicable for forensic work. Pathways to how such a framework might emerge are considered in Chapter 11 of this book.

1.4 Forensic Botany Starts at the Crime Scene

A forensic investigation starts at the scene or location(locus) of a matter of interest to a relevant investigator. Sometimes it will be obvious that there are suspicious circumstances that warrant the matter being treated as a crime scene. For example, a deceased person may be the victim of a crime or may have died from natural causes. A final resolution to this question will most often rely on the outcomes of a post-scene forensic examination, including toxicology and pathology.

It is quite common that bodies in outdoor locations are found by members of the public. Police, relevant emergency services and specialists such as a medical doctor and/or pathologist may be called to the scene as a forensic specialist, such as crime scene examiners (CSEs), and, depending on the circumstances, specialists such as anthropologists, firearms/ballistics examiners, but, as we shall discuss, rarely a botanist.

It will usually fall to the CSE to assess the circumstances at the scene and, along with the investigator, reach an initial decision as to whether there are suspicious circumstances warranting the scene being elevated to a scene of a potential crime. The police and the broader CJS would not be well served if every natural death were treated as a potential murder case. Of course, the other side of the coin is that if the scene is not initially treated as a crime scene, then potentially valuable evidence may be lost. There are too many examples of cases where failure to treat the scene as a crime scene has resulted in loss of potential evidence. The Chamberlain case discussed in Appendix A.1 is one example.

CSEs may not have a high degree of awareness of the potential role of a forensic botanist or ecologist and may not consider involving such a specialist. The early involvement, however, of a specialist in a crime scene team could be the difference between a successful scene examination and a failed scene examination. In order to make informed decisions, the specialist needs to be briefed on *relevant* information about what is known, often very little, at the early stages of an investigation. Often, success or failure will depend on recording and collecting the correct samples at the scene. In the absence of a specialist, this task is likely to fall to the CSE.

With respect to the possible involvement of a specialist botanist, practical circumstances will mean that it is simply not possible for a forensic botanist to always be involved at as early a stage in an investigation. Here, the role of the CSE is critical as they are best placed to advise investigators. They can only do so, however, if they understand how forensic botany could contribute to the whole forensic investigation. Sadly, in many cases, a crime scene will have been compromised as a result of sometimes well-meaning but misguided actions taken at a scene by investigators and even CSEs. Even where an investigation team understands the value of the early engagement of a forensic botanist, it may be difficult to identify a suitable individual or obtain their services in a timely manner. Hence, more often than not, even when a specialist does become involved, it will be some time after the original scene examination.

In the absence of a specialist, how the scene is recorded and what samples are taken by the CSE are critical. From a technical perspective, this really should not be a problem, except that one can only capture images or collect samples of what is recognised as potentially useful. As Wiltshire points out, there is a very poor understanding of just how variable soil and vegetation can be over quite small areas (Wiltshire, 2019). Spencer (2019, at page 60) comments on the '*awful pictures of plants and vegetation in general*' with them being '*either landscape scale and not focused on the vegetation or they are somewhat abstract images of parts of a leaf of flower*'. In my very first appearance as an expert witness in a case in 1976, I was shown just such images, in black and white, and asked if I could identify the plants present.

One of the early tasks for a CSE is to determine possible entry and exit points at a scene that may have been used by a perpetrator and to establish what is called a *common approach path* (CAP). This is a route or path to be used by all personnel to gain access to the crime scene within an established cordon. This 'path' should be marked and searched for potential evidence before being used by others. The purpose of establishing a CAP is to choose a route that is least likely to have been used by a perpetrator and to minimise disturbance at the scene and loss of potential evidence.

Spencer (2019) and Wiltshire (2019) both give many examples where the lack of early engagement of a forensic botanist at a scene, or actions taken at the scene, limited what the forensic botanist was ultimately able to determine. This included cutting down vegetation to make it easier to access scenes! The inclusion of forensic botany into CSE training and into forensic awareness training for investigators and relevant agencies would go some way to improving and enhancing the role of forensic botany.

Equally, it is important that botanists who become involved in forensic work receive formal training in forensic practice rather than having to learn on the run. The specialist botanist needs to operate to the same standards as the CSE with respect to scene procedures; understanding and mitigating contamination; the need to make comprehensive, systematic and contemporaneous records; proper evidence collection and storage of samples. Relevant standards that underpin forensic practice are considered in Chapter 3.

Appendix A.1: Case Studies

A1.1 The Graeme Thorne Case

The 1960 kidnapping and murder of eight-year-old Graeme Thorne in Australia shocked a nation and resulted in a massive police investigation that ultimately led to the arrest and conviction of Stephen Leslie Bradley, but only after an international manhunt! Why was this young boy the target of a kidnapping? Answering this question requires going back to the Australia of 1960 and the construction of the iconic Sydney Opera House. In order to help fund this project, the New South Wales (NSW) Government had instigated Opera House lotteries to raise money. The first prize was A$100,000, equivalent to A$1.7 million today. Bazil and Freda Thorne, the parents of Graeme, were the winners of the first prize in June 1960. In this era, winners had little privacy with their images and personal details splashed across the newspapers.

The Thorne family lived in a rented house in the suburb of Bondi. In the morning, Graeme would walk about 300 meters from his home, where he would be picked up by a family friend, Phyllis Smith, to take him to school with her sons. On the morning of July 7, he left home at 8.30 am as normal, but when she came to collect him at 8.40 am, he was nowhere to be seen. Later that morning, the first ransom demand for $25000 was made in a telephone call to the Thorne household.

It was not until Tuesday, August 16, that the body of Graeme was discovered by two young boys in vacant land in the suburb of Seaforth. His body was wrapped in a blue tartan blanket tied with a string. He had been gagged with a scarf and was still wearing his school uniform. For the reader interested in the detailed investigation and ultimate identification of Bradley, Tedeschi (2015) has written a highly readable account. Here, our focus is on the contribution of botanical materials and examinations to the investigation.

In 1960, there was no forensic laboratory in NSW, with the police having a Scientific Investigation Bureau who were responsible for collecting potential evidence and identifying experts who could assist. At postmortem, it was concluded that Graeme had died as a result of blunt force trauma to the head or asphyxiation, or both. It was important to determine how long Graeme had been dead when his body was discovered. This was the first contribution made by forensic botany when it was established that fungi growing on the shoes worn by the deceased were at least four weeks old. The presence of nematodes in the socks Graeme had been wearing supported a time frame of 35–45 days. Along with other evidence, the Police concluded that Graeme had died shortly after his abduction.

The second role played by botany was the examination of plant fragments recovered from the blanket in which the deceased was wrapped. Initial examination of the locus where the deceased was found showed that these plant fragments could not have originated from plants at that locus. Botanists at the Royal Botanic Gardens in Sydney examined these fragments and identified them as coming from two species of cypress, *Chamaecyparis pisifera* (Sawara cypress) and *Cupressus glabra* (Arizona smooth-barked cypress). The first species was quite common, but the second species was considered to be uncommon. As well as the plant fragments, human and animal hairs were recovered along with soil and what later transpired to be a crumbly mortar. All of these combined led police to conclude that the kidnapper had access to a building with pink and white mortar, near a location with the two types of trees; had a dog and probably lived with a woman with blonde hair who used a red henna rinse (Tedeschi, 2015).

In an attempt to locate possible locations, police asked local postal workers if they came across any buildings with the tree species and other potentially identifying information. Initially, this was disappointing until on Friday, 30 September, when one of the detectives, visiting the home of the parents of one of the boys who had discovered the deceased, had a conversation with the passing postman, who asked him if he had heard from his colleague who thought he had identified a house matching the criteria established by the police. This was the house in which Bradley had lived and had recently vacated. When the police examined the house, they found both tree species at the front of the garden, the presence of pink mortar and, in a low dirt embankment at the rear of the garage under the house, other materials that linked this environment with the materials recovered from the blanket.

Bradley had fled Australia by this point in time and was on a ship travelling to Ceylon (now Sri Lanka), where he was arrested and extradited to Australia, tried and found guilty of the murder of Graeme Thorne. Interestingly, at the time, there was no crime of kidnapping, and this case resulted in that crime being legislated.

This case involved the use of several subject experts from various disciplines and is credited with the further development of forensic science in NSW.

If a similar case were to arise today, would we be better placed from a forensic viewpoint? The answer to this question is yes and no. The yes is that police forensic sciences and forensic laboratories in NSW have made major advances and are now comparable with the best forensic services in the world. In all probability, human DNA would have been found and perhaps identified a suspect at an earlier stage, provided that the offender was on a DNA database. The physical evidence would still be vital, and here is no part of the earlier question: any botanical materials would still have to be sent to specialists at the Royal Botanical Gardens.

A1.2 The Splatt Case

On 3 December 1977, an elderly lady, Rose Amelia Simper, was murdered in her home in Adelaide, South Australia (SA). Her house had been ransacked, and a small number of items were missing. She had been assaulted, and her body was lying on her bed. The Crown theory was that this was a robbery gone wrong. At this time, there was no single forensic laboratory in SA. The SA Police had a technical section that included CSEs who also had the responsibility for initial examination of items collected at a crime scene. When the bed was examined, there were trace materials visible to the naked eye. Macroscopic examination of recovered materials revealed the presence of a complex mix, including what appeared to be paint and metal particles, hairs, fibres and fragments of wood and other plant materials.

The home of the deceased was diagonally across the road from a small paint spraying and metal work factory. Police searched the work clothing of workers in this factory, drawing a link between the paint and metal recovered from the bed and the paint and metal resulting from the activities in the factory. Edward Charles Splatt worked as a spray painter in this factory, and on his work trousers, police found, unsurprisingly, spray paint and metal particles, which they asserted were present in the same proportions as found on the bed. These trousers had turn-ups, popular at this time, and the theory was that, as Splatt was assaulting the victim, materials were transferred from his trousers and the turn-ups, which would have acted as a repository for trace materials. At trial, the Crown case alleged that some 17 types of trace material linked Splatt and his environment with the material recovered from the bed. The Crown case was exclusively circumstantial based on forensic evidence.

The trial Judge even went as far in her summing up to the jury to state

> *you may have a ray of light so feeble that by itself it will do little to elucidate a dark corner, but on the other hand, you may have a number of such rays of light, each of them insufficient in itself to light up that dark corner but, all converging and brought to bear upon the same point, and, when united, producing a body of illumination which will clear the away the darkness.*
>
> (Mann, 2010)

Of course, as was the eventual conclusion of the Royal Commissioner, 17 zeros by 17 zeros amounts to nothing.

Edward Charles Splatt was charged with that murder and convicted on 24 November 1978. As in many cases, which turn out to be wrongful convictions, Splatt maintained his innocence and would probably have remained in jail except for the efforts of a journalist, Stewart Cockburn. It is worth pointing out that this case and the next case, the Chamberlains, were contemporary and indeed featured many of the same actors. There was an atmosphere that there were serious problems with forensic science in Australia at that time. Eventually, the Government of SA agreed to a Royal Commission, which commenced in early 1983 (Anon, 1984).

Two of the 17 trace materials were of a botanical nature, the so-called *seed endosperm* and *wood fragments*. Our focus here will be restricted to only the endosperm component. I became involved in this case in early 1983, instructed by the Crown to review the forensic evidence given by Crown witnesses at trial and evaluate defence reports prepared for the Royal Commission. I eventually spent 7 weeks in Adelaide in September/October of that year, giving evidence to the Royal Commission over a five-day period.

As was the case with the Chamberlain case, the botanist used by the Police was Rex Kuchel, a retired botanist in SA supported by a second botanist, Dr Jenner. Kuchel examined the so-called endosperm fragments based on the microscopic appearance of starch granules and examined a sample of bird seed taken from an aviary at the home of Splatt, identifying seven types of seed: wheat, oats, sorghum, safflower, sunflower, panicum and phalaris. Five of these species were identified in the fragments recovered from the bed. Kuchel further stated that the starch grains in the fragments had not been heated, excluding cooked products as a source of these endosperm fragments. Jenner, in part, supported Kuchel, although only identifying wheat, oats and sorghum. He supported the observation that none of the particles had been cooked, as '*cooking destroys the polarization figure*'.

The evidence given by these two witnesses was problematic in several ways. Firstly, the starting point should have been to examine the 'endosperm fragments' as, from images taken at the time, it would appear that they also included remnants of the seed coat. Secondly, there was a major limitation with the sample of bird seed taken from the aviary being used as a control sample. If the theory was that material had fallen into the trouser turn-ups and this was deposited during the commission of the crime, then the relevant control sample should have been focussed on what was likely to have been available for transfer. Birds break open the seed and leave behind fragments. There seems to have been no criminalistics thinking on the part of Kuchel and Jenner. The Royal Commissioner commented on this saying '*if on the bedsheet only endosperm was found -pure endosperm – with no husk material of any kind, this should surely cause one to pause and ask whether this phenomenon is consistent with the proposition that the fragments originated in the debris and waste left by birds feeding in a home aviary*' (Anon, 1984 at page 87).

Splatt was also aware of this shortcoming, asking why '*other small, microscopic particles involved with my aviary... were missing from my* clothing', drawing attention to missing elements such as small wood fragments, sand from the floor, flaking paint and bird droppings (Mann, 2010). The absence of these other trace materials, including feathers, should have sent a warning signal to the examiners that the aviary may not be the source of the endosperm fragments from the bed and alerted them to consider other possibilities.

When it came to identifying seed endosperm based on starch appearance, in my opinion, the Crown evidence was overstated, giving an appearance of greater accuracy than was realistic. With regard to the effect of heat on starch, they simply were wrong. Dr Jenner at the Commission made two telling remarks. He regretted having appeared as confident at the trial about the identifications and also backed down on his statement regarding the effect of heating on the appearance of starch granules, simply saying that he was not aware of the existence of any cooked foodstuff in which there would be starch granules that had no detectable loss of polarisation.

The key deficiency is that no alternative sources for the 'endosperm fragments' were given serious consideration at the trial and, indeed, they were discounted on the basis of a lack of relevant knowledge. The deceased was known to eat biscuits in bed, which could have been an alternative explanation.

There are several take-home messages from this case. First, accurate identification is essential; any limitations should be clear and not overstated. Second, control samples should represent what may be available for transfer, reflecting a genuine appreciation of criminalistics. Third, there may be elements that are outside of your own knowledge, and great care needs to be taken not to stray beyond your actual expertise, as this may have serious consequences.

Finally, there is a lot to be learned from wrongful convictions, which should be part of any training programme for aspiring forensic experts as cautionary tales. As Dr Jenner put it, '*I was feeling under considerable constrains; and I was feeling apprehensive; and I am not the most articulate of people. And possibly under the circumstances, appeared a little bit overconfident*'. With respect to the effect of heat on starch, his comment was that '*I made an incorrect inference ... all experts can make mistakes, and this was one of mine*' (Mann, 2010).

In drawing attention to these final comments, I do not wish this to be read as unduly criticising Dr Jenner, whose frankness at the Commission should be applauded, but it is a reminder for the inexperienced witness of the responsibility that lies with them in how they give their evidence.

A1.3 The Chamberlain Case

This is without doubt the most famous case in Australian forensic science in which Lindy Chamberlain was convicted of the murder of her infant daughter, Azaria. The conviction was later overturned, and a Royal Commission was held into the convictions of Lindy and her husband, Michael Chamberlain (Anon, 1987).

Much has been written about the diverse elements of the forensic investigation, but the role of forensic botany is perhaps less well known. The Chamberlains had been camping near the famous landmark of Ayers Rock (now called Uluru) when, on 17 August 1980, Azaria went missing with Lindy saying that a dingo had taken her from a tent in which she was sleeping. Despite an extensive search, there was no sign of Azaria, and a body was not, and has not, been found. On August 24, a number of items were located at the base of Ayers Rock, some distance from the campsite across plains and sand dunes. The items found included a jumpsuit and a singlet. Plant fragments and seeds were later recovered from these items.

Similar to the other cases above, at that time in the Northern Territory (NT), there was no forensic laboratory, and the NT Police called for assistance from a number of other jurisdictions in Australia, often from their southern neighbour of SA. In this case that included a consultant botanist with the SA police, Rex Kuchel. On examining the recovered botanical materials, he identified seeds from eight different plants and leaf fragments from some of these species. The largest number of seeds and by far the largest number of other fragments were identified as *Parietaria debilis* (native pellitory), a delicate annual that grows in sheltered and shaded situations among rocks. As it later transpired, critically for the interpretation of how these fragments had been deposited on the clothing, Kuchel was unable to identify some of the plant fragments.

The questions of interest to the Police were how the seeds and fragments had been deposited on the clothing and, specifically, could a dingo have been involved in carrying the items from the campsite to where they were recovered? The police theory was that the plant fragments had been deliberately rubbed upon the jumpsuit and singlet by hand.

At trial, Mr Kuchel tended to give some support to this proposition in that he said that it would be required to agitate the clothing against the plants, and, when shown photographs of where the clothing was recovered, he said that he could not see any disturbance of the plants. Beware relying on crime scene photographs! The Crown, however, relied upon this evidence for the purposes of creating the impression that there had been human intervention and that the plant fragments were not the result of the actions of a dingo. This was further supported by evidence from Kuchel that relied on an interpretation of what plant species might have been expected to be present if the items had been dragged from the campsite to where they were located. The Crown case was that a dingo could not be involved in the transport, as the vegetation recovered did not contain some of the species that would have been expected to be present from the area between the campsite and Ayers Rock, again impugning human intervention.

For the Royal Commission, two botanists, both from the NT, re-examined the fragments and identified the three species that had eluded Kuchel. These pointed to coming from the other environments, the plains and sand dunes, between the campsite and Ayers Rock. These botanists also found a number of inconsistencies in the findings of Kuchel and his interpretation of events. Factors such as the density of vegetation, growth habits of individual plants, and their tendency to fragment and subsequently adhere to clothing all needed to be considered.

The Royal Commissioner concluded that the botanical evidence was not inconsistent with a dingo carrying the clothed baby from the camping area across the plains country to the rock and, further, that there was no substantial support for the contention that it came there by deliberate rubbing of vegetation by human hand (Anon 1987).

This case illustrates the importance of pursuing the accurate identification of all recovered materials, as sometimes subtle differences in the totality of information can alter the overall thrust of the evidence. This case also shows that interpretation is challenging, is best informed by local knowledge and great care needs to be taken to consider alternative explanations (Fiddian & Carroll, 2005).

A1.4 The Gerard Baden-Clay Case

Forensic botany continues to make a rare appearance as evidence in Australia. In Queensland, forensic botany played an important role in the conviction of Gerrard Baden-Clay for the murder (later reduced on appeal to manslaughter) of his wife, Allison Baden-Clay, on or about 19 April 2012. The body of the deceased was discovered lying on its side on an embankment under the

Kholo Creek road bridge at Mount Cosby, some 13 kilometres by road from the Baden-Clay house. Leaves from different species of plants were recovered from the clothing and hair of the deceased at postmortem and sent to Dr Gordon Guymer, Director of the Queensland Herbarium, for examination. He identified six plant species. From extensive visits to the location where the deceased was found, he was not able to find four of the recovered species at that location. These four species were crepe myrtle, cat's claw creeper, fishbone fern and lillypilly. All six species were present at the home of the deceased, with cat's claw creeper growing about the carport. Leaf litter from the crepe myrtle and fern was found in the carport and patio area of the house near its back door.

The Crown case was that the accused was having an affair and, as a result, his marriage was about to break down; that he had argued with his wife on the night she disappeared and that he had killed her and then transported her body to the location at Mount Cosby (Anon, 2015).

Commenting on his involvement in this case, he said

> *I never expected, when I took on the Baden-Clay case, that it would end up occupying a number of months of my life and that I would be so front-and-centre as an expert witness. There was the forensic work to identify the species taken from Allison's hair, visiting the crime scene to identify what was growing there, and then comparing that with plants at the house.*
>
> (Anon, 2022)

These are, of course, the reasons why many experts may be reluctant to become involved in criminal investigations.

Epilogue

In the following chapters, further examples of cases where forensic botany has contributed will be discussed. Two of the cases above were of wrongful convictions where, at least in part, forensic botany made a positive contribution, but where mistakes were made. In both cases, there was a lack of criminalistics thinking with a focus on identification. Of course, accurate identification is a must, but identification that knows its limits as well.

For the reader who is interested in reading about further cases, the books by Spencer and Wiltshire in the references are well worth reading, giving insights and perspectives based on their personal experiences. Also, worth reading is *How Plants Solve Crime* (Gibson, 2022), which brings together numerous examples and case studies written in an easy-to-read style.

References

Anon. 1984. *Royal Commission Report Concerning the Conviction of Edward Charles Splatt*. South Australia: Gov. Printer.

Anon. 1987. *Royal Commission of Inquiry into Chamberlain Convictions*. Report of the Commissioner The Hon. Mr Justice T R Morling. *Parliamentary Paper No 192*. Canberra: Gov. Printer.

Anon. 2015. Supreme Court of Queensland. *R v Baden-Clay* 18, 265.

Anon. 2022. *Connect. The Trials and Triumphs of Botany*. https://www.une.edu.au/connect/news/2022/04/the-trials-and-triumphs-of-botany

Bock, J.H., & Norris, D.O. 2016. *Forensic Plant Science*. Amsterdam: Elsevier.

Coyle, H.M. (ed.). 2005. *Forensic Botany. Principles and Applications to Criminal Casework*. Boca Raton: CRC Press.

Coyle, H.M. (ed.). 2024. *Forensic Botany. Principles and Applications to Criminal Casework*, 2nd ed. Boca Raton: CRC Press.

Coyle, H.M., & Robertson, J. 2023. Forensic Botany. In Houck, M.M. (ed.), *Encyclopedia of Forensic Sciences* (Vol. 1–4, pp. 503–510, 3rd ed.). London: Elsevier.

Fiddian, S. & Carroll, D. 2005. Botany in Forensic Science. In Freckelton, I. & Selby, H. (eds.), *Expert Evidence* (Vol. 5, Chapter 91). Sydney: Thomson Reuters.

Gibson, D.J. 2022. *How Plants Solve Crimes. Planting Clues*. Oxford: Oxford Univ. Press.

Gross, H.L. 1893. *Handbuch fur Untersuchungsrichter als System de Kriminalistik*. Graz: Verlag von Leuschner & Lubenskey.

Hall, D.W., & Byrd, J.H. 2012. *Forensic Botany. A Practical Guide*. Chichester: Wiley-Blackhall.

Kirk, P. 1953. *Crime Investigation: Physical Evidence and the Police Laboratory*. New York: Interscience Pub.

Locard, E. 1931. *Traite de Criminalistique*. Paris: J.Desvigne et ses fils.

Mann, T. 2010. *Flawed Forensics. The Splatt Case and Stweart Cockburn*. Adelaide: Griffen Press.

NAS. 2009. *Strengthening Forensic Science in the United States: A Path Forward*. New York: National Academies Press. https://www.nap.edu/catalog/12589/strenghtening-forensic-science-in-the-united-states-a-path-forward

Robertson, J.M., Damaso, N., & Meiklejohn, K.A. 2021. DNA-Based Analysis of Plant Material in Forensic Investigations. In Dash, H.R., Shrivastava, P., & Lorente, J.A. (eds.), *Handbook of DA Profiling*. Singapore: Springer.

Simpson, M.G. 2019. *Plant Systematics*, 3rd ed. Amsterdam: Elsevier.

Spencer, M. 2019. *Murder Most Florid*. London: Quadrille.

Tedeschi, M. 2015. *Kidnapped. The Crime that Shocked a Nation*. London: Simon & Schuster.

Wiltshire, P. 2019. *Traces*. London: Bonnier Books.

Fundamentals of Botany
An Introduction

2

2.1 Introduction

The purpose of this chapter is to introduce botany, covering plant systematics and taxonomy, plant morphology and plant anatomy as a foundation for understanding the more detailed descriptions in subsequent chapters of this book. This chapter will serve as a revision for the reader who has studied plant science and a foundation for lawyers and investigators who, in all probability, have never studied plant biology. The material is based on lectures delivered by the author when teaching at the University of Strathclyde, brought up to date to reflect the changes that have taken place in the intervening years.

2.2 Plant Systematics and Taxonomy

Systematics is defined as the science that includes and encompasses traditional taxonomy and that has as its primary goal the reconstruction of phylogeny or evolutionary history. Evolutionary history is represented by a *cladogram* or an evolutionary tree (Simpson, 2019).

Taxonomy is defined as a field of science, and a major component of systematics, that encompasses **D**escription, **I**dentification, **N**omenclature and **C**lassification (Simpson, 2019).

As discussed in Chapter 1, the traditional way of defining a plant was by describing and grouping plants based on the common characteristics they possess. Typically, organisms that photosynthesise and have cell walls are considered to be plants. Using this approach, plants included a variety of microorganisms, all of the algae and land-based plants. It is now recognised that organisms that photosynthesise have evolved independently and are not closely related (Simpson, 2019). Hence, there has been a major shift in how organisms are classified to basing this on their evolutionary history.

Organisms are now classified as belonging to one of three major groups or domains. The **Archaea** or Archaea-bacteria, **Bacteria** or Eubacteria and the **Eukarya** or Eukaryotes. Importantly, the eukaryotic domain includes fungi, algae and land plants. Some eukaryotes contain specialised organelles called **chloroplasts** responsible for the process of photosynthesis.

DOI: 10.4324/9781003308508-2

Photosynthesis is the basis of life on our planet, as it is through this process that carbon dioxide is trapped by plants and reacts with water to produce glucose and oxygen. Glucose then becomes the precursor for the production of protein, lipids and nucleic acids. Chloroplasts have their own single-stranded DNA, which differs from plant nuclear DNA. Another vital organelle in plant cells is the **mitochondria**. These are the powerhouses of plant life responsible for the production of energy. Like chloroplasts, mitochondria have their own DNA. Hence, plants have three sources of DNA: nuclear, chloroplast and mitochondrial DNA.

Two groups within the eukaryotes are known as *green plants* and include aquatic green algae and land plants. The latter are what are normally considered to be plants in the field of plant systematics. Land plants (the Embryophytes) include hornworts, liverworts, mosses, cycads, conifers, monocots and dicots.

Mosses (Bryopsida), liverworts (Hepaticophyta) and Hornworts (Anthocerotophyta) belong to the same group, the Bryophytes, with mosses contributing 95% of the species in this group. Bryophytes range in size from barely visible to 20 cm long. The plant body is commonly green but may be reddish or brown. It consists of an undifferentiated thallus or a stem bearing small, delicate leaves. The leaves are always small, rarely more than 12 mm long, and, except for the nerve, usually only one cell thick. They are never stalked. The plant is normally attached to the substrate by fine, colourless or brown threads called *rhizoids*.

Bryophytes possess two distinct stages in their life cycle, a vegetative stage ending in fertilisation and a dependent stage in which spores are produced (see Figure 2.1). Important characteristics used for classification and identification are found in both stages (Simpson, 2019; Watson, 1981).

The remainder of this chapter will focus on land plants and, in particular, monocots and dicots.

2.3 Taxonomy and the Importance of Accurate Naming of Plants

Plant taxonomy starts with the need to accurately describe features or attributes of a **taxon**, plural, **taxa**, defined as delimited groups of organisms and are considered to be a particular **rank**. A higher rank is inclusive of all lower ranks. According to *the International Code of Nomenclature for Algae, Fungi and Plants (ICN)* (Turland et al., 2018), the major taxonomic ranks of land plants are as follows.

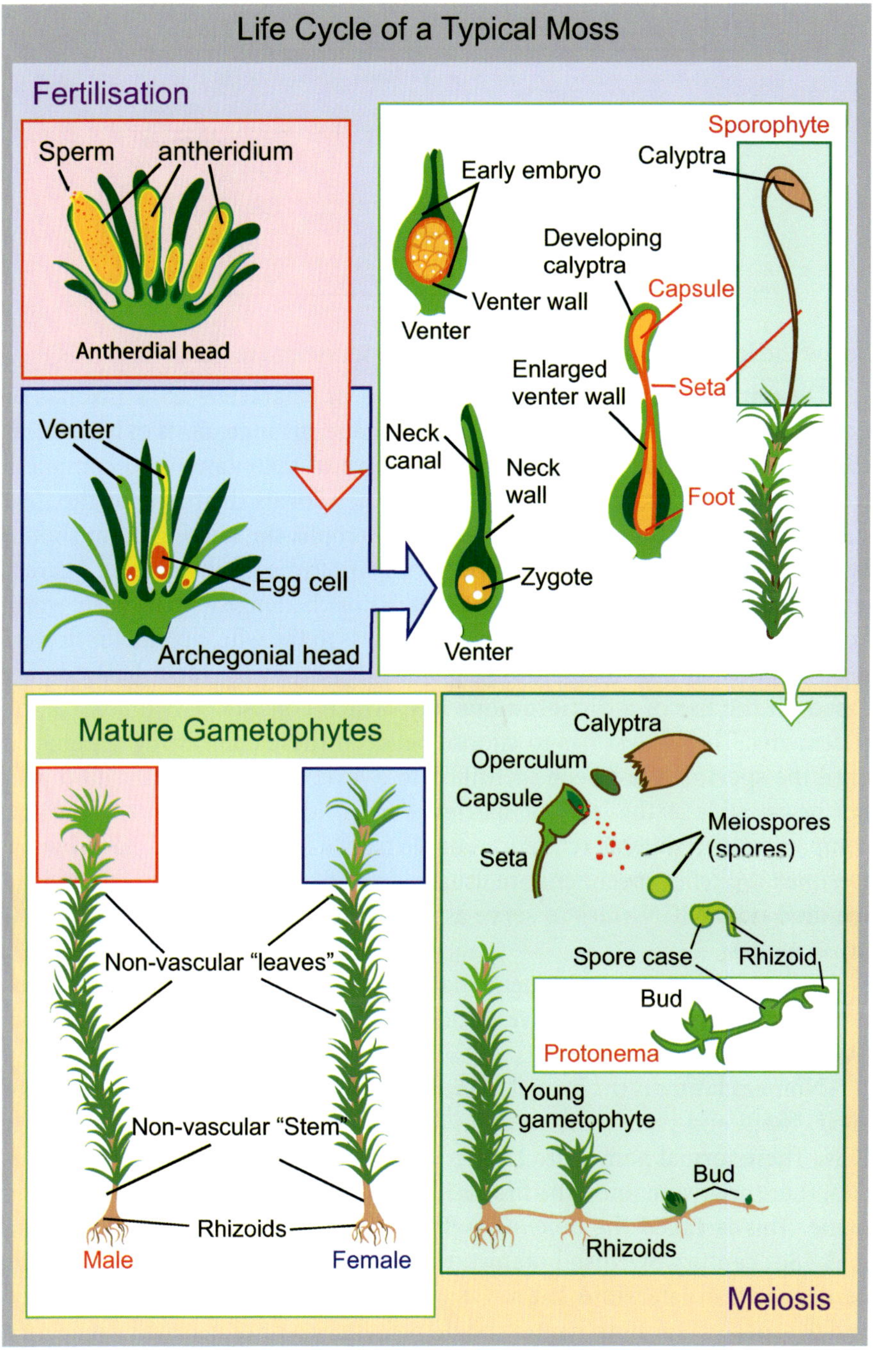

Figure 2.1 Life cycle of a typical moss. Public domain.

(Author: LadyofHats.)

Major Taxonomic Rank	Ending	Taxa
Kingdom	Various	Plantae
Phylum or division	Phyta	Spermatophyta
Class	Opsida	Magnoliopsida
Order	Ales	Urticales
Family	Aceae	Cannabaceae
Genus	Various	*Cannabis*
Species	Various	*Cannabis sativa*

Below the level of species are subspecies, varieties and forms.

Description involves assigning features or characters to a taxon. For example, for leaves this would include leaf shape, arrangements of leaves, leaf edge characters and so on. A major objective of taxonomy is to compile a complete and accurate list of features that then forms the basis for the next stage of **identification**. Given an unknown sample, the starting point should be to note its characteristics and then compare these against known characteristics of other taxa. Herein lies the first major challenge with forensic work, as plant classification depends on having access to the whole plant, for flowering plants usually in their flowering state. Traditionally, identifications are based on the use of a **dichotomous key**, which consists of two contrasting statements. The analyst has to choose between these contrasting statements until the specimen has been identified to a level of taxa. Identification may not be possible at the level of species. Formal identification then relies on comparison to a suitable reference sample from a herbarium, called a *voucher specimen*. Voucher specimens are usually dried herbarium specimens and are required under ICN rules to serve as a *type* in the valid publication of new taxon names.

As we shall see in subsequent chapters, there are different types of taxonomic keys, some of which have distinct advantages from a forensic perspective over dichotomous keys.

Nomenclature is the formal naming of a taxon. As we have seen above for algae, fungi and plants, the rules and regulations are provided through the ICN. These formal names are known as *scientific names*. At the level of species, there are two names, the first being the *genus* and the second the *species* name. This is called the *binomial system of nomenclature*, attributed to the mid-18th century Swedish botanist, Karl Linnaeus. By convention, scientific names are translated into Latin. Convention is that the genus is capitalised and the species not. Both names should be italicised or underlined. Hence, for cannabis, we get *Cannabis sativa* L. The 'L' stands for Linnaeus, who first described this species.

The final component of taxonomy is **classification**. The aim of classification is to provide a system for arranging organisms into some sort of order.

As we have previously discussed, this may be based on the so-called phenetic classification based on overall similarities or a phylogenetic classification based on evolutionary history, but which may or may not reflect similarities.

The overall purpose of plant taxonomy is to provide a system where all organisms have one accepted and correct scientific name, as this forms the basis for accurate communication. The use of common names, however, persists, and this can lead to problems in many areas of life and society, including forensic applications.

For example, the common name *rosewood* is used for a large number of timbers and species. All genuine rosewoods belong to the genus *Dalbergia*, with *Dalbergia nigra* or Brazilian rosewood being the best known. The timber gets its name from the fact that the wood has a strong, sweet smell that persists for years after felling. Other common species of *Dalbergia* include *Dalbergia latifolia*, or East Indian rosewood, and *Dalbergia maritima* or Madagascar rosewood. The term rosewood is, however, used for over 300 other timbers and species. For example, in Australia, *Acacia melanoxylon* and *Acacia excelsa* are sold as rosewood. The issue relevant to forensic investigation is that since 2017, trade restrictions have been placed on rosewood under the CITES (Convention on International Trade in Endangered Species of Wildlife and Flora). Some 300 species are included in these restrictions. According to a recent report, 'African rosewood is by far and away the single largest traded CITES-listed species in the world …it could be as much as 40 percent' (Ong & Carver, 2019). The correct identification and naming of timber products is an obvious prerequisite for any enforcement action.

Another example is Hemlock, the common name used for *Conium maculatum*. Belonging to the carrot family, an extract of this plant is a deadly poison, whereas Hemlock is also used as a common name for a species of *Tsuga*, a coniferous tree.

There are, of course, many other examples where the incorrect identification of plant materials and fungi can have devastating consequences, as we shall see when considering Chinese medicine, poisonous plants and fungi and in Chapter 8.

The naming of plants is made somewhat more complex when it comes to cultivated plants. The *Cultivated Plant Code* covers plants defined as '*distinguishable groups of plants whose origin or selection is primarily due to the intentional actions of mankind*' (Brickell et al., 2009).

Plants named under this code will always have a true botanical name; for example, potatoes are all *Solanum tuberosum*. As a result of plant breeding, there are many identifiable varieties of potatoes differing in appearance and other properties. Formerly, these are called *cultivars* with common names such as King Edward or Maris Piper. A cultivar is defined as '*an assemblage of plants that has been selected for a particular attribute or combination of attributes and that is clearly distinct, uniform and stable in characteristics and that, when propagated by appropriate means, retains these characteristics* '(Spencer et al., 2007).

Adding further complexity to the naming of plants, a number of well-established family names are allowed under the Botanical Code as legitimate alternative names. These are:

Compositae (daisy family): Asteraceae
Cruciferae (cabbage family): Brassicaceae
Gramineae (grass family): Poaceae
Guttiferae (hypericum family): Clusiaceae
Labiatae (mint family): Lamiaceae
Leguminosae (legume family): Fabaceae
Palmae (palm family): Arecaceae
Umbelliferae (carrot family): Apiaceae

Note that family names now all end in 'aceae', meaning 'resemblance'.

Finally, from time to time, names of taxa will change, today often due to emerging new information arising from DNA evidence that results in taxonomic revision. Hence, names used in older texts may not align with current classification terminology.

2.4 Cells and Tissues

2.4.1 Cells

The basic building block for all plants is the *cell*. The term 'cell' is attributed to the 1600s English scientist, Robert Hooke. **Cell theory** postulates that all life is composed of one or more cells, that cells arise only from pre-existing cells and that cells are the units of metabolic processes (Simpson, 2019).

Defined as an organised unit of protoplasm bounded by a plasma membrane, in simple organisms (Prokaryotes) such as bacteria and blue-green algae, the cell has only one cell membrane, with the DNA not separated from the rest of the cell by a membrane. In more complex plants (Eukaryotes), cells have a complex system of membranes that divide the interior of the cell into distinct regions, such as the nucleus and cytoplasm. Other organelles, such as chloroplasts and mitochondria, are also enclosed by membranes.

In land plants, a pectin-rich layer, formed between the plasma membranes of adjacent cells, binds cells together. As cells develop, a cell wall is secreted outside the plasma membrane. This is called the *primary cell wall*. To enable materials to move between cells, microscopic pores may develop, with a group of these being called a *primary pit field*.

Some cell types develop as structural (sclerenchyma) or other functional cells (tracheary elements). In these cells, a secondary cell wall is secreted between the primary cell wall and the plasma membrane. This secondary wall often contains a substance called *lignin*, which imparts strength and rigidity to the cell wall. Movement of materials between cells with secondary cell walls takes place through holes in the secondary wall called *pits*, with pits of adjacent cells forming a *pit pair* (see Figure 2.2). The shape of pits and pit pairs is an important feature in wood identification.

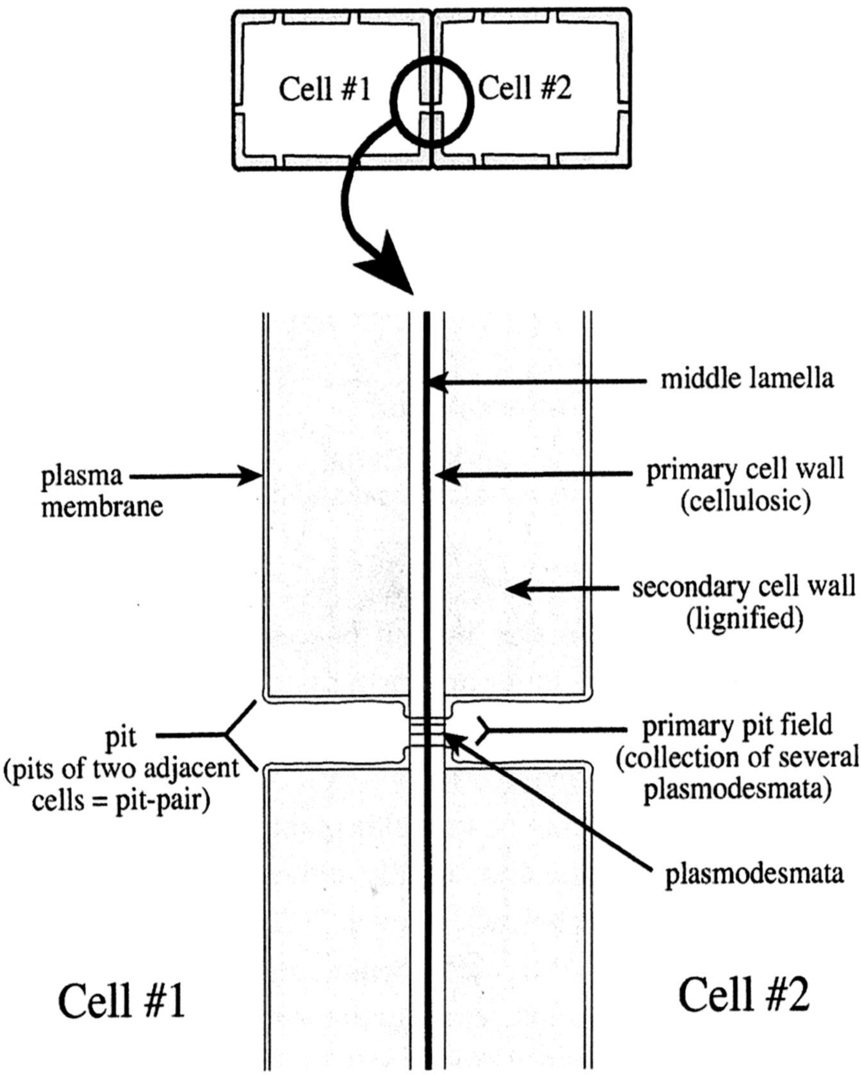

Figure 2.2 Lignified secondary wall of specialised cells of vascular plants.

(Reproduced with permission by Elsevier from Simpson, 2019.)

Cells are always alive at some point, but many plant cells will eventually lose their cell contents and become, in effect, just dead boxes. We still call these cells, although sometimes referring to cells and cell contents.

Plants grow from areas of active cell division, called *meristems*, typically at the apices (root and tip) of plants. Cells expand and change shape, often through elongation and differentiation, into specialised cells.

Cells are organised into groups of *tissues* and then *organs* such as the root, stem, leaves, flowers and so on. Members of a species will show limited variation in these external structures. Members of different species will tend to show greater variations that provide the basis for plant classification.

2.4.2 Tissues

A tissue is defined as a group of cells with a common function or structure. For vascular land plants, there are three tissue types: *ground*, *vascular* and *dermal*.

Ground tissue is the cells inside the dermis, but not vascular. There are three types of ground cells: *parenchyma, collenchyma* and *sclerenchyma*.

Parenchyma cells are generally unspecialised cells with an unthickened primary wall. They are often polygonal in shape but can also be elongated, as in the palisade layer in some leaves. They occupy the bulk of the internal structure of most plants. They are alive at maturity and capable of cell division. For example, they can transform back into a meristem to form new roots.

Collenchyma cells provide mechanical support and are found at the periphery of stems or leaves. They can be stretched during elongation growth and hence are usually elongated. They have a primary wall with uneven thickening that is rich in pectin.

Sclerenchyma cells have a structural role and have thickened secondary walls, which often contain lignin. As previously mentioned, pits are commonly present in these secondary walls. Sclerenchyma cells can be of different shapes and are usually split into two categories, *fibres* and *sclereids*. Fibres are long, narrow cells with tapering ends. Sclereids are isodiametric to irregular in shape and often branched. Examples of fibres and sclereids are shown in Figure 2.3.

Vascular tissue comprises two types of cells, *xylem* and *phloem*.

Xylem is responsible for the movement of water and dissolved minerals from roots to other parts of the plant forming a continuous pipeline the length of the plant and into structures such as leaves and flowers. This requires a series of cells that are connected and structural elements that maintain the structural integrity of the plant. Hence, xylem consists of two major types of cells, *tracheids* and *vessel* members, jointly referred to as *tracheary elements*.

Tracheids are thought to be a primitive form of xylem from which vessels evolved. Like vessels, tracheids are joined end to end to form tube-like structures. However, unlike vessel members, which have perforations between the

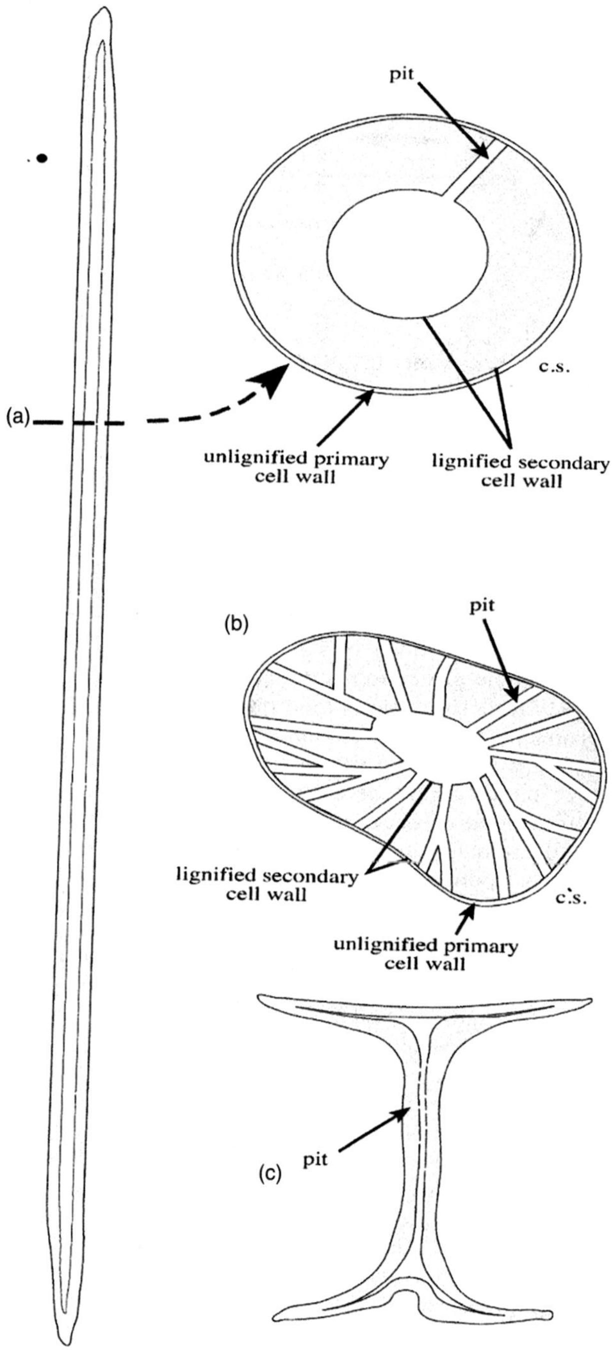

Figure 2.3 Examples of sclerenchyma cells. (a) Fibre cells. (b) Sclereid. (c) Cross-section of a fibre cell with a pit.

(With permission by Elsevier, from Simpson, 2019.)

ends of cells, tracheids are *imperforate*. Water and minerals can still move between adjacent tracheid cells through pit pairs, but not end to end.

The term *vessel* refers to several vessel members joined end to end in a continuous tube. The ends of vessel elements have either a simple single perforation plate or a compound perforation plate. Figure 2.4 shows some examples of tracheary elements and perforation plates. As we shall see in Chapter 7, variation of tracheary elements plays a significant role in wood identification.

Phloem is responsible for the movement of sugars produced by photosynthesis from areas of production to the rest of the plant, supporting its growth, or to areas of storage. Phloem consists of specialised cells called *sieve elements*, but also in association with parenchyma and sclerenchyma, the latter as support cells. In other words, sieve elements do not exist in isolation. Sieve elements are oriented end to end to form a tube-like continuum, as is the case with tracheary elements.

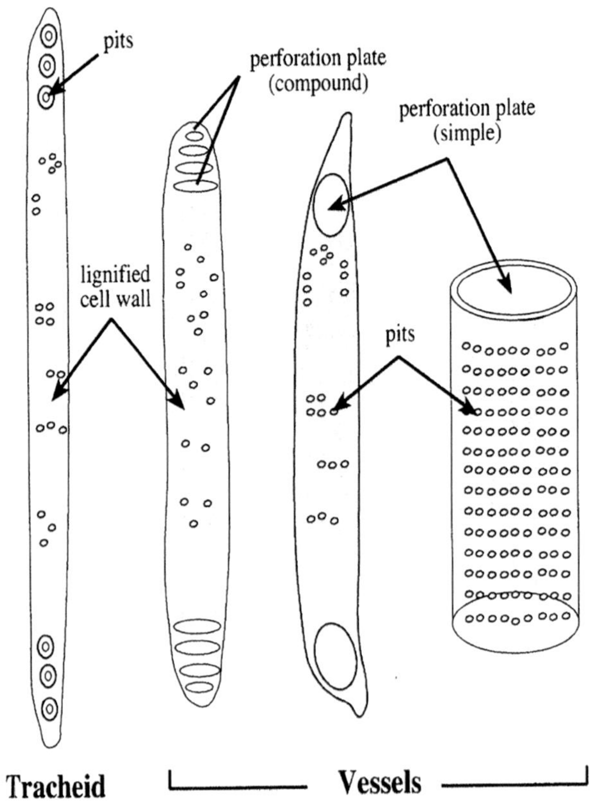

Figure 2.4 Conductive cells of vascular plants: tracheary elements showing perforation plates and pits.

(With permission by Elsevier, from Simpson, 2019.)

There are two types of sieve elements, *sieve cells* and *sieve tube members*. Sieve cells are elongated cells with a primary cell wall and no lignin. At maturity, they lose their nucleus but have other organelles, such as mitochondria. They have specialised pores that together form *sieve areas* or a *sieve plate*. A sieve plate will have multiple individual pores in either a simple or a compound arrangement. Each pore is lined with a substance called callose. Sieve tube members are only found in the flowering plants or angiosperms.

Dermal tissue consists of *epidermis* or *periderm* in woody plants. The epidermis is the primary outermost layer in plants. Epidermal cells vary greatly in shape, size and arrangement. As we will see in Chapter 8, variations in epidermal cells and their arrangement are extremely useful in comparing and identifying fragmented unknown plant samples. As the epidermis provides protection to plants from loss of water and desiccation, they secrete a fatty substance called *cutin* outside the cell wall, forming the cuticle. These plant waxes can form distinct microstructures on the surface of plants, called *epicuticular waxes*. Cutler et al. (1982) provide a comprehensive coverage of all aspects of epicuticular waxes.

The epidermis also has specialised cells, called *stomata*, that enable gaseous exchange between plants and their environment, allowing the capture of carbon dioxide and the emission of oxygen back into the atmosphere. Stomata are further discussed when we return to consider leaf anatomy.

Epidermal cells may also produce a vast array of extensions or hairs ranging from single cells to complex structures. These are called *trichomes* or plant hairs. Trichomes are classified as *covering* or, where they have a secretory function, *glandular*. The particular multiseriate, multicellular glandular trichome found in *Cannabis sativa* is a defining feature used in the identification of that species, as will be discussed in Chapter 4.

Glandular trichomes are only one form of *secretory structures*. Other forms of secretory structures include *resin ducts* or canals in which cells have a lining of cells that secrete resin, oil or mucilage and *laticifers*, cells that store and secrete latex.

2.5 Organs

As we have seen, similar cells that perform a particular function are grouped together to form tissues. The cooperative specialisation of a number of tissues leads to the formation of an *organ*. The interaction of a number of organs is called an *organ system*, and organ systems combine to form the organism. Typical organ systems in higher plants are the root, the stem, leaves and flowers. Figure 2.5 shows a simplified version of the general plant structure.

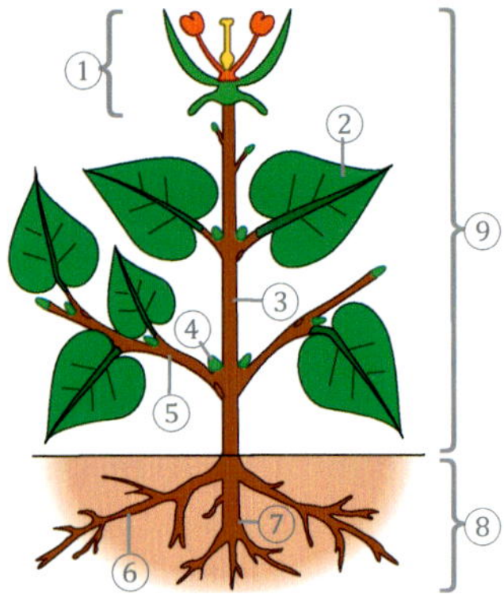

Figure 2.5 Flowering plant schematic. 1. Flower, 2. leaf, 3. stem, 4. bud, 5. side shoot, 6. lateral root, 7. primary root, 8. root and 9. shoot.

2.5.1 Roots

The root of a flowering plant develops from the embryo within the seed. Monocots derive their name from having one (mono) cotyledon, whilst dicots have two (di) cotyledons.

The root develops from the *radicle* and the stem from the *coleoptile*. Roots are protected from damage by the presence of a root cap. Immediately behind this structure is the apical meristem, where cell division takes place. Cells then enter an area of cell elongation and after that an area of maturation or cell differentiation.

The radicle gives way to a *primary root*. If the primary root becomes dominant, it develops into a *taproot*. Many food plants have tap roots that become storage organs and sources of food. Where the taproot does not develop, it is replaced by *adventitious roots* such as those shown in Figure 2.6.

Lateral roots are also a common feature of root systems. These arise from the pericycle layer.

The epidermal cells away from the root tip may also develop short hair-like extensions called *root hairs* that increase the surface area available for water and mineral absorption by the plant.

Most root systems will be below the surface, but in some families, *aerial roots* are found, for example, in many orchids. Roots may adapt in a number of other ways to their growing environment and the plant's habit.

2.5.2 Stems

The stem or shoot of a flowering plant originates from the embryo within the seed, known as the *epicotyl*, which is a continuation of the *hypocotyl*. Stems

Figure 2.6 Root systems schematic. (a) Fibrous root system. (b) Tap root system. **(Creative Commons Attribution – Share Alike 4.0 International license. Author: Liu Kaitlin.)**

provide support, and the internal vascular system conducts water/minerals and sugars between leaves, roots and flowers.

The apex of a growing stem, called the *apical meristem*, is an area of active cell division. Behind the apical meristem are areas of cell elongation and cell differentiation. Figures 2.7 and 2.8 show models of a monocot and dicot stem in cross-section.

In a typical monocot, the vascular bundles are widely spaced and not restricted to a single ring. Monocots do not have secondary growth but can develop large, thickened stems, such as those seen in palms, through primary growth.

In a typical dicot, the vascular bundles are arranged in a distinct ring. Through secondary growth, a new ring is formed each year.

In some stems, a protective tissue of secondary origin will develop, replacing the epidermis. This is called the *periderm*. Periderm is not synonymous with the term bark, which is the outer layer seen in many woody species consisting of all the tissues outside the vascular cambium and including the phloem, cortex and periderm or epidermis when present.

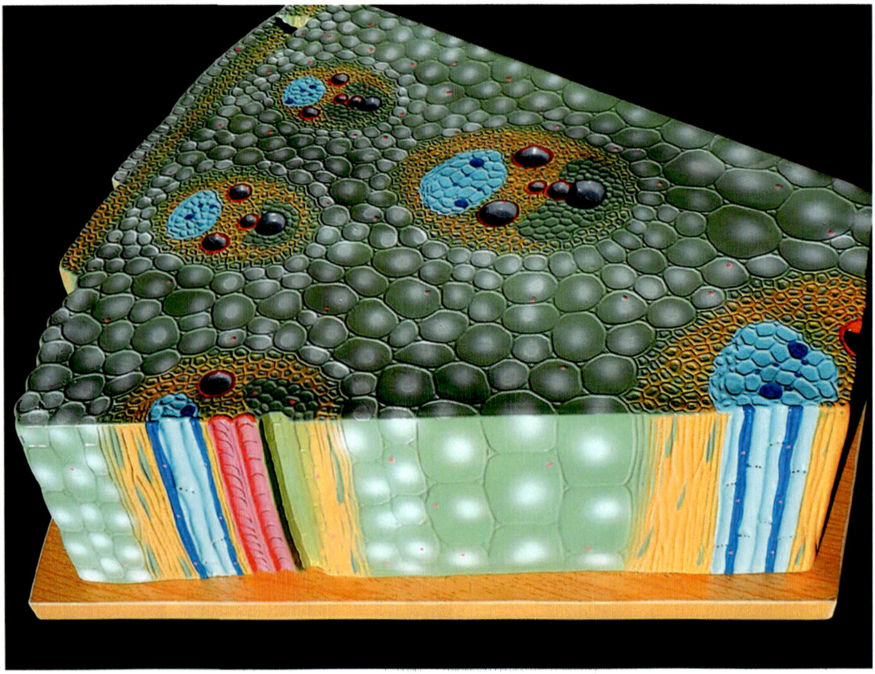

Figure 2.7 Monocot stem model.

(Published under the Creative Commons Attribution 2.0 Generic License. Wikimedia Commons. File: Monocot Stem Model (50669874).jpg. Author: YVC Biology Department. https://www.flickr.com.people/191254970@NO2)

There are numerous modifications of stems adapted for specific purposes. These include the following underground or surface modifications:

- **Bulb**: consists of a small amount of stem tissue and a larger amount of thick, fleshy leaves. An example of a bulb is an onion.
- **Corm**: consists of globose stem tissue surrounded by scale-like leaves. An example is an iris.
- **Rhizome**: a horizontal underground stem with short internodes and scale-like leaves. An example is ginger.
- **Stolen**: is a stem with long internodes that runs just below the surface of the ground and produces new plantlets. An example is strawberry runners.
- **Tuber**: a thick underground storage stem with outer buds but lacking storage leaves or protective scales. An example is a potato.

There are also numerous above-ground modifications of stems. For example, stems in members of the grass and sedge families are called *culms*, and these

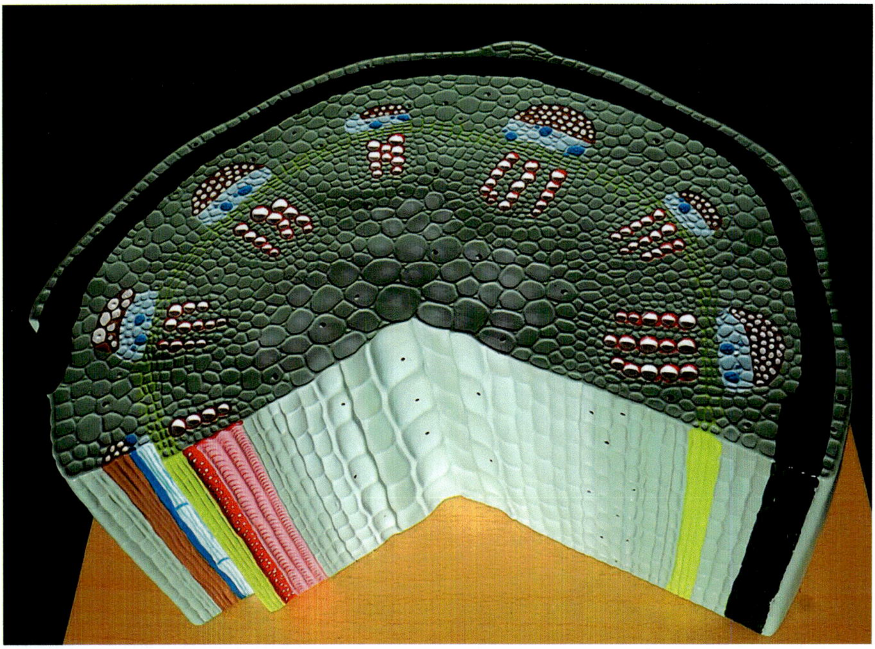

Figure 2.8 Dicot stem model.

will often form clumps growing from axillary buds at the base of the stem to produce *tillers.*

Stems may also be described in terms of their habit and stem branching, with numerous models proposed to describe the arrangement of stems and branch structures (see Simpson, 2019). These are beyond the scope of this introduction.

2.5.3 Leaves

Leaves may develop from either terminal (apical) or lateral (axillary) buds. Leaves are the primary photosynthetic organs of the plant and the main site for transpiration.

As leaves develop, they grow outwards from the base with the final shape and form characteristics of the species. The mature leaf comprises a stalk or *petiole* and a flat portion or *blade.* The leaf may be attached to a stem without a petiole, in which case it is said to be *sessile.* In grasses, leaves are typically attached to the stem with a sheath structure. At the base of the petiole of some leaves are two small structures called *stipules.* The form and structure of the leaf blade may vary within a plant, typically from early leaves to mature leaves, and there is enormous variation in leaves from different species. Not surprisingly, there is a vast number of terms used to capture and describe this variation.

A starting point in describing leaves is whether they are *simple* or *compound*. A simple leaf is where there is a single blade structure. A compound leaf will have from two to several leaflets. Where the leaflets are attached at one point, they are termed compound *palmate*. Where the leaflets are attached along a central axis, called the *rachis*, they are termed compound *pinnate*. Examples of simple and compound leaves are presented in Figure 2.9.

Leaves and leaflets can be described in relation to features such as the shape of the base, the shape of the apex or tip, leaf division, leaf margins, leaf attachment and leaf arrangement. A useful pictorial guide to leaves and leaf terminology is Harris and Harris, 'Plant Identification Terminology. An Illustrated Glossary' (2019). Hickey and King (2014) also have excellent diagrams of leaf features. For example, Hickey and King provide diagrams for 15 (×15) types of leaf margin, including entire (smooth edge), crenate (rounded edge) and serrate (spiked edge). Stearn (1995), in his encyclopaedic work on botanical Latin, lists some 505 terms to describe leaves!

The vascular bundles in leaves are seen as veins and form distinct patterns that can also be described and classified. Like leaves, the vein patterns are described as palmate, pinnate or parallel. Typically, parallel venation is seen in monocot leaves such as grasses. Leaf venation and oil gland patterns are used in the identification of Eucalypts (Brooker & Dean, 2013).

Figure 2.10 shows some of the features used to describe leaves.

Figure 2.9 Examples of (a) a simple palmate leaf and (b) a compound (pinnate) leaf structure.

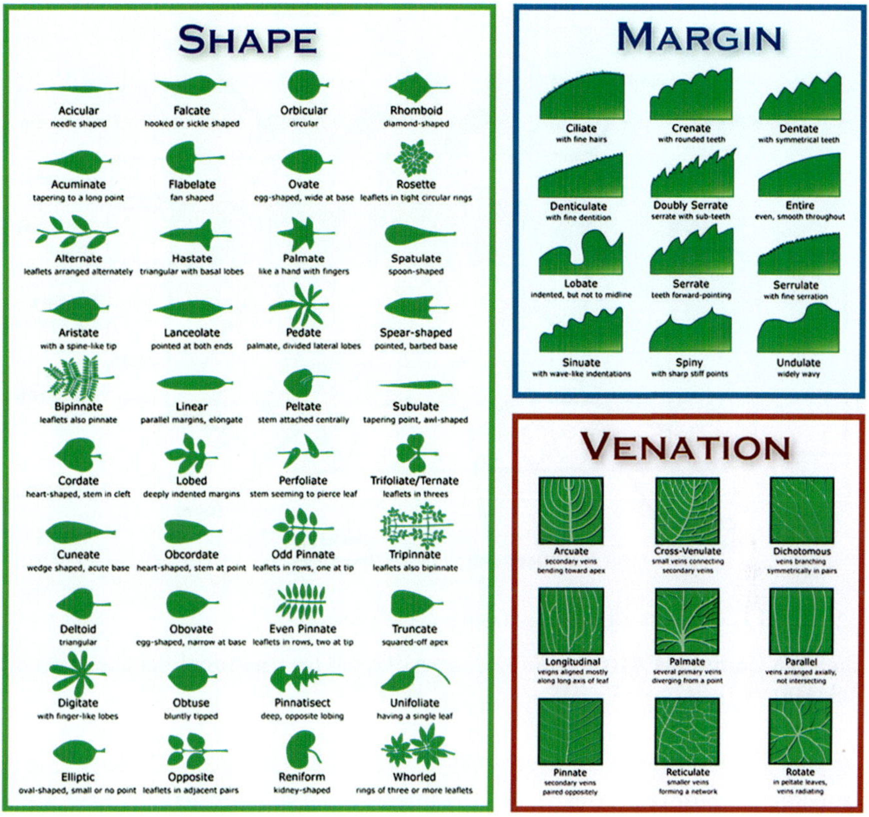

Figure 2.10 Leaf morphology.

Figure 2.11 shows cross-sections of typical dicot leaf anatomy. The upper and lower epidermis is protected by a cuticle. Between the epidermis lie layers of parenchyma cells called *mesophyll*. When these cells contain chloroplasts, they are called *chlorenchyma*. In dicots, the mesophyll can be further differentiated into well-ordered elongated cells called the *palisade layer* and less organised cells called the *spongy layer*. Where a leaf has a palisade layer below the upper epidermis and a spongy layer above the lower epidermis, this is called a *dorsiventral* leaf. If a leaf is structurally the same on both sides, then it is called an *isobilateral* leaf. The mesophyll is traversed by numerous vascular bundles (veins).

Veins in monocot leaves typically have more mechanical tissue and xylem fibres than the veins of a dicot – this has practical significance as we shall see when looking at vegetable fibres.

As we have discussed earlier, the epidermis also has specialised cells called *stomata*, which are responsible for gaseous exchange with the

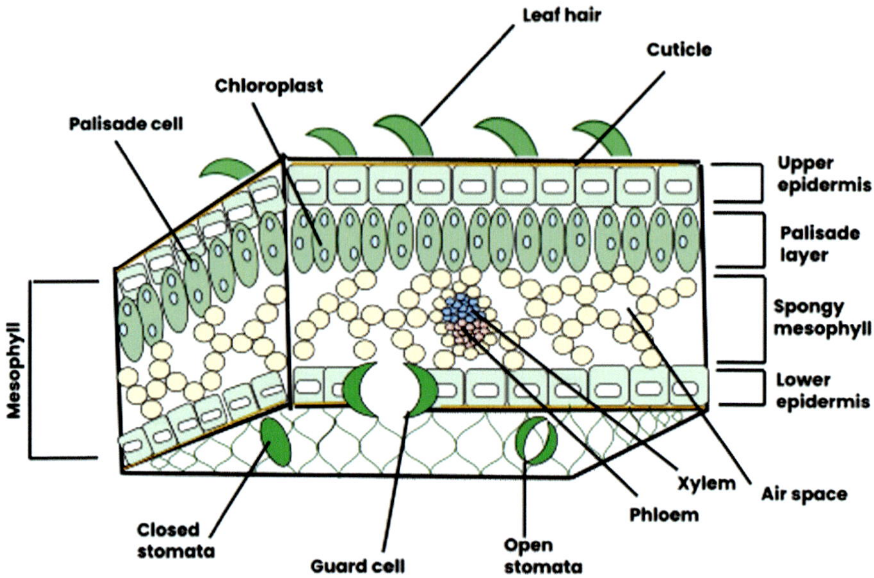

Figure 2.11 Leaf anatomy schematic – cross section.

atmosphere. Stomata consist of two guard cells surrounded by epidermal cells. The epidermal cells surrounding the stomata can be classified based on the number and arrangement of epidermal cells into a number of distinct types – see Figure 2.12. In dicots, four basic patterns are recognised as follows:

- *Anomocytic* – the guard cells are surrounded by a variable number of cells with no specific pattern. The pattern appears random or lacking a defined structure.
- *Anisocytic* – the guard cells are surrounded by three cells, where one is distinctly smaller than the other two cells.
- *Diacytic* – the guard cells are surrounded by two cells that are oriented perpendicular to the long axis of the stomatal pore and guard cells.
- *Paracytic* – the guard cells are surrounded by two cells that are oriented parallel to the long axis of the stomatal pore and guard cells.

A fifth type is sometimes recognised where multiple cells surround the guard cells in a circle of radiating cells resembling the spokes of a wheel. This type is called a radiate-celled type or *actinocytic* stomata.

Stomata in monocots, such as the grasses, have an entirely different appearance, which will be considered in Chapter 10.

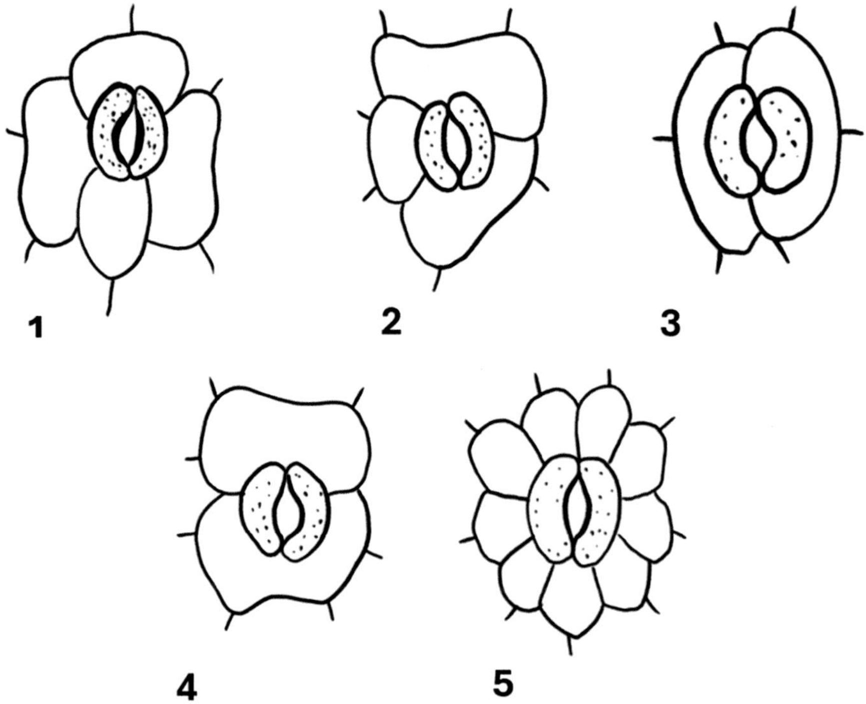

Figure 2.12 Classification of types of dicot stomata based on the arrangement of subsidiary epidermal cells around the guard cells of stomata. 1. Anomocytic, 2. anisocytic, 3. paracytic, 4. diacytic and 5. actinocytic.

The epidermis may also give rise to a variety of appendages (hairs) called *trichomes*. These can vary in shape and size from a single cell to multicellular structures and may be covering and secretory (glandular). The range of shapes and sizes is truly amazing, and trichomes are one of the most valuable features of leaves in differentiating plants.

Finally, leaves show a wide variety of adaptations. Of special interest are modified leaves that are usually smaller than the major leaves called *bracts*. In angiosperms, bracts are typically associated with flowers or the axis of inflorescences (groups of flowers). Smaller secondary bracts are called *bracteoles*. We shall return to look at bracts and bracteoles when considering the structure and morphology of Cannabis plants in Chapter 4. Grasses also have specialised bracts called *glumes* (lemmas and paleas). These will be described in more detail in Chapter 10.

Another common adaptation is a *tendril*. A tendril is a slender, twining organ used to grasp support in climbing plants, for example, many forms of beans and peas. A tendril can also be a modified stem. Of fascination to many are carnivorous plants, where the *pitcher* is a modified leaf shaped like a container.

2.5.4 Flowers and Inflorescences

In the Angiosperms, flowers are a major diagnostic feature. Flowering plants produce stems with modified leaves that form the parts of a flower. Flowers may be subtended by a modified leaf or *bract*. As discussed above, smaller or secondary bracts, called *bracteoles*, may also be present.

A complete flower has four groups of parts arranged symmetrically and in whorls at the end of an enlarged stem called the *receptacle*. When the stem bears a single flower, it is known as a *peduncle*. Where there are groups of flowers, each stalk is called a *pedicel*. These four groups are the *sepals, petals, stamens* and the *pistil*.

The presence of four circles of parts makes up a complete flower, as shown in Figure 2.13.

The outer circle comprises a number of modified leaf structures called *sepals*, which together form the *calyx*. In most instances, sepals are green in colour. The next circle of parts comprises the *petals* collectively called the *corolla*. The corolla shows enormous variation in shape and colours related to function. In some plants, the corolla is shaped to attract bees in the process of insect pollination. Taken together, the calyx and corolla are called the *perianth*.

The third whorl of modified leaves consists of the *stamens*, which collectively are known as the *androecium*. Each stamen has two parts: a stalk or *filament*, which supports at its end, and an *anther* containing two pollen sacs. There is a plethora of technical terms to describe stamen type, attachment, insertion and fusion, as well as anther parts, type and attachment (see Simpson, 2019).

Within the pollen sacs at maturity are pollen grains that show enormous variation in shape and size between species. The study of pollen is called

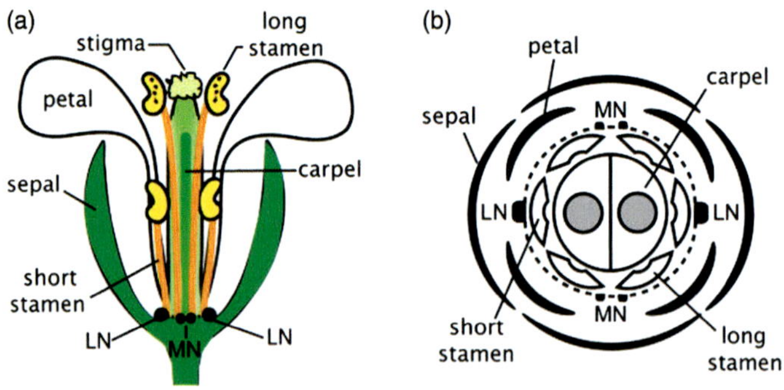

Figure 2.13 Schematic of *Arabidopsis thaliana* showing typical floral parts in (a) cross-section and (b) as a floral diagram. In addition, this flower has two lateral (LN) and two median (MN) nectaries at the base of the short stamen.

palynology. In Chapter 8, we shall consider the forensic applications of palynology.

The fourth circle of parts forms the *pistil*. The base of the pistil is enlarged to form an *ovary* from which extends a slender filament called the *style*. Inside the ovary are a number of hollow spaces called *locules* in which one or more *ovules* will develop, giving rise eventually to seeds. At the end of the style is the *stigma*, whose function is to catch and hold pollen. The stigma is typically either sticky and/or roughened.

There is enormous variation and adaptation in flowers, and a nomenclature that matches. It is these variations in morphology that are important in distinguishing between species. Hence, a deep knowledge of the variation and understanding of the nomenclature is necessary if one is to accurately identify a plant.

Floral structures and the arrangement of the floral parts can be captured by the use of two conventions, the *floral diagram* and the *floral formula*. The floral diagram shows the floral features depicted as a number of concentric circles representing the whorls of a flower. The floral formula is a shorthand way of showing the number of members composing the flower. Judd et al. (2008) cover in detail how to 'read' a floral formula.

Thus far, only single flowers have been considered. The production of a single flower on an individual plant is quite rare. An example is a tulip. In this case, the *inflorescence* is said to be *solitary*. Where there are multiple flowers in clusters, these are firstly grouped into two classes: indeterminate or *racemose* and determinate or *cymose*. Where the inflorescence is indeterminate, the first flowers to open up are at the base or the outside of the cluster. In a determinate inflorescence, the first flowers to open up are at the tip or centre of the inflorescence.

Finally, with respect to flowers, they can also be classified based on the relationship of the ovary to other floral parts and by the type of *placentation*. There are three basic arrangements known as *epigynous*, *hypogynous* and *perigynous*. In the hypogynous arrangement, the ovary is above the other floral parts and is said to be superior. Conversely, in an epigynous arrangement, the ovary is inferior or below the other floral parts. In a perigynous arrangement, the ovary and the floral parts are fused and arranged around and above a free ovary. Simpson (2019) provides an in-depth treatment of ovary attachment and position and further detailed descriptions of placentation and ovule types.

The ovule is attached to the placenta of the ovary wall by a short stalk called a *funiculus*. If the placenta is on the ovary wall, this is termed *parietal*; if on a central axis, the placentation is called *axial* and if it is on a short stalk that extends from the base of the ovary into the cavity, it is called *central*.

The pistil can also be classed as *simple* or *compound*. The former is where there is only one carpel, and the latter is where two or more carpels are fused. A compound pistil may have more than one style or stigma.

2.5.5 Fruits and Seeds

In everyday use, fruit conjures up an image of an apple, an orange, a strawberry or other typical fruit. In botanical terms, a fruit is defined as a ripened ovary and any other structures that are attached and ripen with it (Harris & Harris, 2019). Fruits contain one or more seeds.

During the development of a fruit, the ovary wall (*pericarp*) usually thickens and becomes differentiated into three layers, the outer *exocarp*, middle *mesocarp* and inner *endocarp*. In some fruits, other structures may contribute to the formation of a fruit. For example, in strawberries, individual fruits, called *achenes*, are borne on a fleshy receptacle. A strawberry is called either an *accessory fruit* or an *aggregate fruit*. Other examples of accessory fruits are apples, pears and quince, where the edible portion is not the mature ovaries but a modified stem and floral tubes in which the ovaries are embedded.

Fruits can be classified as *simple*, *aggregate* or *multiple*. A simple fruit is the product of a single ovary of a single pistil. The ovary may be superior or inferior (see discussion above). Where the ovary is superior, the resulting fruit will comprise only the ovary, whereas if the ovary is inferior, the resulting fruit will contain contributions from other parts of the flower.

A compound fruit is the product of an *apocarpous* (meaning separate carpels) gynoecium, where each of its separate ovaries remains distinct, forming a fruit by itself.

An aggregate fruit is formed where the fruit is the product of a whole inflorescence. This is also called an *infructescence*, an example of which is the pineapple.

Fruits may then be subdivided based on whether or not the fruits open at maturity to release their seeds in the process of seed dispersal. Where the fruit opens to release its seeds, it is called a *dehiscent* fruit. The opposite is an *indehiscent* fruit. Obviously, all simple fruits are the latter, as they have no need to release their seed.

As fruits reach maturity, the pericarp may become very dry, or it may become a bulky tissue, such as seen in tomatoes. This gives a third means of classifying fruits as either *dry* or *succulent*. The most common dry indehiscent fruit is called an *achene*. A special type of achene is found in grasses, where it is termed a *grain* or *caryopsis*. Achenes with a hard coat are called a *nut*. A classic example of a dry dehiscent fruit is a *legume* or *pod* found in the Leguminosae family.

Harris and Harris (2019) provide a key to common fruit types based on the features discussed above.

A *seed* is defined as an embryo, which is in an immature diploid sporophyte that develops from the zygote, surrounded by nutritive tissue and enveloped by a seed coat (Simpson, 2019).

Seeds show almost infinite variation in shape, size and structure. For example, Sweedman and Merritt (2006), in their book on Australian seeds, catalogue more than 1,200 coloured images of seeds, and these are only representative of different genera.

With relevant experience and expertise, the external morphology of a seed can lead to identification to at least the level of genus. Useful features can be the presence of an obvious *hilum*, a scar that is left where the seed was attached to the mother plant by the funiculus and the *raphe*, a ridge on the seed coat leading away from the hilum.

In this section, our primary interest does not lie in the embryo, although clearly this is the reason seeds exist, as once a seed has germinated, it is the embryo that produces a young plant.

From a forensic viewpoint, we need to know something about the protective seed coat and its anatomy. The seed coat is called the *testa* and may vary from being a thin membrane to a fleshy structure to a hard external coat. The testa is derived from the outer integument, which may also be subdivided into inner, middle and outer layers. As we shall see in Chapter 7, the seed coats of some food plants are very useful in identifying fragments of plant material.

To this point, most of this chapter has dealt with Angiosperms. Gymnosperms also produce seeds. The term gymnosperm is derived from the Greek words *gymnous*, meaning naked, and *sperma*, meaning seed. Hence, gymnosperm seeds are considered naked, lacking the structures derived from floral parts found in angiosperms. Gymnosperm seeds, however, are protected in many instances as the seeds are attached to the inside of scales found in structures called *cones*.

There are less than 1,000 gymnosperms compared to 250,000+ angiosperms. Hence, it is not surprising that there is less diversity in the seed coats of gymnosperms. Nonetheless, there are significant differences in the seeds of the four major groups of gymnosperms, Conifers, Ginkgo, Gnetales and Cycads. The terminology used to describe gymnosperm seeds differs from that used to describe angiosperms The typical Gymnosperm seed coat comprises three layers: an outer fleshy parenchymatous layer (*sarcotesta*), a middle sclerenchymatous layer (*sclerotesta*) and an inner parenchymatous layer (*endotesta*).

The mechanisms by which seeds are dispersed are of forensic interest. Fruits and seeds have developed many modifications that aid in their dispersal. These include dispersal by wind (*anemochory*) and self-dispersal (*autochory*). In the latter category, there are many forms of self-dispersal, including dispersal by explosive dehiscence, dispersal by gravity or by geocarpy where the plant pushes fruits into the ground, dispersal by water, dispersal by ants and dispersal by animals. The latter may involve the fruit or seed becoming attached to and carried away by an animal, or where a fruit or seed is eaten and excreted in faeces unharmed. We are all familiar with the experience of picking up burrs on our clothing whilst walking in the countryside. Figure 2.14 shows some examples of seed and fruit dispersal.

Figure 2.14 Examples of seed and fruit dispersal. (a) *Acer pseudoplatanus*, windborne. (b) *Cirsium arvense*, airborne. (c) *Pisum sativum*, pops when dry. (d) *Xanthium strumarium*, animal.

References and Additional Literature Resources

Arnett, R.A. & Bazinet, G.F. 1977. *Plant Biology. A Concise Introduction*, 4th ed. Saint Louis: CV Mosby.

Black, M., Bewley, J.D., & Halmer, P. (eds.). 2006. *The Encyclopaedia of Seeds. Science, Technology and Uses*. Trowbridge: Cromwell Press.

Brickell, C.D., Alexander, C., David, J.C., Hetterscheid, W.L.A., Leslie, A.C., Malecot, V., Jin, X., & Cubey, J.J. (eds.). 2009. *International Code of Nomenclature for Cultivated Plants*, 8th ed. Vienna: International Association for Plant Taxonomy.

Bridson, D. & Forman, L. (eds.). 1998. *The Herbarium Handbook*, 3rd ed. Kew: The Royal Botanic Gardens Pub.

Brooker, I., & Dean, N. 2013. *Atlas of Leaf Venation and Oil Gland Patterns in the Eucalypts*. https://www.pulish.csiro.au/bbk/7083

Cutler, D.F., Alvin K.L., & Price C.E. (eds.). 1982. *The Plant Cuticle. Papers Presented at an International Symposium*, September 1980. London: Academic Press.

Esau, K. 1960. *Anatomy of Seed Plants*. New York: John Wiley and Sons Pub.

Evert, R.F. 2006. *Esau's Plant Anatomy. Meristems, Cells, and Tissues of the Plant Body: Their Structure, Function, and Development*, 3rd ed. Hoboken: John Wiley and Sons Pub.

Fahn, A. 1982. *Plant Anatomy*, 3rd ed. Oxford: Pergamon Press.

Harris, J.G., & Harris, M.W. 2019. *Plant Identification Terminology: An Illustrated Glossary*, 2nd ed. Spring Lake: Spring Lake Pub.

Hickey, M., & King, C. 2014. *The Cambridge Illustrated Glossary of Botanical Terms*. Cambridge: Cambridge University Press.

Judd, W.S., Campbell, C.S., Kellogg, E.A., Stevens, P.F., & Donoghue, M.J. 2008. *Plant Systematics. A Phylogenetic Approach*, 3rd ed. Sunderland: Sinauer Associates Pub.

Ong, S., & Carver, E. 2019. *The Rosewood Trade: An Illicit Trail from Forest to Furniture*. https://e360.yale.edu/features/the-rosewood-trade-the-illicit-trail-from-forest-to-furniture

Simpson, M.G. 2019. *Plant Systematics*, 3rd ed. Amsterdam: Elsevier Pub.

Spencer, R., Cross, R., & Lumley, P. 2007. *Plant Names. A Guide to Botanical Nomenclature*, 3rd ed. Collingwood: CSIRO Publishing.

Stearn, W.T. 1995. *Botanical Latin, History, Grammar, Syntax, Terminology and Vocabulary*, 4th ed. Newton Abbot: David and Charles Pub.

Sweedman, L., & Merritt, D. (eds.). 2006. *Australian Seeds. A Guide to Their Collection, Identification and Biology*. Collingwood: CSIRO Pub.

Turland, N.J., Wiersema, J.H., Greuter, W., et al. 2018. *International Code of Nomenclature for Algae, Fungi, and Plants (Shenzhen Code). Adopted by the Nineteenth International Congress*, Shenzhen, China, July 2017.

Watson, E.V. 1981. *British Mosses and Liverworts*. 3rd ed. London: Cambridge University Press.

Forensic Processes and the Role of Standards

3

3.1 Introduction

> I had to learn quickly about how criminal investigations were managed and structured. I soon discovered that there is a complex ecology of people working with different mindsets and power infrastructures… my experience of entering forensics rather later in my career has been fascinating, but some formal training would have eased the transition.
>
> (Spencer, 2019)

As commented on in Chapter 1, there are very few forensic botany specialists employed in mainstream forensic organisations. Hence, the experience of Mark Spencer and others, such as Patricia Wiltshire (Wiltshire, 2019), is not uncommon coming to the world of forensic investigation as almost 'accidental' forensic scientists. The nature of some forensic work, dealing with death and decay, is not for everyone, nor is the cut and thrust of the legal process. For some, engaging with this world may be a one-off experience! For those who stay the distance, they need to quickly learn the fundamentals of forensic science and what is required for forensic evidence to stand up to the rigours of the legal system.

As will become obvious in the following chapters, not all forensic botany will involve the forensic botanist visiting a scene. In practice, however, **all** forensic science starts at the scene of what is deemed to be suspicious or an alleged criminal activity. Like Mark Spencer, my first encounter with forensic botany as an 'expert' was being presented in court with a series of black and white photographs of vegetation at an outdoor scene and being asked if I could identify the plants that were present. As Mark Spencer puts it:

> I've yet to meet a botanically inclined police officer, and they invariably take awful pictures of plants and vegetation … the images are either landscape scale (as was the case in my experience!) and not focused on the vegetation or they are somewhat abstract images of parts of a leaf or a flower.

Today, one might expect a more informed and technically competent approach from qualified crime scene examiners (CSEs). As Wiltshire, however, records in her book 'Traces' (Wiltshire, 2019), when arriving at a scene, a smiling CSE greeted her by telling her, 'I've cut all of the grass and bushes down for you'. Not only had he cut all the vegetation, but he had dumped it on the only path that might have been taken by an offender. As Spencer, Wilshire and others

DOI: 10.4324/9781003308508-3

have commented on with some degree of exasperation, a forensic botanist may not be involved in the initial crime scene process, and, all too often, the possibility of botany contributing is an afterthought. I will come back to how this situation might be improved in the final chapter of this book.

Where scene attendance is required, for the individual who, as a 'one off', does become involved assisting the police in an investigation, they will obviously be reliant on the advice and assistance of CSEs with regard to recording and recovering materials of potential forensic significance. In some jurisdictions, evidence can only be collected by police. As, however, already discussed, there may be little to no appreciation of what should be collected, and this *evidence recognition component* is where the specialist can play an important role, as well as advising on how best to preserve and store such materials for subsequent examination.

As Spencer (2019) comments:

> generally we botanists are used to looking at the whole plant, or, if a portion of a plant, such as a flowering branch of a tree or a seed, it is normally relatively pristine… this is frequently not the case in the world of forensics, as evidence gathered at a crime scene is often fragmentary and far from ideal for identification.

The forensic botanist should always be conscious that whilst identifying botanical materials is probably why they have been asked to be involved, this is only one aspect of the potential value of forensic botany.

As discussed in Chapter 1, most forensic work requires attempting to answer the other investigative questions of when, where, what and why. Along with the 'who', these are referred to as the **five W's**.

WHO

- was the victim
- was the perpetrator
- witnessed the crime

WHEN

- did the crime occur

WHERE

- did key events take place – in the case of a body found in suspicious circumstances, did the person die at the scene or elsewhere
- did people involved enter and, if applicable, leave the crime scene
- were people involved in the crime located at the time of its commission, standing, kneeling, etc.
- are there inanimate objects in the crime scene, and were they introduced to the scene during the commission of a crime

WHAT

- was the sequence of events that may have occurred during the commission of the crime
- was the motive
- was the modus operandi of the perpetrator
- was added to or removed from the scene by the perpetrator

WHY

- did the crime happen where it did
- did the crime happen when it did

From a forensic perspective, attempting to provide useful input to answering these questions requires the application of a **criminalistics** approach or thinking. How can forensic science and, specifically, forensic botany help provide answers to some of these questions and assist investigators? Wiltshire (2019) states that

> the complexity and variation is pivotally useful when one square metre of land is different from the next, and that, it becomes possible, by profiling the biological trace evidence, to identify the **when** and **where** of a crime, to pinpoint where somebody has been, when they were there, and perhaps even, by the traces left on their clothes, **what** they were doing.

Other disciplines of forensic science are better placed to provide answers to the **who** question, and the **why** question is one for the investigators.

Botanical materials recovered from a scene or from items of interest, such as clothing or from a vehicle, will usually be visible to the naked eye but may be small enough to be considered as a **trace material**. A trace material can be defined as a very small amount of substance, which, because of its size or texture, may transfer from one location to another and persist there for some period of time. They can be thought of as *the surviving evidence of a former occurrence of an action of some event or agent*. This, of course, is the often-quoted Locard's Exchange Principle, simply stated as 'every contact leaves a trace'.

It should be noted that there is a very limited literature dealing with the transfer and persistence of botanical trace materials, and care should be taken in drawing parallels to other trace materials such as fibres for which there is an extensive literature (Robertson et al., 2018). In recovering small botanical materials, particular attention should be paid to factors that might influence their retention. In general, the major factor influencing the redistribution and loss of trace materials is movement.

Hence, a botanist undertaking any aspect of forensic work needs to understand the basic principles of forensic science and investigations, including evidence *recognition*, *recording* and *recovery* techniques and requirements. Whilst it is reasonable to assume that all scientists operate in a professional environment, forensic science has its own framework of standards that need to be followed and complied with, even by sole practitioners. These will be considered in this chapter.

The primary goal of this chapter is to provide an introductory framework addressing the 'some formal training' remark of Spencer (2019) aimed at the botanist who accepts the challenge to become involved in this fascinating application of their primary discipline.

3.2 An Introduction to Crime Scene Considerations – Recognition, Recording and Recovery of Potential Evidentiary Material

This section draws heavily on Australian Standard (AS) 5388-1 Forensic Analysis. Part 1: Recognition, recording, recovery, transport and storage of material.

Primary responsibility for management of the scene of an alleged crime lies with the investigating agency, with CSEs playing a central role. All scenes present their own challenges, with outdoor scenes often being subject to inclement weather and/or presenting safety issues. Depending on the nature of the scene and seriousness of the alleged crime, a team may be assembled that includes experts from a number of disciplines. It is important that each person understands their role at the scene and works as a team – there is no place for individual egos.

In the early stages of an investigation, some specialists, such as a botanist, may not get called to a scene. Unfortunately, even when investigators and CSEs have thought about forensic botany, they may still be poorly informed as to what will be important to the forensic botanist, and a botanist will have to rely on what has been recorded and/or collected by a CSE.

In an ideal world, when a forensic botanist has been called to a scene, several steps will have been taken before the forensic botanist enters the scene. After an initial assessment of the scene, the CSE will have considered safety issues, identified possible routes into and out of the scene that may have been used by a perpetrator, identified a **common approach path** (CAP) and will have established an outer cordon or perimeter to secure the scene.

In choosing a CAP, the CSE will have considered a route that is least likely to have been used by an offender and one that minimises disturbance of the scene and any potential evidence. The focal point of the scene will usually be the location of a body, a grave or where the injured person or other piece of potential evidence was located. An inner cordon will then be established.

Between these cordons, a staging post may have been established for equipment and personnel working on the crime scene.

Police should have established a formal logbook to record anyone entering the scene. Where an individual is injured and alive, paramedics are likely to be amongst the first people to enter a scene. Preservation of life is the primary consideration whilst attempting as far as possible to preserve or 'freeze' the scene to minimise loss or contamination of potential evidence. In cases involving a deceased person, a forensic medical officer or a forensic pathologist will be amongst the first people to enter the scene.

In the real world, of course, the resources allocated to a scene will depend on the type of crime (volume or major crime), the nature of the scene, specific police protocols, legal considerations and the experience of the CSE.

In a complex scene, CSEs will develop an overall examination plan considering factors such as:

- what human resources will be needed
- whether or not specialist advice or assistance will be required
- the types of physical materials that may need to be collected and any specific requirements for collection
- the search approach and methodology appropriate for the scene
- review of the existing cordons and establishing a new CAP if necessary

The search approach and methodology will be determined by the nature of the scene. The five typical search approaches are the **zone**, **grid**, **line**, **wheel** or **ray** and the **spiral** methods. In many instances, a mix of these approaches will be appropriate. The important point to make is that the search should be thorough and systematic. More details about these methods can be found in any introductory textbook on crime scene examination (Fisher, 1992; Houck et al., 2018).

Prioritisation will normally be given to:

- collection or protection of any material at risk of deterioration or loss
- collection of materials that will enable safe access to any deceased remains or other critical areas of an incident scene along entry and exit paths
- any critical areas that may render the most relevant material, once processed, enable the removal of deceased remains
- assessment of any area or material that may provide an early indication of the identity of an offender(s)

Recognition of the potential for **contamination** at a crime scene is critical and has been magnified by the sensitivity of contemporary human DNA analysis, leading to much stricter protocols regarding the wearing of **personal**

Figure 3.1 Image showing the use of a tent to protect the scene.

(Creative Commons CCO 1.0 Universal Public Domain Dedication. Author: Zboralski.)

protective equipment (PPE). Hence, a specialist at a scene should expect to wear appropriate PPE, which would normally include a Tyvek suit, protective footwear and gloves and, where appropriate, a mask.

As crime scenes are often exposed to inclement weather, the use of crime scene tents is now common (see Figure 3.1). In suitable circumstances, protective footplates can be deployed.

One of the most important roles of the CSE is that of **evidence recognition**, put simply, deciding what is relevant and should be collected at the scene. This may involve subsampling at the scene and/or removal of larger items for subsequent laboratory examinations. Before anything is collected/recovered, the scene needs to be fully recorded. Although this sounds straightforward, it is not, however, just a technical issue involving various means of recording. You can only record what you identify as being potentially important. Evidence recognition is not just about individual items and has to capture the scene in a holistic way. In a sense, the skill of a good CSE is their visual acuity and ability to see patterns and understand what may be important in contributing to answering the five 'W's'.

At all stages in the evidence recovery phase, and before materials are collected, the scene should be **recorded** using a mix of notes, sketches, scene plans and still and video photography capturing who examined; what, when and how items were packaged and transported and then where they were stored.

Each organisation will have its own formal procedures and protocols for record keeping. Whatever the details of these procedures, records should:

- be contemporaneous – for practical reasons, it may not be possible to actually write notes as examinations are conducted but there should be no significant time lapse or reliance on memory. Notes may be made by an assistant, or they may be dictated
- uniquely identify the item and record what was collected, and where, by whom and when
- explain why items were collected – if a CSE cannot explain why an item has been collected, then why collect it
- be sufficiently detailed, comprehensive and intelligible that they can be understood by a peer

It is beyond the scope of this brief introduction to cover in detail the techniques used to record crime scenes. The overall aim is to produce a record that:

- accurately represents the physical facts
- establishes the precise location and relationships of objects at the scene
- attempts to relate the sequence of events at the scene
- creates a mental picture of the scene for those not present (investigators and eventually jurors)
- images should be a fair and accurate representation of the scene – these will usually include *overall*, *midrange* and *close-up* images

Some basic principles to follow with regard to evidence recovery are:

- as far as possible follow the *GIFT principle*, 'get it first time'
- prioritise collection from transient/fragile to more stable to avoid potential loss of evidence
- wear appropriate PPE both for safety and to minimise the potential for contamination
- consider how collecting one type of material may impact other types of potential evidence
- move from less intrusive to more intrusive techniques

Typical techniques employed for the recovery depend greatly on the nature of the material being collected. Where an item is visible, then it should be collected by **handpicking** with or without the use of forceps. For botanical materials, it may not be possible to collect larger specimens, and parts may have to be **cut or scraped**. At some scenes, the material may be collected by

sweeping, and for difficult-to-access areas, such as car boots, **vacuuming** may be used. Trace materials may be recovered with the use of tweezers. Where the trace material is essentially not visible to the naked eye, typically forensic scientists will use **tape lifts**. There are specific techniques used to recover pollen, and these are covered in Chapter 9.

Items recovered at a scene may include materials that may have evidential value, *recovered or questioned samples*. Forensic examinations often involve comparison of recovered 'evidence' materials with *known* and/or *reference* materials. Although the two terms known and referenced are often used as though they were interchangeable, in my view, they are not.

A *known sample* simply implies that its origin is known, such as samples taken from an individual, or, in the context of forensic botany, samples from a scene or other known source. The key challenge in taking known samples is to ensure that they are *representative* of variation, and there is *sufficient* material for subsequent examinations, potentially for access to both prosecution and defence parties.

The term *reference sample* is better reserved for reference collections, with the ultimate botanical reference collection being voucher specimens in herbaria.

As already highlighted, the potential problem, and often actual problem, for the use of botany at a scene is more often than not the lack of any specialist botanical knowledge on the part of investigators and CSEs, so that even with the best of intentions, botanical materials may not be adequately recorded and collected. If called to a scene at an early stage, one of the most important contributions a specialist can make is to work with the CSE and other specialists, to ensure important elements of the scene are captured, materials collected and then *packaged* to protect their integrity.

Packaging of materials will depend on the size and nature of the items being collected. Some general guidelines are:

- packaging should protect potential evidence against alteration, contamination or loss
- packaging should not inhibit or make more difficult subsequent examinations
- packaging should meet occupational health and safety requirements
- packaging should be suitable for transport and storage
- packaging should be *sealed* as soon as possible using a *tamper-proof evidence tape*
- packaging should be labelled with a unique identifier to establish the start of a chain of custody

Figure 3.2 shows examples of appropriate packaging appropriately sealed with a tamper-proof evidence tape.

Figure 3.2 Example of packaging with appropriate labelling and a tamper-proof seal.

For handpicked small items, the use of a *druggist's fold*, also known as a paper boat, is recommended. These can then be placed in snap-seal plastic bags – see Figure 3.3.

The use of plastic containers is in general to be avoided as it can be difficult to remove items for examination. A paper boat can simply be folded out and placed under a stereo microscope for initial examinations.

For larger plant samples, these can be placed in paper bags and/or cardboard boxes, provided they are properly sealed. Plastic bags should be avoided, as even visually dry plant material can quickly disintegrate during storage.

The *Herbarium Handbook* details quite specific criteria for the collection of plant specimens intended for herbaria (Bridson & Forman, 1999). Guidelines for the collection and preservation of specimens include what and how much to collect, depending on the nature of the plant material. Typically, a plant press is used to ensure specimens are flat as they dry. At least some of

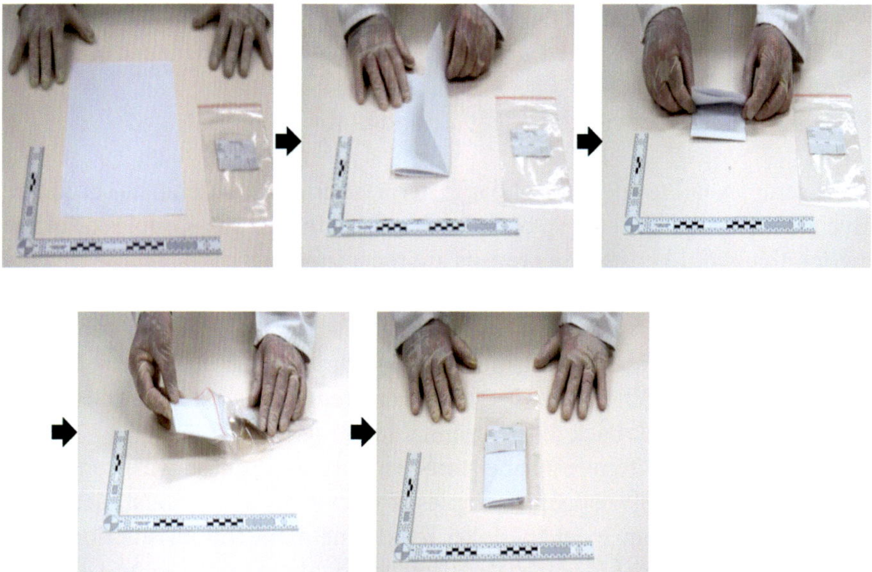

Figure 3.3 Creating a 'druggist's fold' for the collection of trace materials.

the information in the Herbarium Handbook can help guide appropriate procedures for forensic application.

3.3 Laboratory Examinations

The focus in Section 3.2 was on the role of a forensic botanist, either attending a scene or advising on what samples should be collected. This would, in most circumstances, be followed up by examinations carried out in a suitable laboratory. Samples may also be received directly into the laboratory for examination with no prior field involvement. AS 5388.2 – 2012 offers guidance on standard practices for the forensic examination and analysis of physical materials. A practitioner working in an accredited forensic laboratory would be expected to meet the requirements of this standard or similar standards and/or guidelines. The focus in this section is to offer guidance to those individuals who find themselves involved in an investigation requiring their specific botanical expertise, who may not be familiar with forensic requirements and practices.

All samples received for examination should meet basic standards for acceptance into the laboratory, including that:

- the material is packaged appropriately such that its integrity is not compromised
- the material is appropriate for the testing requested

- continuity of the material has not been compromised
- the examinations/analysis requested is likely to yield evidence of probative value
- any material provided is adequate and fit for purpose

If the samples fail to meet the above criteria, serious consideration should be given to rejecting the request, or, at least, requesting further samples or information that would ensure any results are meaningful.

Information provided by the instructing agency should be limited to that required to focus examinations, and that is *task-relevant*. All items submitted must have a unique identifier, which, as far as practical, should be used as the primary identifier. Where subsamples are created during examinations, there has to be a direct link back to the original identifier.

When received, items must be stored in a secure location with controlled access. The storage of suspected samples of illicit drug plants must meet storage requirements that meet the legal requirements of the jurisdiction. This would usually require formal legal approval to hold such samples and a secure safe or vault.

Questioned/recovered items should be kept separated from any reference/known comparison samples. When items are being examined, they should not be left unattended without appropriate measures to protect against contamination and to maintain the integrity of the items.

The United States–supported Scientific Working Group for Materials Analysis has produced *Trace Evidence Recovery Guidelines*, which state that 'the examination of questioned and known items for trace evidence must be conducted separately in different locations, at different times, **or both** (my emphasis), to prevent contamination.' Where items can only be examined in the same laboratory due to physical limitations of the facility, then separation in time is essential, as are rigorous anti-contamination procedures. Due to the potential for contextual bias, it is recommended that questioned/recovered materials be examined before known and/or reference materials, although, for practical purposes, this is not always possible.

Where physical material is destroyed during an examination, the reasons for such destruction should be recorded in case notes or covered by documented procedures. A guiding principle should be to ensure sufficient material is available for re-examination by an opposing expert.

In the real world of forensic organisations, cases will most often require a series of forensic examinations, tests and subsequent analysis. Many laboratories have instigated a *case triage and management* approach led by a *case manager* to determine the sequence of examinations based on a series of case management considerations. The role of a triage approach includes the following:

- The case manager develops a broad examination plan based on relevant information that gives appropriate weight to investigator questions, keeping an open mind to alternative theories that may emerge over time as more information becomes known
- Depending on the complexity of the case, decisions will be made as to what items are examined, prioritisation of items and the sequence of testing, considering possible forensic specialisations
- In many laboratories, specialists will work together in the initial examination of items to maximise the collection of materials/samples in the most efficient sequence
- In many cases, an early focus will be on providing information that identifies a person or persons who may have been involved
- Effective case management aims to minimise compromise between sample collection for identity purposes and collection that helps answer other forensic questions, such as event reconstruction

Once an examination is complete, items should be returned to their original packaging or, if necessary, into new packaging, sealed and returned to the designated storage area.

As covered in Section 3.2, case notes should be contemporaneous and should include the following:

- continuity details
- a description of any packaging
- datc and location of examination
- a description of the material received for examination and any general observations
- the condition of the material received
- any visual examinations and results, including any presumptive tests
- sampling techniques used and, if relevant, the location of sampling and the reasons why it was sampled
- any procedures used
- details of the person involved in the examination and testing

Examination notes may include a mixture of written notes, dictated notes, drawings/sketches, photography and video. Where electronic recording is used, it should comply with relevant digital evidence guidelines.

AS 5388-2-2012 provides general guidance on sampling procedures.

Forensic botany may use a range of analytical and comparative techniques and, often, a combination of techniques. All techniques used should be *validated* as described in *International Organization for Standardization* (ISO) 17025. In brief, validation is the process that ensures an instrument, technique or procedure is working according to specification. Validation involves having a plan that is *appropriate* and *proportionate*. Validation experiments typically examine the *reliability*, *reproducibility* and *robustness* of techniques, processes and methods.

A forensic laboratory seeking accreditation has to retain records of its validation procedures.

> **Reliability**: a reliable method is one that ensures the obtained results are *accurate* and correctly reflect the sample being tested.
>
> **Reproducibility**: a reproducible method means that the same or a very similar result is obtained each time a sample is tested.
>
> **Robustness**: a robust method is one that ensures acceptable results are obtained a high percentage of the time and few samples need to be repeated.

Section 14 of AS5388-2 deals with the identification of physical materials using *comparative examinations*. The single specific reference to forensic botany is the identification of cannabis by microscopy. Visible light microscopy, polarised light microscopy, scanning electron microscopy and macro-scale observations are included as techniques that may be applied for enhanced visualisation. DNA typing by gel or capillary electrophoresis is considered as a comparison aided by an analytical instrument.

In Chapters 4–10, the various types of physical materials encountered in forensic botany applications are considered, and the methods and approaches used in each category are covered in detail. Finally, how analytical results are transformed into meaningful information through interpretation is considered in Chapter 11.

3.4 Creating a Case File

Regardless of what support may or may not be offered by an instructing agency, often none, the specialist is responsible for creating their own case file.

A case file should be a full and transparent record of what has been done, when, where and by whom. All records are open to *disclosure* and *discovery* and should not be viewed as personal property. Figure 3.4 gives an example of a case file.

Typically, a case file is started with a request by an instructing agency, police or defence. The information provided should be limited to that which

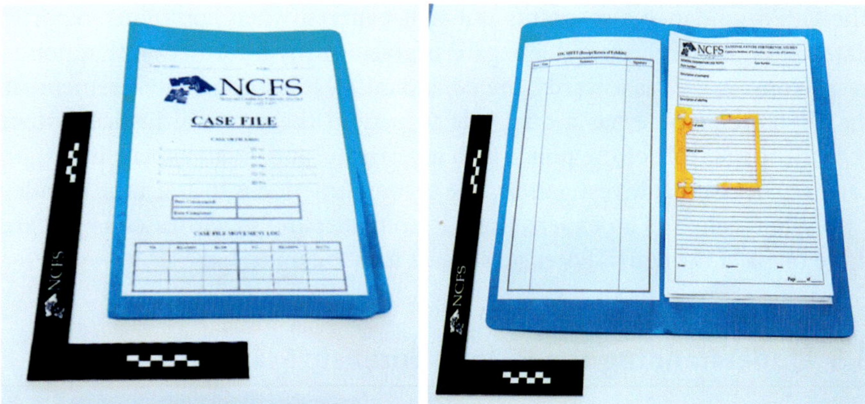

Figure 3.4 Examples of a case file.

is relevant to the task, and should not include information that is task-irrelevant and has the potential to create bias.

It is important to record any conversation with investigators or lawyers. A summary of what has been discussed should be recorded and formalised through an email to ensure there is an agreed and shared understanding by both parties to the conversation. This minimises the potential for any mis-communication leading to a later disagreement between parties as to what was agreed.

Typically, a case file will include:

- contemporaneous notes, signed and dated – notes should be permanent, with pencil only used for diagrams
- where electronic records are created, these should comply with relevant standards
- images, diagrams and any analytical results
- exhibit continuity information
- interim and final reports

The UK Crown Prosecution Service (CCPS) offers the following advice on the obligations of an expert under the title of **Retain-Record-Reveal**.

- Retain: you should retain everything, including physical, written and electronically captured material
- Record: the requirement for you to commence recording begins at the time you receive instructions and continues for the whole of the time you are involved
- Reveal: you are required to reveal everything you have recorded

The CPPS guidance also points out that experts, when compiling reports/statements, should ensure due regard is given to any information that points away from, as well as towards, the defendant(s). It is important to remember that forensic science has a dual role to play in the criminal justice system (CJS). The first role of the police is to investigate, and, as an expert, it is quite proper to assist that investigation. The second function is to assist the broader CJS. When called as an expert, there is no ownership in witnesses as you are there to assist the court. Forensic science must be impartial.

3.5 Understanding the Role of Forensic Standards

Many disciplines in science operate under a framework that includes professional practice and responsibilities, often under the umbrella of a professional body, and, at an organisational level, formal standards.

For forensic science, there are many professional societies around the world. Many of these groups have *codes of conduct* in their rules with which members are expected to comply. Only a few of these organisations, however, have professional membership categories with more stringent requirements for this level of membership.

The reality is that there are no forensic member organisations that are regulated in such a way that entry into the profession is controlled and ongoing membership is monitored to ensure members maintain their expertise and competency, including a requirement for continuous professional development. Whilst scientists should be encouraged to be members of relevant professional bodies, it is important to understand this does not equate to a qualification, nor does it guarantee competence.

In the absence of a truly professional framework (Robertson, 2011), forensic science is heavily reliant on organisations adopting relevant standards in their operations. The relevant international body setting such standards is the ISO. Established in 1947, ISO is an independent, nongovernmental, worldwide federation of national standards organisations or member bodies. International standards are prepared through ISO technical committees (TCs). Standards may also be developed at the level of recognised standards bodies at the country level – these bodies are collectively termed *Standard Development Organizations* (SDOs).

Doyle (2019) has written a comprehensive text on quality management in forensic science with a fuller background to how standards for forensic science have evolved, developed and been adopted around the world.

Below the level of a formal standard, a variety of bodies have developed guidelines for a broad range of forensic activities.

At the ISO level, two standards are used for different aspects of forensic science: ISO 17025 (General Requirements for the Competence of Testing and Calibration Laboratories) and ISO 17020 (Conformity Assessment – Requirements for the Operation of Various Bodies Performing Inspections). Neither was developed specifically for forensic purposes.

In most developed countries, it would be expected that laboratories conducting forensic testing would have adopted ISO 17025 as the relevant high-level standard. It has been argued that ISO 17020 is a better fit for organisations involved in delivering crime scene examinations; however, this standard has not been widely adopted (Doyle, 2019).

In Australia, with the exception of one police provider, all police organisations have adopted ISO 17025 to cover their crime scene and other forensic disciplines, as have all forensic laboratories. Worldwide, it is a more mixed story. In general, organisations providing crime scene and other classic police forensic disciplines have not adopted formal standards. This is not to say they do not have standard operating procedures, which may meet some, if not all, of the requirements of an ISO standard.

Given the considerable resources involved in meeting the requirements of ISO standards, organisations will typically then seek to have this formally recognised through third-party *accreditation*. Countries generally will have their own accreditation bodies. In Australia, the relevant body is the National Association of Testing Authorities (NATA). Established in 1947, it is the oldest accreditation agency in the world. In the United Kingdom, the relevant body is UK Accreditation Agency, and in the United States, ANSI/ASQ National Accreditation Board, amongst other organisations.

As ISO 17025 and ISO 17020 are generic standards, they are not always a great fit for the work of a forensic organisation, and typically, more detailed requirements have been developed to supplement these standards and make them more forensic-specific and relevant. In Australia, a series of forensic-specific standards, AS 5388, was developed to cover all aspects of forensic analysis with the view to removing the need for supplementary documentation. This standard is in four parts. AS 5388.1 covers recognition, recording, recovery, transport and storage of material; AS 5388.2 analysis and examination of material; AS 5388.3 interpretation and AS 5388.4 reporting.

A similar process was underway in Europe under the auspices of CEN (European Committee for Standardization) until ISO PC 272 assumed responsibility for developing an ISO standard based on AS 5388 (Wilson-Wilde, 2018). Currently, two standards have been developed by ISO based on AS 5388. These are ISO 21043-1 Forensic Sciences. Part 1: terms and definitions and ISO 21043-2 Forensic Sciences. Part 2: recognition, recording, collecting, transport and storage of items.

A comparison of ISO 21043-2 with AS 5388-1, both dealing with recognition, recording, collecting, transport and storage of items, shows that the ISO standard is a much shorter document; some might say concise, but certainly a less detailed document than the AS standards from which it was developed.

At a meeting in October 2024, the TC reached final consensus on the content of ISO 21043 parts 1, 3, 4 and 5 and approved these as standards. These standards have been released in 2026.

The title of ISO 21043-2 was also changed to 'General requirements for the forensic process and recovery and management of items' and will be subject to review to align this part with the new intent and other standards (Wilson-Wilde & Neville, 2024). When ISO 21043 is completed, there will be a truly international set of standards covering forensic science.

As stated above, below the level of ISO recognised standards, there are numerous guideline documents covering forensic disciplines. For example, under the umbrella of *Scientific Working Groups* (SWGs) in the United States, a series of guidelines has been developed, including the previously mentioned Trace Evidence Recovery Guidelines that have relevance to the recovery of botanical trace materials. Some of the SWG guidelines are in the process of being recognised as formal standards. There are also a number of 'best practice' guidelines, some published by the *European Network of Forensic Science Institutes* (ENSFI).

The only specific guideline for forensic botany has been published under the auspices of the *Society for Wildlife Forensic Science* in conjunction with members of the United Nations Office of Drugs and Crime (UNODC) Expert Group on Timber Identification. This document states that this standard and guideline are meant to augment and not replace current laboratory quality management systems and that consultation with relevant experts and authorities in applicable jurisdictions is recommended to ensure that all local requirements are also met (SWFS, 2016/2017). This guideline includes sections on DNA analysis and on wood anatomy standards, as well as some general guidance similar to the scope of ISO standards for laboratory analysis. Section 3.4 covers reference materials and collections. Notably, 3.4.5 states that

> the identity of a botanical reference specimen must be verified before the material is used in case work... verification of specimens is made with reference to other previously verified specimens at hand, to voucher specimens in an herbarium and/or xylarium.

Wiltshire (2015) has published protocols for forensic palynology covering the spectrum of activity from first encounter with the client to final court appearance.

Central to all standards is the requirement to have documented procedures in place to cover all aspects required by the standard. Whether formal accreditation is sought or not, compliance against the standard can be assessed through audit processes to test whether or not an individual or organisation is compliant. Essentially, a quality system is about '*saying what you do and doing what you say*' and being able to demonstrate this through controlled documentation. Although this basic principle sounds simple, it can be far from simple for even a sizeable organisation to develop the documentation required to meet relevant standards, seek accreditation and then manage the implementation and maintenance of an ongoing quality system.

3.6 Implications of Forensic Standards for a Forensic Botanist?

If you are part of an organisation that is accredited, then your individual responsibilities will fall under the organisation's documented procedures. If you are not part of a forensic organisation, you may still meet some of the requirements if your organisation has ISO 17025 accreditation. If, however, you are an individual coming from an organisation such as an academic department, a natural history museum, or a botanical garden, or you are a sole practitioner, then it is unlikely that you will have accreditation against a relevant standard.

ISO 17025 recognises that accredited laboratories may need to use an *external provider*. This does not remove the responsibility of the laboratory if it intends to incorporate results from an external provider into an accredited report. Specifically, Section 6.6.1 states that

the laboratory shall ensure that only suitable externally provided products and services that affect laboratory activities are used, when such products and services:

- are intended for incorporation into the laboratory's own activities
- are provide, in part or in full, directly to the customer by the laboratory, as received from the external provider
- are used to support the operation of the laboratory.

Section 6.6.2 deals with the requirements of the laboratory to ensure any externally provided service conforms to the relevant requirements of ISO 17025 before they are used or provided directly to the customer. Finally, Section 6.6.3 states that the laboratory is responsible for communicating its requirements to an external provider, including the competence and required qualifications of personnel. Section 7 further emphasises *that externally provided activities can occur when the laboratory does not have the resources or competence to perform the activities.*

As very few forensic organisations have in-house employed forensic botanists, ISO 17025 allows the use of external providers, BUT if the instruction comes through that forensic provider, they have quite onerous requirements to ensure that the provider meets the relevant requirements of ISO 17025.

The United Kingdom has taken a somewhat different approach in that they have regulated forensic science. In addition to the various ISO standards, the *Forensic Regulator* has issued a *Code of Practice*. This is a lengthy document running to 362 pages. Part E recognises that some disciplines will be **infrequently commissioned**. Evidence provided by an infrequently commissioned expert 'shall not be of a type which can ordinarily be obtained from a forensic science unit' and the provider 'shall not be a member of staff of a forensic unit ... nor represent themselves as a forensic practitioner'. An infrequently commissioned expert can also only be called once in a 12-month period.

Section 46.1.1 of the above code of practice recognises that it is impractical to require compliance with all of the provisions of this code or compliance with the means of demonstrating compliance (e.g. accreditation). However, the infrequently commissioned expert still has quite onerous requirements under Section 46.2. In addition to general obligations of expert witnesses and the requirements for the contents of reports, experts have:

- retention, recording, revelation and disclosure obligations
- to declare the use of any reference collections and databases that have been used
- a requirement to use validated methods or procedures based on sound scientific principles and methodology
- to demonstrate competence in using these methods or procedures and evaluating the result obtained objectively and impartially, and according to the established scientific and statistical methodology
- to consider the impact that confirmation/cognitive bias can have at different stages and use appropriate avoidance strategies
- to include a declaration required in the Criminal Practice Directions V 19B (34)

Forensic botany is not specifically mentioned in the code but falls into the category FSA-BIO 202, Section 92 of Non-human biological examination and analysis: plants, microbes and invertebrates, which are excluded from the code.

These activities are defined as 'examination and analysis to determine species and/or the potential source of plant, microbial, and/or invertebrate material'. Subactivities include morphological examination and macroscopic and microscopic examinations.

One of the most challenging requirements for a forensic botanist is demonstrating competence, as, in forensic practice, this is usually demonstrated by the use of *proficiency testing*. There are few, if any, meaningful commercially available botany proficiency tests. In the absence of external proficiency tests, laboratories, either singly or working as a group, have developed in-house proficiency tests. A recent example of this approach is the ENSFI

APST CE 2023 Botany collaborative exercise. This test aims to test routine sample identification of botany traces, including pollen, spores, seeds, whole plant and wood chips.

In Australia, NATA recognises the identification of Cannabis as an evidence type. In most jurisdictions in Australia, Cannabis identification is performed by police forensic officers or by forensic chemists who will normally be authorised analysts for that purpose under the relevant legislation. An exception in Australia is the State of Victoria, where the Police forensic laboratory has a dedicated botany section. Internal proficiency trials shared between laboratories are organised through the National Institute of Forensic Sciences Illicit Drug Specialist Advisory Group.

In summary, given the specific requirements of forensic science, it is perhaps not surprising that there is a paucity of individuals worldwide who undertake forensic botany as individual practitioners.

References

Bridson, D., & Forman, L. (eds.). 1999. *The Herbarium Handbook*, 3rd ed. Kew: Royal Botanic Gardens.

Doyle, S. 2019. *Quality Management in Forensic Science*. London: Academic Press.

Fisher, B.A.J. 1992. *Techniques of Crime Scene Investigation*, 5th ed. New York: Elsevier.

Houck, M.M., Crispino, F., & McAdam, T. 2018. *The Science of Crime Scenes*, 2nd ed. London: Academic Press.

Robertson, J. 2011. Forensic Science – A True Profession? *Australian Journal of Forensic Sciences*, 43, 105–122.

Robertson, J., Roux, C., & Wiggins, K.G. (eds.). 2018. *Forensic Examination of Fibres*, 3rd ed. Boca Raton: CRC Press.

Spencer, M. 2019. *Murder Most Florid*. London: Hardie Grant Pub.

Wilson-Wilde, L. 2018. The International Development of Forensic Science Standards - A Review. *Forensic Science International*, 288, 1–9.

Wilson-Wilde, L. & Neville, D. 2024. Standards Update. *The Forensic Exhibit*, 7, 12–13.

Wiltshire, P. 2015. Protocols for Forensic Palynology. *Palynology*, 40(1), 4–24. https://doi.org/10.1080/01916122.2015.1091138

Wiltshire, P. 2019. *Traces*. London: Bonnier Books UK.

Standards and Guidelines

ISO 17025:2018. General Requirements for the Competence of Testing and Calibration Laboratories.

AS 5388.1-2012. Forensic Analysis. Part 1: Recognition, Recording, Recovery, Transport and Storage of Material.

AS 5388.2-2012. Forensic Analysis. Part 2: Analysis and Examination of Material.

AS 5388.3-2012. Forensic Analysis. Part 3: Interpretation.

AS 5388.4-2012. Forensic Analysis. Part 4: Reporting.

European Network of Forensic Science Institutes (ENFSI). Best Practice Manuals and Forensic Guidelines. https://ensfi.eu/about-ensfi/structure/orking-groups/documents-page/documents/best-practice-manuals
European Network of Forensic Science Institutes (ENFSI). Best Practice Manual for Scene of Crime Examination. https://ensfi.en/wp-content/uploads/2022/021/BPM-SOC-01-v.2022014finalV2.PDF
Forensic Science Regulator: Code of Practice. Version 1. 2023. https://www.gov.uk/government/publications/forensic-science-activities-statutory-code-of-practice-version-2
ISO 21043-1. Forensic Sciences – Part 1: Terms and Definitions.
ISO 21043-2. Forensic Sciences – Part 2: Recognition, Recording, Collecting, Transport and Storage of Items.
Scientific Working Group on Materials Analysis (SWGMAT). 1998. Trace Evidence Recovery Guidelines. https://www.ojp.gov/ncjrs/virtuallibrary/abstracts/trace-evidnec-quality-assurance-guidelines-scietific-working-group
Society for Wildlife Forensic Science (SWFS). Standards and Guidelines for Forensic Botany Identification. https://www.wildlifeforensicscience.org/wp-content/uploads/2016/07/annex-18-standards-and-guidelines-for-forensic-botany-identification-2.pdf
(Note: there are no specific ENFSI best practice guidelines for forensic botany.)

Cannabis

4

4.1 Introduction

The focus of this chapter is the forensic examination and identification of the cannabis plant and its constituent parts. Some knowledge of other plants is required to ensure that a misidentification does not occur. A number of other plants, including poppies and plant products, such as khat and kava, which are either regulated or illegal in some countries, will be considered in Chapter 5.

Surprisingly, cannabis receives no more than a brief mention in a number of texts on forensic botany, with no detail on the structure of cannabis plants or the microscopic features of cannabis (Bock and Norris, 2016; Coyle, 2005; Hall and Byrd, 2012), with a default to the use of DNA testing. This chapter aims to fill this gap and provide practical advice and guidance to those tasked with identifying cannabis seized by law enforcement agencies.

4.2 History and Legal Status

The term 'cannabis' is used in a broad and generic sense to include all aspects of the plant and its uses. For example, in addition to the obvious focus on cannabis due to its drug properties, cannabis seeds are used as a source of oil and fibre-producing varieties of cannabis produce hemp fibre used in the production of a variety of products.

Iversen (2018) provides a brief history of cannabis from its earliest use by humans, dating back to 10,000 BC in Taiwan, for its fibre properties. Its herbal properties are first recorded in the 1st–2nd century AD, in an ancient Chinese herbal, *Pen-ts'ao*. The term *ma* was used, a pun for 'chaotic'. Ma mixed with wine, called *ma-yao*, was used for its pain-relieving properties. Hence, its stupefying and hallucinogenic properties were well understood (Iversen, 2018).

Fast forward to the 19th century, from 'The Indian Hemp Drugs Commission Report of 1894', it is clear that cannabis and its products had become incorporated into Indian life and culture for its psychoactive properties (Anon, 1894). This seems to have been the origin of the use of cannabis in Western medicine, moving first to Britain and then the rest of Europe and to the United States. Cannabis was included in Pharmacopoeias of the time and in textbooks of pharmacognosy (the study of powdered vegetable drugs), for

DOI: 10.4324/9781003308508-4

example, in classic texts such as Greenish and Collin (1904) and Wallis (1960). 'Tincture of Cannabis' was only removed from the British Pharmacopoeia in the early 1970s, as, due to the growing recreational use, cannabis was made illegal in the UK Misuse of Drugs Act 1971 (Iversen, 2018). Even in 1937, there were no less than 28 different pharmaceutical preparations in the United States, which included cannabis in some form.

However, 1937 marked a watershed for cannabis with the passing of the Marijuana Tax Act in the United States. Patton (2020) explains how this act came into being and how it effectively ended the use of cannabis for medicinal purposes and medical research in the United States. To understand the origins of this act, one has to backtrack to 1934, and a case called *Sonzinsky v United States*. As Patton describes, in response to gun violence (some things never change in the United States) associated with alcohol prohibition, Congress passed the National Firearms Act, which required gun sellers to pay a $200 tax for every gun sold. Sonzinsky was convicted of violating this act and appealed his conviction. The Supreme Court upheld the law as a *prohibitive tax*, that is, a tax primarily designed to prohibit conduct and not to raise revenue. This ruling endorsed the legality of prohibitive taxes. Hence, the real purpose of the Marijuana Tax Act of 1937 was not to raise taxes but was aimed at prohibiting the use of cannabis using Federal legislation. This act remained in effect until 1969, when it was declared unconstitutional on Fifth Amendment grounds (self-incrimination) (Patton, 2020).

At an international level, cannabis and cannabis resin are included in Schedules 1 and IV of the 1961 United Nations' *Single Convention on Narcotic Drugs*. Schedule 1 includes substances that are considered highly addictive and liable to substance use disorders, whilst Schedule IV further qualifies some drugs from Schedule 1 as not only being highly addictive but also highly liable to abuse and rarely used in medical practice (Anon, 1961). Following a scientific assessment in 2020 by the World Health Organisation, cannabis and cannabis resin were deleted from Schedule IV of the 1961 Single Convention, but they remain in Schedule 1.

The last decade or so has seen dramatic changes in attitudes to cannabis and its legal status. Whilst cannabis remains illegal in many countries, an increasing number of countries have moved to decriminalise personal use of cannabis, allowing small amounts of cannabis product and/or plants to be possessed for personal use. Canada has been the most radical in its relaxation of laws, pertaining to the possession and use of cannabis, passing laws in 2018 that legalised cannabis in Canada for personal use and opening up an industry to grow commercial quantities for the production of medicinal cannabis. Following California in 2014, in the United States, more than 30 States permit the provision of medical cannabis. This trend has been mirrored across countries and continents with a concomitant growth in research on naturally occurring cannabinoids. Iversen (2018) reports that from around 200 scientific papers in 2000, this has grown to over 16,000 citations for cannabis in PubMed statistics in 2017.

4.3 Classification, Nomenclature Issues and the Importance of Legal Definitions

Cannabis has a long and ancient history with a subculture that uses a nomenclature that often does not conform strictly to accurate use of botanical terminology. For example, cannabis has a multitude of common names such as ganja, Bhang, hashish, kennab and marijuana, the term *marijuana* being a Spanish Mexican word originally used to describe tobacco but now used to describe cannabis.

Cannabis is defined in the Single Convention on Narcotic Drugs 1961 (Anon, 1961) as 'the flowering or fruiting tops of the cannabis plant (excluding the seeds and leaves when not accompanied by the tops) from which the resin has not been extracted, by whatever name may be designated'.

'Cannabis plant' means any plant of the genus *Cannabis*.

'Cannabis resin' means the separated resin, whether crude or purified, obtained from the cannabis plant.

Provided relevant jurisdictional legislation is based on the above definitions and refers only to the genus *Cannabis*, there is no valid basis on which an identification that meets the requirements of the Single Convention should be capable of being challenged in a legal setting or a court of law.

Where an issue has historically arisen is when legislation has been more specific and named *Cannabis sativa*. For example, in US Federal law, under Title 21, US Code, Section 802, *C. sativa* is defined as being 'marihuana'. Title 21, US Code, Section 209, recognises a separate species of *Cannabis indica* as being a poison (Lowry and Garriott, 1975). Notwithstanding, Lowry and Garriott (1975) cite seven cases in which courts in the United States have ruled that *C. sativa* refers to all forms of cannabis plant material on the basis of *legislative intent*. Some US States have redefined 'marijuana' to overcome any debate as to the number of species by including all species and variants. However, in the 1970s, US courts were faced with a semantic problem and relied largely on expert witness testimony. Specifically, a defence was raised in 1971 based on testimony by several botanists that there was more than one species of cannabis (Small, 2017).

So, is cannabis a single species or are there several species of cannabis? The primary advocates for there being several species were Emboden (1974) and Schultes et al. (1974), who argued that there were three species, *C. sativa*, *C. indica* and *C. ruderalis*. Lowry and Garriott (1975) concluded that 'whether the genus *Cannabis* is monotypic or polytypic is as yet uncertain'. Certainly, anyone who has examined cannabis grown outdoors, or even greenhouse-grown cannabis, will be well aware that cannabis plants can be highly variable in appearance and stature, even when grown in the same environmental conditions.

Small and Cronquist (1976) provide an in-depth consideration of the taxonomy of cannabis. They cite as evidence for a single species that all cannabis is diploid ($n = 10$) and that there appear to be no barriers to successful hybridisation within the genus, concluding that the variation seen today is in large part due to the influence of man. Indeed, there is no truly wild progenitor for cannabis. They accept that there are two widespread classes of plant, a group of generally northern plants with relatively limited intoxicant potential, influenced by selection for fibre and oil agronomic qualities, and a group of generally southern plants with considerable intoxicant potential, influenced by selection for the latter quality. These two groups are then treated, respectively, as subspecies *sativa* and *indica*, with the only species of the genus *Cannabis* being *C. sativa* L.

Small (2017) offers the following identification key that goes further and recognises varieties below the level of subspecies based on the level of tetrahydrocannabinol (THC) and wild or domesticated traits.

- *C. sativa* subsp. *sativa* var. *sativa* (low THC with domestication traits)
- *C. sativa* subsp. *sativa* var. *spontanea* (low THC, wild-type traits)
- *C. sativa* subsp. *indica* var. *indica* (high THC, domesticated traits)
- *C. sativa* subsp. *indica* var. *kafiristanica* (high THC, wild-type traits)

In a recent review of cannabis systematics, at the level of family, genus and species, McPartland (2018) offered further support for the Small classification, confirming that this has not been replaced.

Contrary to what is sometimes quoted in the literature, *Cannabis* is not the sole genus in the *Cannabaceae*. The family *Cannabaceae* currently consists of *Cannabis* and *Humulus* (hops) and eight genera formally in the *Celtidaceae* (Sytsma et al., 2002).

It should be noted that there are a number of publications, such as 'The Cannabis Encyclopaedia' and numerous magazines, such as 'High Times' or 'Sinsemilla Tips', that in general still talk about *C. sativa*, *C. indica* and *C. ruderalis* as separate species (Cervantes, 2015). Information in this type of publication should be treated with caution, as it is unlikely to meet what would be expected of scientific research. McPartland (2018) refers to the use of '*indica*' and '*sativa*' in this type of literature as *vernacular taxonomy*, and that ubiquitous interbreeding and hybridisation have rendered their distinctions almost meaningless.

Finally, over the past few decades, there has been a move to indoor cultivation of cannabis, often using some form of soil-free growing medium or true hydroponics cultivation. There are a number of suppliers of cannabis seeds, to

Figure 4.1 Example of a commercial catalogue.

name but two, 'Dutch Passion 'and Sensi Seed Bank', both based in the Netherlands. The latter produces a glossy catalogue where they describe an image and a description of the 'variety' (see Figure 4.1). For example, 'Shiva shanti' is described as being 'Afghani with penetrating indica aroma', or 'Super Skunk' is described as being mostly indica and a cross of the best Skunks of Afghani origin. Another 'variety' claims to be of 'ruderalis' origin. Whatever the truth of these claims and descriptions, what it does illustrate is the emergence of the so-called *skunk* varieties, which are largely used for indoor hydroponic cultivation using cloned cuttings. This move has certainly changed the nature of cannabis material presented to law enforcement and forensic scientists.

4.4 Botanical Description of Cannabis

4.4.1 Morphology

Cannabis is possibly the most photographed plant in the world, with its leaf an iconic structure. The physical height and shape of the plant depend on whether it is the 'drug' subspecies or the 'fibre' subspecies and on the growing conditions. For example, the modern trend to hydroponics and plants grown from cuttings produces a quite different plant from large outdoor grown cannabis. Furthermore, plants grown for fibres are produced in an agricultural setting to produce tall, thin plants with minimal branching (see Figure 4.2). Hence, whilst a general description of the overall morphology can be given, this needs to be interpreted in the context of the aforementioned factors.

Cannabis is an annual, although plants can survive beyond one year's growth. The shoot or stem architecture depends to some extent on how the plants are grown. As mentioned above, fibre cannabis varieties are grown in a crowded, high-density planting pattern that suppresses branching. A typical outdoor-grown cannabis 'drug' plant, however, produces a highly branched structure with a robust central stem with four ridges lengthwise (see Figure 4.3). Outdoor-grown cannabis can reach a height of several metres with a similar girth.

Small (2017) discusses shoot architecture and how this can be manipulated through growing techniques to produce plants suitable for their intended end use for fibre, oil seeds or drug varieties. For example, for an indoor hydroponic setting, growers may remove tops to limit the height of the plants and

Figure 4.2 Field-grown crop of hemp.

(Published under Creative Commons Attribution 4.0 International. Creator: Gary Halvorson, Oregon State Archives. Wikimedia Commons. File: Case Road-DPLA-f5962afae6112f87be8c3437c72282d5.jpg.)

Figure 4.3 Example of a large outdoor-grown cannabis plant.

promote branching. Other methods, such as 'Sea of Green' and 'Screen of Green', developed in the Netherlands, aim to grow cannabis where the 'buds' (female inflorescences) mature and can be harvested whilst the plants are short and well-branched. According to Clarke and Merlin (2013), breeders are also developing early maturing and high-yielding varieties that are short and compact for indoor grow rooms to avoid detection outdoors.

The first leaves are unifoliolate and are located at the branch apices. The larger leaves are multifoliolate or have several leaflets. Lower leaves are opposite, becoming alternate in upper leaves. Cannabis leaves are palmately compound with from three to 13 leaflets, usually between seven and nine. There may be some difference in the leaves of sativa and indica subspecies, with indica leaves being somewhat larger and darker green in colour; however, this may be simply due to indica-type plants being grown in a crowded and therefore shaded environment (Small, 2017). Leaflets are narrowly lanceolate with coarsely serrated to toothed margins. A typical mature cannabis leaf can be seen in Figure 4.4.

When cannabis is grown outdoors in a natural environment, it will produce vegetative growth for several months before producing flowers. Flowering is under photoperiodic control with shortening days (actually, it is

Figure 4.4 Example of a classical cannabis leaf.

(Author: Dohduhdah. Public domain. Wikimedia Commons. File: Potleaf.jpg.)

the length of night or darkness that is critical), inducing flowering. Floral primordia are initiated mid-summer and grow from the axis of leaves, first near the tip of the plant and then along branches.

Cannabis in nature is mostly *dioecious*, meaning that individual plants will either have female flowers (*pistillate*) or male flowers (*staminate*). Before flowers develop, it is not possible to tell if a plant will produce female or male flowers. Cannabis can produce plants that are *monoecious*, having both female and male flowers on the same plant. Cultivars produced for hemp have been bred for this characteristic. Monoecious plants are rarely seen in drug varieties of cannabis.

Sexual development in cannabis is influenced by a number of factors, including temperature and the light regime or by treatment with plant hormones (Small, 2017). As female plants have far higher levels of THC than male plants (Clarke and Merlin, 2013), growers of drug varieties select for female plants by physically removing male plants before they release pollen.

Flowers in both female and male plants do not exist as single flowers but rather are found in groups or *inflorescences*. Technically, the female inflorescence is a *spicate cyme*. Visually, these inflorescences are tightly packed, dense structures. A typical female inflorescence is shown in Figure 4.5a and a harvested inflorescence or bud in Figure 4.5b.

Figure 4.5 Mature female flowering cannabis. (a) Whole inflorescence. (b) Harvested cannabis 'bud'.

(a. Published under Creative Commons Attribution – Share Alike 3.0 unported license. b. Public domain. Author: Evan-Amos.)

Figure 4.6 shows cannabis inflorescences and flowers in a diagrammatic format.

The cannabis flower is a simple structure and consists of a single ovary with a single ovule surrounded by a thin membranous coat called a *perigone*. The perigone is all that remains of the perianth or sepals and petals. Emerging from the flower are two feathery *stigmas*. These are covered in short hair-like structures called *papillae*. As the female flower matures, these stigmas turn a brown colour, as can be seen in Figure 4.5a.

Each flower produces a single seed, technically called an *achene*. Cannabis seeds vary in size and have a distinctive reticulate patterned surface (see Figure 4.7).

The individual flowers are held in the axils of two *bracteoles* subtending a bract.

All parts of the plant bear numerous hairs or *trichomes*, these being denser on the bracts and bracteoles of the female flower. These trichomes will be considered in detail in Section 4.4.2.

By comparison, male flowers are loose, axillary and relatively diffuse. Technically, the male inflorescence is called a *cymose panicle*. A typical male inflorescence is shown in Figure 4.8.

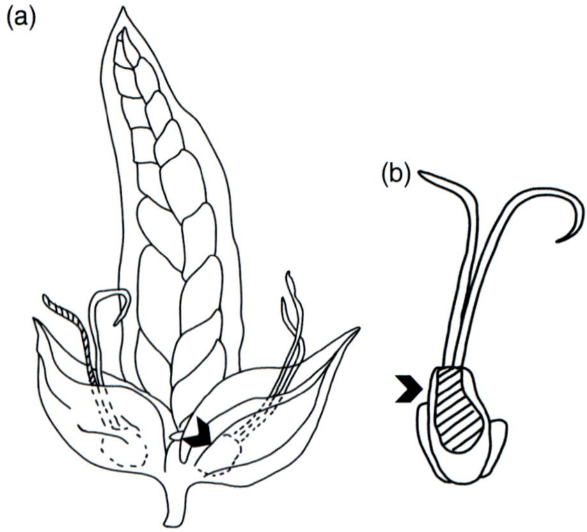

Figure 4.6 Schematic of the component parts of a female flowering cannabis inflorescence. (a) Bract with a pair of bracteoles, each having a flower in its axil. (b) Pistillate flower with two feathery stigmata.

(Original drawing Dino Todorovic.)

Figure 4.7 Examples of cannabis seeds showing variation in size.

(Author: Topi Pigula.)

Figure 4.8 Cannabis male flowers.

Individual male flowers have a short stalk or *pedicel*, with five greenish or white *tepals* (derived from a combination of sepals and petals) and five stamens with five flaccid filaments. At maturity, a single stamen can produce about 350,000 pollen grains (Faegri et al., 1989). A single male plant may have several hundred flowers! Flowering male cannabis plants are taller than female plants and also come into flower from one to three weeks before female plants have receptive stigmas. This promotes cross-pollination and fertilisation in nature. Cannabis is primarily wind pollinated, although insects may play a role in transferring pollen. Although the vast majority of cannabis pollen will likely fall quite close to its origin, there have been reports of cannabis pollen travelling considerable distances (Cabezudo et al., 1997).

4.4.2 Anatomy and Microscopic Features

The following draws on the descriptions from Greenish and Collin (1904) and from Jackson and Snowdon (1968). Figure 4.9 depicts the microscopic features of cannabis.

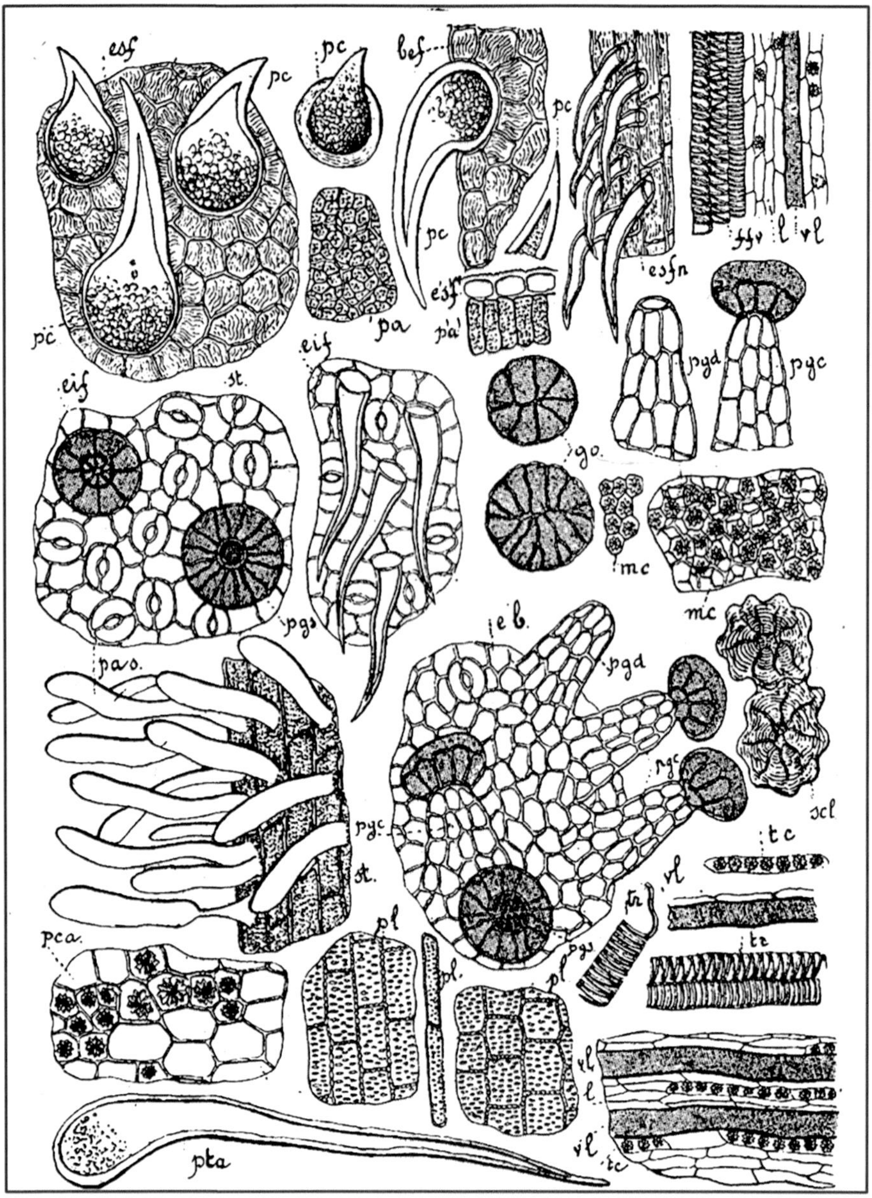

Figure 4.9 Powdered Indian Hemp – Microscopic detail for powdered female flowering cannabis.

(From Greenish and Collin (1904), page 101, figure 53.)

(*Continued*)

bef, Epidermis from margin of bract.
eb, Epidermis of bracteole.
eif, Lower epidermis of bract.
esf, Upper epidermis.
e's'f', Upper epidermis of bract, with palisade, **p'a'**, in profile.
esfn, Neural epidermis of bract.
ffv, Portion of fibro-vascular bundle from bract.
go, Oil-gland.
l, Bast, rich in calcium oxalate.
mc, Mesophyll of bract, with crystals.
pa, p'a', Palisade.
pas, Papillæ of stigma.
pc, Cystolithic hairs.
pca, Cortical parenchyma of tem with crystals.
pgc, Stalked glands.
pgd, Stalk deprived of gland.
pgs, Sessile glands.
pta, Simple hair from stem.
pl, Lignified parenchyma of stem.
st, Stigma with papillæ,
tc, Crystal tubes,
tr, Vessels, etc.
vl, Laticiferous vessels.

x240

Figure 4.9 (Continued)

The identification of cannabis relies on the presence of large glandular trichomes and, to a lesser extent, other types of trichomes. These trichomes can be seen when suspected cannabis plant material is examined using a stereo microscope with epi-illumination (see Figure 4.10).

4.4.2.1 Trichomes

There are two broad categories of trichomes, *covering trichomes* and *glandular trichomes*.

Covering trichomes include unicellular trichomes, which may or may not be cystolithic.

Cystoliths are deposits of calcium carbonate in the cells of the epidermis that have given rise to the trichome. The cystolithic trichomes are either very short with an enlarged base or large and elongated, and have a distinct *warty wall.*

The non-cystolithic trichomes are either short with a slightly enlarged base and an abruptly tapering apex or much longer with a narrow base and a gradually tapering apex.

Glandular trichomes are very distinct and characteristic.

Less common are small glandular trichomes with a uniseriate stalk composed of one or two cells and a spherical head with from one to four or occasionally eight cells.

The most abundant glandular trichomes are composed of a multicellular, multiseriate stalk with a multicellular head containing from eight to 12 radiating cells. The stalk is cylindrical, three to five cells in diameter, and the

Figure 4.10 Glandular trichomes on a cannabis bracteole as seen under stereo microscopy.

(Author: Psychonaught.)

constituent cells are elongated with slightly thickened walls. The stalk is constricted at the apex so that the head may be attached by only two cells. The cells of the head portion are thin-walled and have a common cuticle, which is raised to form a bladder-like covering. In dry powdered cannabis, the head portion is frequently found detached.

Covering trichomes are found on leaves, bracts and bracteoles and are fairly abundant. Large glandular trichomes are most abundant on the outer surface of bracteoles, less so on the inner surface of bracteoles or on the epidermal surfaces of bracts and even less so on leaf surfaces.

Most of the other features described can only be visualised if the plant material has been cleared, placed in a suitable mountant and examined at higher magnifications using transmitted light microscopy (TLM).

The characteristic microscopic features of cannabis are as follows.

4.4.2.2 Bracts

The upper epidermis is composed of straight-walled polygonal cells with a faintly striated cuticle. Short, conical unicellular cystolythic trichomes with an enlarged base, few small and large glandular trichomes. Underlying palisade cells with occasional cluster crystals of calcium oxalate.

The lower epidermis is composed of smaller cells than the upper epidermis with distinct sinuous walls. Numerous anomocytic stomata. Frequent small glandular trichomes.

4.4.2.3 *Bracteoles*

The upper epidermis is composed of polygonal cells with unevenly thickened and beaded walls. Calcium oxalate crystals are present in the underlying mesophyll cells. Numerous large glandular trichomes are present.

The lower epidermis is composed of smaller cells than the upper epidermis, with walls that are more sinuous and only slightly thickened and beaded. Anomocytic stomata are present. Numerous short, conical unicellular covering trichomes with a wide base and tapering to a point at the apex. Cluster crystals of calcium oxalate are present in the underlying mesophyll cells.

4.4.2.4 *Stigmas*

Fragments of stigmas are fairly abundant. In mature flowers, they are orange to reddish brown in colour. The surface of stigmas is covered in short, thin-walled cells called *papillae.*

Jackson and Snowdon (1968) also describe some other features, including features of stem fragments.

The key microscopic features that enable the bract and bracteole to be differentiated are summarised in Table 4.1.

4.4.2.5 *Seeds*

Technically, the seeds of cannabis are fruits, called *achenes*. For convenience, however, the fruits of cannabis are referred to as seeds. An achene is a small, dry, indehiscent fruit with a single locule and a single seed (ovule), attached to the ovary wall at a single point.

According to Small (2017), the seeds of domesticated cannabis are larger than the seeds of wild plants. They are ellipsoidal to oval in shape, brown in colour, with a characteristic reticulate or mosaic (Tortoise shell) pattern on

Table 4.1 Microscopic Features of Bracts and Bracteoles

	Upper Epidermis	Lower Epidermis
Bract	Straight-walled polygonal cells, faint striations	Sinuous walls with smaller cells than on the upper surface
	Short conical unicellular trichomes often with well-defined deposits of calcium oxalate	Numerous anomocytic stomata
	A small number of glandular trichomes	Quite frequent glandular trichomes
Bracteole	Beaded-walled polygonal cells	Slightly beaded cell walls than on the upper epidermis
	Numerous glandular and covering trichomes	Numerous anomocytic stomata
	Unicellular and multicellular glandular trichomes	Numerous short, conical unicellular trichomes

the surface. Typically, seeds may be 2.5–3.5 mm wide and 4–8 mm long – see Figure 4.7. At the point where the seed is attached to the plant and becomes unattached, a small scar is left at the base of the seed.

4.5 Forensic Considerations

4.5.1 Who Identifies Cannabis – The Role of the Forensic Botanist?

The role of a forensic botanist with respect to cannabis will vary depending on the jurisdiction and their approach to dealing with cannabis. The police, or other relevant law enforcement or investigative agency, will normally have primary responsibility for seizures of cannabis. This may involve small (personal use) to much larger dealer or commercial quantities of cannabis plant material and/or plants. The latter may be from indoor and outdoor crops of varying sizes. Cannabis may also be present in the form of cannabis resin and, rarely today, as cannabis oil.

For simple and small seizures, these will be examined either 'in-house' or by a specialist agency. 'In-house' examinations may be conducted by police forensic officers who have been trained in cannabis identification. In some jurisdictions, cannabis identification is conducted by forensic chemists who have also received training in cannabis identification.

To the best of my knowledge, there are very few forensic organisations that have specific forensic botany sections – one such organisation is the Victoria Police in Australia. In some jurisdictions, cannabis identification will be conducted by botanists in botanical institutions such as Kew Gardens in the United Kingdom and other botanical institutes and gardens. Such institutes may, however, be reluctant to become involved in legal matters that may end up in court.

Wherever identification work is conducted, it may be mandated that the organisation holds appropriate accreditation under ISO 17025. Regardless of formal accreditation, the work has to meet the normal standards for forensic work, which have been considered in Chapter 3.

In many jurisdictions, there will be a legal requirement for analysts to be formally authorised to report the results of cannabis identification. Reports will also usually be in the form of a certificate.

4.5.2 Crime Scene Considerations

Larger seizures and/or crops are *crime scenes*, and it is advisable for qualified crime scene examiners to be involved, as there may be many other aspects with potential forensic significance in addition to the examination of 'alleged' cannabis.

In Australia, large outdoor-grown crops of cannabis were once the norm. It was also quite common to find cannabis being grown in market garden environments in greenhouses (see Figure 4.11 for examples).

At least in Australia, we are still seeing large greenhouse cultivations. In a recent seizure in Queensland, police located a 1,970-metre-long greenhouse containing almost 3,000 cannabis plants, more than 1,500 seedlings in hydroponic growth rooms and 33 kg of dried cannabis.

Figure 4.11 Examples of cannabis crops. (a) Outdoor setting. (b) Greenhouse setting.

It is also still quite common to have small to medium cultivations in remote areas and dense bushland. These are often left unattended for much of the time, with growers setting up booby traps, which can include rigged firearms and razor blades in the plant stems that can cause significant injuries to the unaware. Hence, forensic specialists should not be in the first responder team as these scenes may pose significant danger.

With the increasing prevalence of indoor-grown cannabis, crime scenes will have multiple cables, often illegal access to electricity and in an environment with water – never a good mix (see Figure 4.12).

In my previous employment in South Australia, I was a member of a small team who attended major crops, ran a twice weekly across the counter identification service and travelled throughout the state and conducted an identification service to minimise the need to transport cannabis seizures back to Adelaide. One of the questions that was frequently asked by the Judiciary was to estimate the likely yield of a cannabis crop, assuming it reached maturity. Our group compiled data from crops we attended, but sadly, these data were never published.

Knight et al. (2010) report a doubling of indoor cannabis crops in New Zealand in one year, from 7% in 2006 to 15% in 2007. The main purpose of the Knight study was to estimate cannabis yields from indoor cultivations. A conservative estimate was given as an average of 881 g of dry female

Figure 4.12 Example of a typical indoor hydroponic growing set-up.

flowering head per plant, although this did vary depending on the phenotypic appearance of individual plants.

It is out of scope in this chapter to go into any detail about scene examination. All normal protocols should be followed with a holistic approach to gathering forensic evidence. Some aspects of crime scene work have been covered in Chapter 3. At some point, crime scene examinations will involve recording the plant material and making decisions about whether to identify cannabis *in situ* or to sample and transport samples back to a laboratory for identification. These decisions will be determined by a mixture of practical considerations, such as how to transport samples back to a laboratory, legal requirements around destruction on site, and any requirement for 'best evidence'.

Regardless of whether or not an expert attends at the scene, it is critical that the scene is recorded, including overall, mid-range and close-up images in addition to professional-quality video and/or other digital recording that meets all of the quality requirements for digital evidence. It is sadly still not uncommon at a later stage for a forensic specialist to be shown images and asked to comment on some evidential aspect when the images have not captured adequate detail. Where the decision is made to take samples at the scene, these need to be representative of any variation in plants.

Plant material should never be placed in plastic bags, as even partially dried material can rapidly turn into unidentifiable 'mush'. Plant material should be placed in paper evidence bags appropriately sealed and labelled such that it is suitable for transport and, at least, short-term storage. As far as is practical, examination of seized plant material should be expedited to avoid issues with storage and to facilitate subsequent destruction.

4.5.3 What Constitutes an 'Identification'?

As previously discussed, cannabis remains in the 1961 Single Convention on Narcotic Drugs under which it is defined as 'the flowering or fruiting tops of the cannabis plant (excluding the seeds and leaves when not accompanied by the flowering tops)'. In addition, it stipulates that 'the resin has not been extracted'. Leaves of cannabis are not listed in the schedules and are therefore not a 'drug' in the meaning of the Single Convention. Notwithstanding the status of cannabis in the Single Convention, cannabis may be defined in jurisdictional legislation in different terms, and an analyst will need to be clear as to how cannabis is defined and, hence, the requirements with regard to what may need to be proven under their legislation.

Relevant to what constitutes an identification are the United Nations Recommended Methods for Testing Cannabis (Anon, 1987), recognising that these are only guidelines and not mandatory requirements. These guidelines state that 'because cannabis and cannabis resin are plant material it is mandatory that the analyst includes macroscopic and/or microscopic examination of the material as part of the testing protocol'. Furthermore, it is suggested that in order to establish the identity of any controlled substance, the criteria should include 'at least two independent analytical parameters' and 'where possible, three entirely different analytical techniques should be used' (Anon, 1987). In addition to microscopic examination, these techniques may include colour tests and any two chromatographic techniques such as thin-layer chromatography, gas-liquid chromatography (GLC) or high-pressure liquid chromatography.

In many laboratories, mass spectrometry (MS) has been added to chromatography to identify the presence of cannabinoids and, specifically, the presence of THC.

In practice, plant material suspected of being cannabis may be identified based solely on its botanical features, with the UN guidelines commenting that 'the very abundant trichomes which are present on the surface of the fruiting and flowering tops of cannabis being the most characteristic features' (Anon 1987), with evidence for this conclusion drawing on Jackson and Snowdon (1968).

In the field (indoor or outdoor crop), identification of cannabis relies on observing morphological characteristics, including the highly definitive leaf shape for non-flowering juvenile plants, the characteristic inflorescences found in female flowering plants and, when present, the equally definitive appearance of male flowering plants. It would be unusual for a botanist attending a crop to have a field microscope with which they would be able to recognise the characteristic glandular trichomes (multicellular, multiseriate stem with a multicellular glandular head). However, in many instances, a field-based identification would be subsequently confirmed on samples taken and examined in a laboratory setting.

In ideal circumstances, identification should involve direct comparison with a voucher herbarium specimen. This may take place where the expert works in, or has access to, a herbarium collection that includes *C. sativa*. A minimum requirement for organisations that provide a cannabis identification service is to have representative samples of cannabis plant material and seeds, which have been identified by a recognised expert as reference materials. Such samples can then form the basis for training programmes and internal proficiency tests.

In the laboratory, samples, suspected of being cannabis, will be examined using LPM (low power microscopy) with epi-illumination and magnifications of up to ×40. At this level of examination, the major features seen will be the glandular and covering trichomes and, in female inflorescences, the two feathery stigmata, which in mature flowers will be of a light brown colour. Seeds may also be present.

Where plant material lacks morphological features, such as the female inflorescence or 'bud', as is seen with lower quality seizures, the examiner may be presented with essentially fragmented plant material. The minimum requirement for a botanical identification is the presence of large multicellular glandular trichomes. It is not sufficient to only observe covering trichomes, although, of course, these should also be seen in cannabis leaves and bracts.

Many publications refer to the presence of cystolythic covering trichomes as an identifying feature of cannabis. These trichomes are characterised by having deposits of calcium carbonate at the base of the trichome. In cannabis, broad-based cystolythic trichomes are seen with both smooth walls and warty walls (Jackson and Snowdon, 1968). It is important to understand that these deposits can only be seen when plant material is cleared of internal pigment with the use of TLM. Hence, where only LPM is used, it is more accurate just to call such trichomes covering trichomes with a broad base and either smooth or warty walls.

Of course, both covering and glandular trichomes are widely seen in plant material. Thornton and Nakamura (1972) surveyed some 600 dicotyledonous plant species known to have cystolythic trichomes and concluded that cystolythic trichomes cannot be used as a sole criterion to identify cannabis. One species not examined by Thornton and Nakamura was the mulberry, *Morus nigra*, from the family Moraceae. According to Maloney and Thornton (1982), mulberry shares with cannabis the arrangement of cystolythic trichomes only on the upper surface of the leaf and covering trichomes only on the underside of the leaf, concluding, however, that the microscopic appearance of the cystolithic trichomes is sufficiently similar to those found in cannabis to warrant careful scrutiny. Maloney and Thornton (1982) make the point that mulberry should be incorporated into training programmes for persons being trained to identify cannabis. I would add that any training programme should include a range of commonly available herbs and spices, along with any local plants, that may be used as a substitute for cannabis or as an additive to a cannabis sample.

The closest relative of cannabis is the hop, *Humulus lupulus*. Hops are generally dioecious with female flowers in cone-like catkins (called strobiles) or inflorescences (see Figure 4.13). Similar to cannabis, a bract encloses a pair of bracteoles with each bracteole containing a small flower with a cup-shaped perianth and two large stigmas. The bracts and, especially, the bracteoles are covered in small glandular trichomes called 'lupulin glands' (Langer and Hill, 1982). There should be no confusion with the glandular trichomes found in cannabis.

Cannabis may also be mixed with tobacco, *Nicotiana tabacum* – see Figure 4.14. Hence, it is important to be able to distinguish tobacco from cannabis. It is outside the scope of this chapter to examine the botany of tobacco plants or the industrial process to produce tobacco leaf used in cigarette

Figure 4.13 Flowering inflorescence of hop, *Humulus lupulus*.

(Creative Commons Attribution 2.0 generic license. Author: Luc, T.)

manufacture – for the interested reader, Wind and Wind (2008) and Langer and Hill (1982) provide excellent descriptions of the botany and cultivation of tobacco. Chopped tobacco leaf found in cigarettes should provide no challenge to differentiate from cannabis. Tobacco leaf has two types of glandular trichome, a tall multicellular, uniseriate stalk with a unicellular or multicellular head and a short unicellular stalk with a multicellular head (see Figure 4.14). An interesting feature of these trichomes is the presence of deposits of calcium oxalate in the glandular heads.

The forensic examiner, working at the LPM level of examination, needs to take a cautious approach when examining fragmented plant material and to be conscious that many plants have covering trichomes. For example, glandular and covering trichomes are common in many culinary herbs. Jackson and Snowdon add descriptions of an additional 20 herbs and spices to their earlier work (Jackson and Snowdon, 1990).

To reiterate, the presence of covering trichomes, cystolythic or not, cannot, in isolation, be used as a definitive feature of cannabis.

There have also been examples where synthetic cannabinoids have been added to non-cannabis plant material. Elkins et al. (2017) report on the use of a real-time polymerase chain reaction (PCR) multiplex high-resolution melt assay as an inexpensive assay to quickly detect and identify trace biological material where there is a mixture and where analysts lack expertise in plant morphology. In this study, the authors simultaneously identified a mixture of plants found in 'herbal highs', morning glory, jimson weed, Hawaiian wood-rose and cannabis.

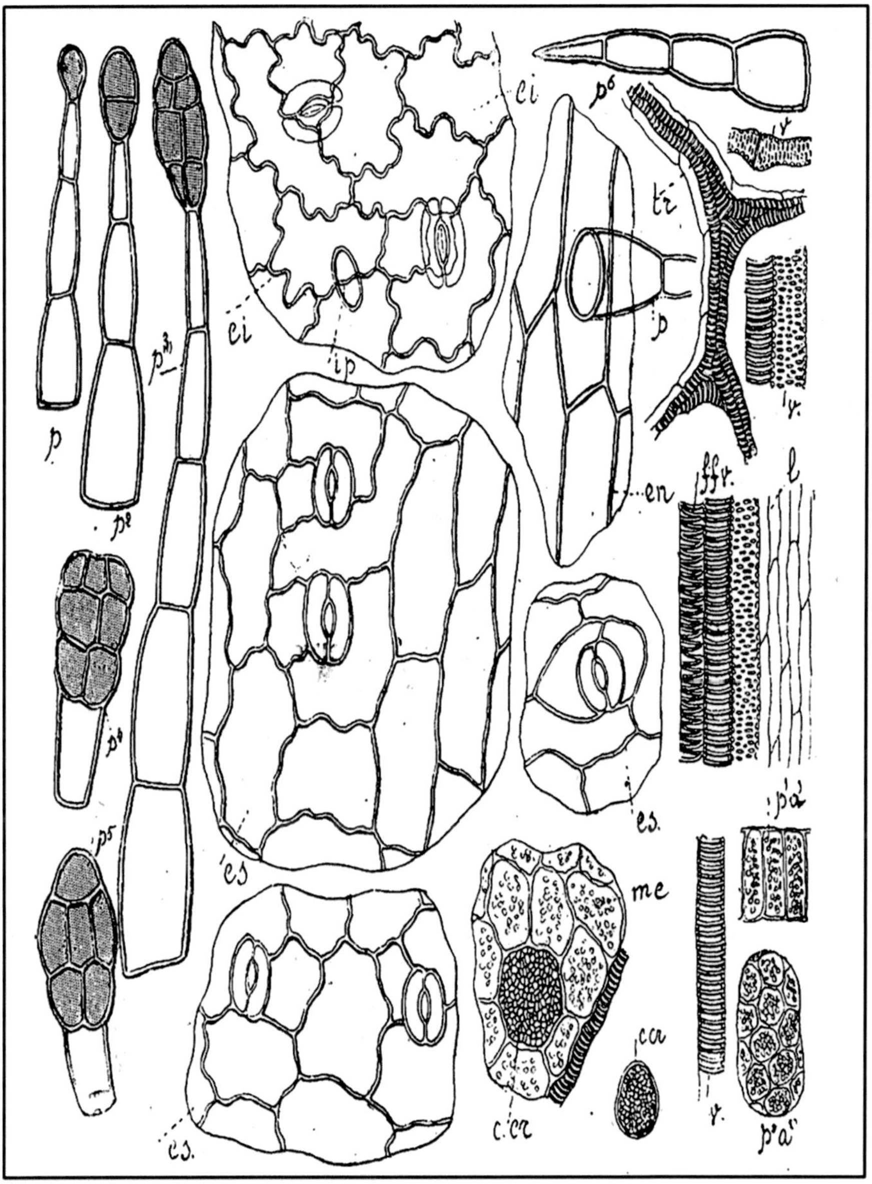

Figure 4.14 Powdered tobacco leaves – Microscopic detail for powdered tobacco leaf.

(From Greenish and Collin (1904), page 91, figure 49.)

(*Continued*)

ccr, Cells with sandy crystals.

ei, Lower epidermis.

en, Neural epidermis.

es, Upper epidermis with stomata.

ffv, Fragments of fibrovascular bundle.

me, Cells of spongy parenchyma.

p, p^2, p^3, Long pluricellular hairs with unicellular or pluricellular glands.

p^4, p^5, Short glandular hairs.

p^6, Conical, pluricellular simple hair

pa, p'a', palisade cells.

st, Stomata surrounded by three cells, one of which is smaller than the others.

tr, Tracheids

v, Vessels.

X240

Figure 4.14 (Continued)

4.6 Cannabis Resin

Cannabis resin is defined in the Single Convention as 'the separated resin, whether crude or purified, obtained from the cannabis plant'. With the emergence of Skunk and high-resin-containing forms of cannabis, there is less of a commercial need to produce cannabis resin. Notwithstanding, cannabis resin may still appear at street level or as very large seizures. The largest seizure made during my time with the Australian Federal Police involved the export of a very large roller to Fiji and its re-importation back to Australia – see Figure 4.15. The roller was never collected, but was full of cannabis resin!

One aspect not to be overlooked is that larger blocks or pieces of cannabis resin are sometimes broken up for sale or to be shared. I was quite often asked during my time in South Australia in the late 1980s to compare seizures of cannabis resin to determine if they could have a common source. Figure 4.16 shows an example where a physical fit provided definitive proof of a common source.

Regrettably, I do not have any images of the most significant case in South Australia involving a cannabis physical fit. This involved the arrest of the then officer in charge of the South Australia Police drug squad, who was charged with alerting criminal gangs that their cannabis cultivations were to be raided. He was also responsible for drug destruction, and it was alleged that he was

Figure 4.15 Image of a large roller used in cannabis resin importation.

Figure 4.16 Example of cannabis resin case involving a physical fit.

placing drugs back onto the street and not having them destroyed. This latter aspect was difficult to prove until I was asked to compare two pieces of cannabis resin. One piece of resin block had been retained by the Police for training purposes, and the second piece, which was supposed to have been destroyed, had been seized from a small-time dealer. When I examined the two pieces, they formed an exact physical fit down to scrape marks on the surface that aligned like rifling marks on firearms projectiles. At that time, we took Polaroid images of seizures. I had examined the original complete block and later the piece of a block relating to the small -time dealer. The images confirmed that the latter piece was from the original block. The next day, the accused changed his plea to guilty.

From a botany viewpoint, to prove the material is cannabis resin requires examination of the questioned material and, usually, an extract of the material with TLM. Figure 4.17 shows a typical example of what can be expected from such an extract when the material is indeed cannabis. The presence of a glandular trichome and a covering trichome confirms that the sample is botanically cannabis. Note, however, that it is common for the head portion of glandular trichomes to be lost and that the stalk on its own would not be sufficient to identify the material as cannabis.

For cannabis resin and cannabis plant material, a botanical identification does not prove the presence of resin as required by Schedule 1. The next

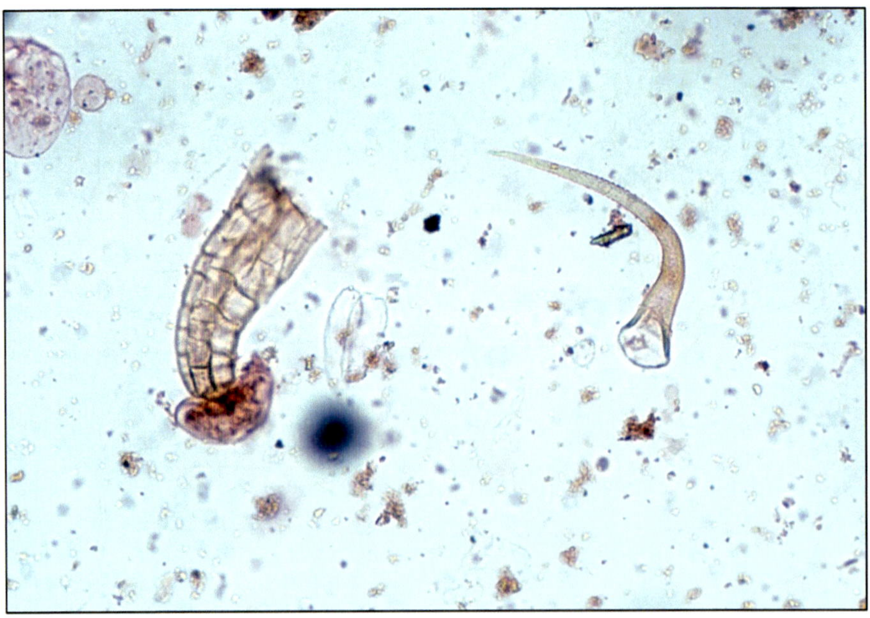

Figure 4.17 Microscopic detail from cannabis resin showing covering and glandular trichomes.

section deals with how resin can be detected. In Section 4.8, the role of DNA analysis for cannabis is considered. El Alaoui et al. (2013) have shown that high-quality DNA can be extracted from cannabis resin.

4.7 Rapid Tests for the Presence of Cannabinoids – The Duquenois-Levine (D-L) Test

A botanical identification is not in itself sufficient to meet the definition of cannabis in the 1961 Single Convention, as it is necessary to show that 'the resin has not been extracted'.

The UN Recommended Methods for Testing Cannabis includes the use of the Rapid Duquenois Test (Duquenois-Levine (D-L) test) (Anon, 1987) whilst stressing that a positive result is only a presumptive indication of the possible presence of cannabis products. The UN Scientific and Technical Note, SCITEC/6, provides details of this test (Kovar & Laudszun, 1989).

The method is rapid and simple. The Duquenois reagent consists of 2 grams of vanillin and 2.5 ml of acetaldehyde made up to 100 ml of ethanol. The test can be performed in a cavity slide where there is very limited material (such as debris from a pipe or bong) or in a small test tube. In a test tube, 10 drops of Duquenois reagent are added to about 10–20 mg of the questioned material. After shaking, 10 drops of concentrated hydrochloric acid are added, and the tube is shaken again. In the presence of cannabinoids, a purple colour will develop. The Levine addition involves then adding 20 drops of chloroform, vortexing the tube and allowing it to settle, during which time two layers will form. A positive result is where the colour moves into the chloroform layer.

The value of this test has been questioned as it is not specific for cannabis, and false positives may result. Reliance on this test has been placed on the study by Thornton and Nakamura (1972). In this study, the authors found that of 82 species of plant, from 600 screened, where there could be some doubt based on the presence of cystolythic trichomes (note, not glandular trichomes!), none yielded a positive D-L test result. A contemporary study by Bailey (1979) reported on 400 herbally derived products and 200 organic compounds, and concluded that there is 'no obviously botanical material, apart from cannabis, that gives a positive D-L test. Apart from cannabinoids, some resorcinol derivatives give a positive test'.

Note, a positive D-L test does not equate to THC being present. Fairbairn et al. (1976), as long ago as 1976, reported the loss of THC from resin over a period of up to 2 years, with exposure to light and air being the most important factors in the loss of THC. More recently, Zamango et al. (2019) reported similar results when cannabis and cannabis products were stored over a four-year period, with exposure to light and temperature being the most important factors in the loss of THC.

The D-L is what it says it is, a presumptive test based on the presence of cannabinoids. Used as an exclusionary test, based on a negative result and along with a lack of botanical features, it is safe not to conduct any further test. In laboratories that now conduct more advanced analytical procedures such as GLC and/or MS, the D-L has fallen out of favour. In my view, the D-L has value as a simple, inexpensive and quick test that shows the presence of cannabinoids (and resin) meeting the requirements of Article 1.

4.8 Is There a Role for DNA Testing of Cannabis?

4.8.1 The Early Days

If one were to rely on a recent review of DNA-based analysis of plant materials in forensic investigations (Robertson et al., 2022), the answer might be a resounding 'no' as there is no mention of DNA testing for Cannabis! This is a surprising omission in what is otherwise an informative and comprehensive review. The reality is that there is an extensive literature dealing with DNA analysis of cannabis from early work in the 1990s, using techniques such as RAPD (random amplified polymorphic DNA) and AFLP (amplified fragment length polymorphism), to the use of metabarcoding in the 2020s.

The first papers that looked at the potential for DNA analysis for cannabis involved the use of RAPD (Cole et al., 1995; Faeti et al., 1996; Jagadish et al., 1996). At that time, a major advantage of the RAPD approach was that no prior knowledge of the DNA sequence was required. However, it was already apparent that there were concerns as to the reproducibility and reliability of data using this method (Fowler & Kijas, 1994; Jones et al., 1997).

Around the same time, an alternative approach was being developed, called AFLP. Coyle et al. (2003a) point out that the strength of AFLP is that they are particularly useful for separating closely related individuals from inbred genetic lines, as might be expected for cannabis. Coyle and co-workers have validated the AFLP method for cannabis, including samples of cloned cannabis, where they have shown the profiles to be highly reproducible. In contrast, profiles from unrelated cannabis plants were easily distinguished (Coyle et al., 2002, 2003b). Datwyler and Weiblen (2006) have also reported on the use of AFLP to distinguish between fibre and drug plants.

Both RAPD and AFLP have two significant potential drawbacks for use with real-world cannabis seizures, as they require a relatively large amount of high molecular weight DNA. Neither approach can differentiate a single source from mixed samples (Carita, 2005).

A later study by Rotherham and Harbison (2011) used a different approach to differentiating drug and non-drug cannabis, focussing on seeds, pollen and degraded material unsuitable for chemical analysis of THC content. Using a single-nucleotide polymorphism (SNP) assay, these authors demonstrated that the presence of a copy of the active form of the tetrahydrocannabinolic acid synthesis gene is only found in the drug cannabis phenotype.

Shirley et al. (2013), using a single hexanucleotide short tandem repeat (STR) locus, investigated the distribution of genotypes of cannabis seeds. This marker was highly specific and highly individualising even amongst seeds of known common origin.

Another line of research involved cloning and sequencing the nuclear ribosomal DNA internal transcribed spacer (ITS) regions for species identification (Siniscalco, 1998, 1999). These studies established that the cannabis ITS I and ITS 2 sequences were unique to cannabis, with its closest relative, hops, showing a 90% match.

Hsieh et al. (2004) further examined the potential for the IGS (intergenic spacer) locus to provide a possible source of variation in establishing the population data of different accessions or origins of cannabis. Mostly, however, the utility of the IGS loci has been directed to a simple method to identify samples of alleged cannabis where morphological examination has been compromised.

Linacre and Thorpe (1998) identified a nucleotide sequence within the IGS between the *trn*L 3' exon and *trn*F gene of cannabis chloroplast DNA, which is cannabis specific and can be used to identify cannabis.

4.8.2 The Move to Short Tandem Repeats (STRs)

As early as 2003, papers started to appear reporting on the development of STR markers, with Coyle predicting that 'STR loci identified in cannabis should be very useful in the future for establishing forensic linkages between source and evidentiary samples' (Coyle et al., 2003a).

STR markers are DNA sequences of six or fewer bases that are repeated multiple times in tandem arrays and are flanked by unique sequences, allowing species-specific primers for PCR amplification. They are at the core of human DNA typing systems (Butler, 2010). In general, STRs provide an excellent method to assess genetic variation due to their high information content, relative ease of genotyping, codominance and reproducibility (Mendoza et al., 2009).

In 2003, Hsieh et al. (2003) claimed the first STR locus to be isolated from cannabis. This locus was highly polymorphic when screened against a population of 108 samples of cannabis. Alghanim and Almirall (2003) reported on the development of 11 STR markers, which were able to identify 27 unique profiles of the 41 cannabis samples studied. The same group later reported on the development of a one reaction six STR multiplex, which they used to examine 98 cannabis samples, including fibre and drug samples (Mendoza et al., 2009). This study could not distinguish between fibre and drug samples.

In 2003, Gilmore and Peakall et al. (2003) developed a five-locus multiplex and examined 93 samples of cannabis sourced from accessions around the world, including both fibre and drug cultivars. All but four individuals from a single drug-type accession had a unique multilocus genotype. The authors, however, could not readily distinguish fibre and drug samples. The need to develop a robust STR genotype database was stressed to enable the full potential of these genetic markers to be realised. These same loci were used in a study of cannabis grown in an indoor hydroponic setting in New Zealand (Knight et al., 2010). DNA analysis distinguished distinct groups in general agreement with phenotypic variation.

A series of publications from Gilmore and co-researchers continued to lay the foundations necessary if cannabis DNA typing was to meet the requirements for use in the criminal justice system. Building on earlier studies by Gilmore et al. (2007), Howard et al. (2008a) demonstrated consistent amplification of 10 STR loci in four multiplex reactions and showed that air-dried leaf tissue, typical of what is seen in drug seizure samples, was particularly suitable as a source of DNA. With these now validated loci, the next step was to establish a genetic database. Howard et al. (2009) examined 510 cannabis plants representing drug plant seizures across Australia and 57 fibre/hemp samples. A total of 106 alleles and 314 different genotypes were detected. All fibre samples had unique genotypes, whilst 55% of the 'drug' samples shared a genotype with one or more samples. Genotype sharing was shown to be a consequence of clonal propagation rather than a lack of genetic resolution. The database in this study was the first database in the world to be produced for validated STR profiles of cannabis. Finally, this group conducted a validation study based on the recommendations of the Scientific Working Group on DNA Analysis Methods (SWGDAM) (Howard et al., 2008b). DNA was extracted from dried root, stem and leaf sources, and from fresh, frozen and dried leaf with a template DNA range of 10.0–0.01 ng. The loci were amplified and scored consistently for all DNA sources within this template range. Cross-species amplification was seen in hops (*H. lupulus*) for three loci, but there was no allelic overlap. The authors acknowledge that some SWGDAM requirements could not be met by this study, as these guidelines were specifically developed for human DNA. For example, these guidelines recommend that inheritance and chromosomal mapping studies be completed. Due to legal restrictions, at the time of this study, it was not possible to conduct breeding experiments that would provide this type of information. With the emergence of medical cannabis, it is likely that we will see breeding experiments to further develop cultivars with desirable qualities.

Interest in developing STR-based DNA approaches to cannabis analysis, however, has not stopped elsewhere. Kohnemann et al. (2012) reported on a 15 STR multiplex for cannabis, claiming that it was 'validated according to common laboratory quality standards'. Although highly informative, these authors pointed out that it is not possible to apply the type of biostatistics that is used to interpret human DNA due to the common use of clonal propagation for cannabis. The same group (Valverde et al., 2014) pointed out that there has been little use made of the various cannabis STR-based systems in the published literature, arguing that one reason for this has been the lack of standardised methods, which makes it impossible to compare results between studies. Through sequencing the 15 STRs from their previous study, these authors propose a nomenclature that can be applied for standardising results and supporting comparisons between laboratories based on the use of allelic ladders.

Houston et al. (2016) developed a 13-loci autosomal STR multiplex system for cannabis and validated this following International Society for Forensic Genetics (ISFG)/SWGDAM guidelines, including the use of a sequenced allelic ladder for accurate designation of alleles and the potential for interlaboratory STR profile sharing. These authors drew particular attention to the potential for biogeographic origin determination and intelligence purposes. Houston et al. (2018) extended their work on autosomal (nuclear) DNA to include data for five chloroplast DNA and two mitochondrial DNA (mtDNA) markers. For the autosomal typing, 356 unique profiles were generated from 425 samples, with 25 identical genotypes within the seizures. For chloroplast and mtDNA, the authors built on previous work by Gilmore et al. (2007), developing a pentaplex STR and a triplex SNP assay with one chloroplast marker shared by both methods for quality control. Complete haplotypes (STR and SNPs) were observed for 127 samples from a subsample of 134 seizures. The same haplotype was seen in 38 samples, with two unique haplotypes detected. The authors concluded that both autosomal and organelle markers could discern population substructure and may be useful in classifying seized cannabis samples. This study also established a US DNA database of cannabis samples for nuclear, chloroplast and mtDNA.

Roman et al. (2019) investigated the potential for chloroplast markers to determine biogeographic origin and crop type for cannabis, developing assays to genotype two regions of the chloroplast genome (*trn*S-*trn*G and *rpl*132-*trn*L), which had four and five polymorphic loci, respectively. Perhaps disappointingly, this study found these markers to have limited use in determining geographic origin and crop type.

Roman and Houston (2020) and Cheng and Houston (2021) further explored the value of chloroplast regions to differentiate crop type and biogeographic origin of cannabis, but again, concluded that it was essential to have a broader sample database to evaluate the true potential of chloroplast markers, suggesting that a massively parallel sequencing (MPS) assay could be developed to more easily and rapidly genotype samples at hundreds of polymorphic sites (Cheng and Houston, 2021).

Roman et al. (2022), in the latest research published by this group, reported on MPS of chloroplast hotspots for forensic typing of cannabis. The advantage of MPS, also called next-generation sequencing (NGS), is the high throughput, which allows the collection of DNA sequence data from multiple targets and samples in parallel. In this study, a total of 40 polymorphisms were observed, 16 of which had not previously been reported. Whilst this study provided some additional information and discrimination to previous studies using capillary electrophoresis (CE), the authors again pointed out the need to develop a more comprehensive sample set from around the world to provide investigative leads for law enforcement agencies.

Adamek et al. (2023) have also used an NGS or genotyping-by-sequencing approach to identify over 9,000 variants in cannabis clones, revealing genetic diversity within clonal lines, demonstrating the limitations in using SSRs or similar markers to evaluate genetic fidelity in clonal populations.

4.8.3 Current Status and Future Potential

A major limitation for cannabis DNA analysis, compared to human DNA, is that there is no commercially available cannabis DNA kit that could then be easily accessed and used in different countries. A promising development, however, has been a European collaborative study based on the 13 locus STR multiplex developed by Houston et al. (2016). Twenty-one laboratories from three European countries performed STR typing of two cannabis samples using the multiplex supplied by the University of Barcelona and San Houston State University (SHSU) in the United States (Di Nunzio et al., 2021). The results were very promising, with an overall genotyping success rate of 96% and only minor artefacts observed, demonstrating the robustness of this system. Two limitations of this study were that it was conducted on DNA extracts and not on cannabis plant material, and that this STR-based system cannot differentiate hemp from drug forms of cannabis. For the latter, chloroplast markers are more successful (Roman et al., 2019).

Fett et al. (2019) and Ribeiro et al. (2020) have also evaluated two 13-loci STR multiplex systems developed by the SHSU group (Houston et al., 2016, 2018) for Brazilian seized samples of cannabis. Both panels were effective in individualising and origin discriminating all samples with the 2016 system outperforming the later 2017 system.

Finally, at the time of writing this chapter, the latest report on DNA in cannabis is the development of a novel and fast 19-plex STR typing system (Xia et al., 2022). This panel includes 15 autosomal STRs, two X-chromosomal STRs and one sex-determining marker and claims to be the first to be able to determine sex. Validation studies showed this panel to be robust, accurate and sensitive. The authors claim to be the first to also report on the chromosomal location of the STRs, pointing out that a limitation of current systems is that several STRs are on the same chromosome and that it would be more informative to include markers that cover all of the chromosomes of cannabis. A further limitation of having loci on the same chromosome is the potential for genetic linkage, which means that the product rule cannot be used to calculate likelihood ratios. These authors flag that they are currently developing STR loci on different chromosomes to address this limitation.

Howard et al. (2009) optimistically predicted that with the establishment of a cannabis DNA database and validation that DNA typing could be handed to an established forensic laboratory for case work. The demand for this type of information existed. The research was in part funded in Australia by the National Drug Law Enforcement Research Fund. Following the publication of a major report on our cannabis DNA work (Howard et al., 2008b), I was approached on a number of occasions by investigators interested in having cannabis seizures compared to try to establish some common links. Sadly, no forensic laboratory has operationalised a cannabis DNA service in Australia.

In writing this short review of the history of the development of DNA testing for cannabis, it is therefore encouraging to see that much of the research that followed the work of the group with which I was involved has drawn on that research and further evolved and developed from that base. Whilst it may still be too optimistic to claim that there is a universal cannabis DNA system available to law enforcement, the work of the SHSU group has certainly laid a strong foundation. The STR-based system has great advantages in that it can be performed on the existing analytical platforms in mainstream forensic biology laboratories. The lack of a commercial supplier of kits remains an obstacle to its widespread adoption. Finally, although a cannabis STR system may look like that used for human DNA, the comparison stops there! The interpretation of plant DNA is different from that of human DNA and requires experts with a knowledge of plant DNA.

References

Adamek, K., Grainger, C., Jones, A.M.P., & Torkamaneh, D. 2023. Genotyping-by-Sequencing (HBS) Reveals Greater Somatic Mutations Than Simple Sequence Repeats (SSRs) in Micropropagated Cannabis Plants. *In Vitro Cellular & Developmental Biology: Plant*. https://doi.org/10.1007/s11627-023-10377-x

Alghanim, H.J., & Almirall, J.R. 2003. Development of Microsatellite Markers in *Cannabis sativa* for DNA Typing and Genetic Relatedness Analysis. *Analytical and Bioanalytical Chemistry*, 376, 1225–1233.

Anon. 1894. *Indian Hemp Drugs Commission Report*. Simla, India: Government Central Printing Office.

Anon. 1961. *Single Convention on Narcotic Drugs*. New York: United Nations. https://www.unodc.org/pdf/convention_196_en.pdf
Anon. 1987. *Recommended Methods for Testing Cannabis. Manual for Use by National Narcotics Laboratories*. New York: United Nations.
Bailey, K. 1979. The Value of the Duquenois Test for Cannabis – A Survey. *Journal of Forensic Sciences*, 24, 817–841.
Bock, J.H., & Norris, D.O. 2016. *Forensic Plant Science*. London: Academic Press.
Butler, J.M. 2010. *Fundamentals of Forensic DNA Typing*. Amsterdam: Elsevier.
Cabezudo, B., Recio, M., Sanchez-Laulhe, J.M., Del Mar Trigo, M., Toro, F.J., & Polvorinos, F. 1997. Atmospheric Transportation of Marijuana Pollen from North Africa to the Southwest of Europe. *Atmospheric Environment*, 31, 3323–3328.
Carita, E.J. 2005. Classical and Future DNA Typing Technologies for Plants. In Coyle, H.M. (ed.), *Forensic Botany. Principles and Applications to Criminal Case Work* (Chapter 15, pp. 253–274). Boca Raton: CRC Press.
Cervantes, J. 2015. *The Cannabis Encyclopedia. The Definitive Guide to the Cultivation and Consumption of Medical Marijuana*, Clifton Park: Van Patten Pub.
Cheng, Y., & Houston, R., 2021, Evaluation of the trnK-matK-trnK, ycf3, and accD-psal Chloroplast Regions to Differentiate Crop Type and Biogeographic Origin of *Cannabis sativa*. *International Journal of Legal Medicine*, 135, 1235–1244.
Clarke, R.C., & Merlin, M.D. 2013 *Cannabis Evolution and Ethnobotany*. Berkeley: University of California Press.
Cole, R., Linacre, A., Thorpe, J.W., & Watson, N.D. 1995. Comparison of *Cannabis sativa* by Random Amplification of Polymorphic DNA (RAPD) and HPLC of Cannabinoids: A Preliminary Study. *Science & Justice*, 35, 169–177.
Coyle, H.M. (ed.) 2005. *Forensic Botany. Principles and Applications to Criminal Casework*. Boca Raton: CRC Press.
Coyle, H.M., Germano-Presby, J., Ladd, C., Palmbach, T., & Lee, H.C. 2002. *Tracking Clonal Marijuana Using Amplified Fragment Length Polymorphism (AFLP) Analysis: An Overview. Proceedings of the 13th International Symposium on Human Identification*, Phoenix, AZ. https://www.promega.com
Coyle, H.M., Palmbach, T., Juliano, N., Ladd, C., & Lee, H.C. 2003a, 'An Overview of DNA Methods for the Identification and Individualisation of Marijuana. *Croatian Medical Journal*, 44, 315–321.
Coyle, H.M., Shutler, G., Abrams, S., Hanniman, J., Nelson, S., Ladd C., Palmbach, T., & Lee, H.C. 2003b. A Simple DNA Extraction (AFLP) Analysis Method for Marijuana Samples Used in Amplified Fragment Length Polymorphism. *Journal of Forensic Sciences*, 48, 343–347.
Datwyler, S.L., & Weiblen, D. 2006. Genetic Variation in Hemp and Marijuana (*Cannabis sativa* L) According to Amplified Fragment Length Polymorphisms. *Journal of Forensic Sciences*, 51, 371–375.
Di Nunzio, M., Agostini, V., Alessandrini, C., et al. 2021. A Ge.F. I-ISFG European Collaborative Study on DNA Identification of *Cannabis sativa* Samples Using a 13-locus Multiplex STR Method. *Forensic Science International*, 329, 111053.
El Alaoui, M.A., Melloul, M., Alaoui Amine, S., et al. 2013. Extraction of High-Quality DNA from Seizes Moroccan Cannabis Resin (Hashish). *PLoS One*, 8, e74714.
Elkins, K.M., Perez, A.C.U., & Quinn, A.A. 2017. Simultaneous Identification of Four Legal High Plant Species in a Multiplex PCR High-Resolution Melt Assay. *Journal of Forensic Sciences*, 62, 593–601.

Emboden, W.A. 1974. A Botanical History of the Genus *Cannabis* with Reference to Legislation. *California Attorneys for Criminal Justice Forum*, 5(Suppl), 1–5.

Faegri, K., Iverson, J., Kaland, P.E., & Krzywinski, K. 1989. *Textbook of Pollen Analysis*. New York: John Wiley & Sons.

Faeti, V., Mandolino, G., & Ranalli, P. 1996. Genetic Diversity of *Cannabis sativa* L. Germplasm Based on RAPD Markers. *Plant Breeding*, 115, 367–370.

Fairbairn, J.W., Liebmann, J.A., & Rowan, M.G. 1976. The Stability of Cannabis and Its Preparations on Storage. *Journal of Pharmacy and Pharmacology*, 28, 1–7.

Fett, M.S., Mariot, R.F., Avila, E., Alho, C.S., Stefenon, V.L., & De Oliveira Camargo, F.A. 2019. 13-loci STR Multiplex System for Brazilian Seized Samples of Marijuana: Individualisation and Origin Differentiation. *International Journal of Legal Medicine*, 133, 373–384.

Fowler, J.C.S., & Kijas, J.M.K. 1994. Current Molecular Methods for Plant Genome Identification. *Australasian Biotechnology*, 4, 153–157.

Gilmore, S. & Peakall, R. 2003. Isolation of Microsatellite Markers in *Cannabis sativa* L (Marijuana). *Molecular Ecology Notes*, 3, 105–107.

Gilmore, S., Peakall, R., & Robertson, J. 2007. Organelle DNA Haplotypes Reflect Crop-use Characteristics and Geographic Origins of *Cannabis sativa*. *Forensic Science International*, 172, 179–190.

Greenish, H.G., & Collin, E. 1904. *An Anatomical Atlas of Vegetable Powders*. London: J&A Churchill.

Hall, D.W., & Byrd, J.H. 2012. *Forensic Botany. A Practical Guide*. Chichester: John Wiley & Sons.

Houston, R., Birck, M., Hughes-Stamm, S., & Gangitano, D. 2016. 'Evaluation of a 13-loci STR Multiplex System for *Cannabis sativa* Genetic Identification. *International Journal of Legal Medicine*, 130, 635–647.

Houston, R., Birck, M., LaRue, B., Hughes-Stamm, S., & Gangitano, D. 2018. Nuclear, Chloroplast, and Mitochondrial Data of a US Cannabis DNA Database. *International Journal of Legal Medicine*, 132, 713–725.

Howard, C., Gilmore, S., Robertson, J. & Peakall, R. 2008a. Developmental Validation of a *Cannabis sativa* STR Multiplex System for Forensic Analysis. *Journal of Forensic Sciences*, 53, 1061–1063.

Howard, C., Gilmore, S., Robertson, J., & Peakall, R. 2008b. Application of new DNA Markers for Forensic Examination of *Cannabis sativa* Seizures – Developmental Validation of Protocols and a Genetic Database. *Monograph Series No. 29*, National Drug Law Enforcement Research Fund (NDLERF): Commonwealth of Australia.

Howard, C., Gilmore, S., Robertson, J., & Peakall, R. 2009. A *Cannabis sativa* STR Genotype Database for Australia Seizures: Forensic Applications and Limitations. *Journal of Forensic Sciences*, 54, 556–563.

Hsieh, H., Hou, R., Chen, K., Tsai, L., Liu, S., Liu, K., Linacre, A., & Lee, J. 2004. Establishing the rDNA IGS Structure of *Cannabis sativa*. *Journal of Forensic Sciences*, 49, 477–480.

Hsieh, H., Hou, R., Tsai, L., Wie, C., Liu, S., Huang, L., Kuo, Y., Linacre, A., & Chun, J. 2003. A Highly Polymorphic STR Locus in *Cannabis sativa*. *Forensic Science International*, 131, 53–58.

Iversen, L.L. 2018. *The Science of Marijuana*, 3rd ed. Oxford: Oxford University Press.

Jackson, B.P., & Snowdon, D.W. 1968. *Powdered Vegetable Drugs. An Atlas of Microscopy for Use in the Identification and Authentication of Some Plant Materials Employed as Medicinal Agents*. London: J&A Churchill Ltd.

Jackson, B.P., & Snowdon, D.W. 1990. *Atlas of Microscopy of Medicinal Plants, Culinary Herbs and Spices*. New Delhi: CBS Pub.

Jagadish, V., Robertson, J., & Gibbs, A. 1996. RAPD Analysis Distinguishes *Cannabis sativa* Samples from Different Sources. *Forensic Science International*, 79, 113–121.

Jones, C.L., Edwards, K.J., Castiglione, S., et al. 1997. Reproducibility Testing of RAPD, AFLP and SSR Markers in Plants by a Network of European Laboratories. *Molecular Breeding*, 3, 381–390.

Knight, G., Hansen, S., Conner, M., Poulsen, H., McGovern, C., & Stacey, J. 2010. The Results of an Experimental Indoor Hydroponic Growing Study, Using the 'Screen of Green' (ScrOG) Method – Yield, Tetrahydrocannabinol (THC) and DNA Analysis. *Forensic Science International*, 202, 36–44.

Kohnemann, S., Nedele, J., Schwotzer, D., Morzfeld, J., & Pfeiffer, H. 2012. The Validation of a 15 STR Multiplex PCR for *Cannabis* Species. *International Journal of Legal Medicine*, 126, 601–606.

Kovar, K., & Laudszun, M. 1989. Chemistry and Reaction Mechanisms of Rapid Tests for Drugs of Abuse and Precursor Chemicals, United Nations, Scientific and Technical Notes, SCITEC/6.

Langer, R.H.M., & Hill, G.D. 1982. *Agricultural Plants*. Cambridge: Cambridge University Press.

Linacre, A., & Thorpe, J. 1998. Detection and Identification of Cannabis by DNA. *Forensic Science International*, 91, 71–76.

Lowry, W.T., & Garriott, J.C. 1975. On the Legality of Cannabis: The Responsibility of the Expert Witness. *Journal of Forensic Sciences*, 20, 624–629.

Maloney, R.S., & Thornton, J.L. 1982. Trichomes of the Mulberry, *Morus nigra*. *Microgram*, 15, 78–79.

McPartland, J.M. 2018. *Cannabis* Systematics at the Level of Family, Genus, and Species. *Cannabis and Cannabinoid Research*, 3, 203–212.

Mendoza, M.A., Mills D.K., Lata, H., Chandra, S., ElSohly, M.A., & Almirall, J.R. 2009 Genetic Individualisation of *Cannabis sativa* by a Short Tandem Repeat Multiplex System. *Analytical and Bioanalytical Chemistry*, 393, 719–726.

Patton, C.V. 2020. A History of United States Cannabis Law. *Journal of Law & Health*, 34, 1–30.

Ribeiro, L.O.P., Avila, E., Mariot, R. F., Fett, M. S., de Oliveira Camargo, F. A., & Alho, C. S. 2020. Evaluation of Two 13-loci STR Multiplex System Regarding Identification and Origin Discrimination of Brazilian *Cannabis sativa* Samples. *International Journal of Legal Medicine*, 134, 1603–1612.

Robertson, J.M., Damaso, N., & Meikelejohn, K.A. 2022. DNA – Based Analysis of Plant Materials in Forensic Investigations. In Dash, H.R., Shrivastava, P., & Lorente, J.A. (eds.), *Handbook of DNA Profiling*. Singapore: Springer. https://doi.org/10.1007/978-981-16-4318-7_59

Roman, M.G., Gangitano, D., & Houston, R. 2019 Characterization of New Chloroplast Markers to Determine Biogeographic Origin and Crop Type of *Cannabis sativa*. *International Journal of Legal Medicine*, 133, 1721–1732.

Roman, M.G., Gutierrez, R., & Houston, R. 2022 Massively Parallel Sequencing of *Cannabis sativa* Chloroplast Hotspots for Forensic Typing. *Journal of Cancer Research*, 4, 13–23.

Roman, M.G., & Houston, R. 2020. Investigation of Chloroplast Regions rps16 and clpP for Determination of *Cannabis sativa* Crop Type and Biogeographic Origin. *Legal Medicine*, 47, 101759.

Rotherham, D., & Harbison, S.A. 2011. Differentiation od Drug and Non-drug *Cannabis* Using a Single Nucleotide Polymorphism (SNP) Assay. *Forensic Science International*, 207, 193–197.

Schultes, R.E., Klein, W.M., Plowman, T., & Lockwood, T.E. 1974. *Cannabis*: An Example of Taxonomic Neglect. *Botanical Museum Leaflets Harvard University*, 23, 337–367.

Shirley, N., Allgeier, L., LaNier, T., & Coyle, H.M. 2013. Analysis of the NM101 Marker for a Population of Database of *Cannabis* Seeds. *Journal of Forensic Sciences*, 58, s176–s182.

Siniscalco, G.S. 1998. Identification of *Cannabis sativa* L. (Cannabaceae) Using Restriction Profiles of Internal Transcribed Spacer II (ITS2). *Science & Justice*, 38, 225–230.

Siniscalco, G.S. 1999. Preliminary Data on the Usefulness of Internal Transcribed Spacer I (ITSI) Sequence in *Cannabis sativa* L. Identification. *Journal of Forensic Sciences*, 44, 475–477.

Small, E. 2017. *Cannabis. A Complete Guide*. Boca Raton: CRC Press.

Small, E., & Cronquist, A. 1976. A Practical and Natural Taxonomy for Cannabis. *Taxon*, 25, 405–435.

Sytsma, K.J., Morawetz, J.C., Pires, M., Nepokroeff, M., Conti, M., Zjhra, M., Hall, J.C., & Chase, M.W. 2002. Urticalean Rosids, Circumscription, Rosid Ancestry, and Phylogenetics Based on *rbcl*, *trnL-F*, and *ndhf* Sequences. *American Journal of Botany*, 89, 1531–1546.

Thornton, J.I., & Nakamura, G.R. 1972. The Identification of Marijuana. *Journal of Forensic Sciences*, 12, 461–519.

Valverde, L., Lischka, C., Scheiper, S., Needle, J., Challis, R., De Pancorbo, M.M., Pfeiffer, H., & Kohnemann, S. 2014. Characterization of 15 STR Cannabis Loci; Nomenclature Proposal and SNPSTR Haplotypes. *Forensic Science International: Genetics*, 9, 61–65.

Wallis, T.E. 1960. *Textbook of Pharmacognosy*. London: J&A Churchill.

Wind, T.E., & Wind, K. 2008. *Guide to Cultivated Plants*. Wallingford: CABI International.

Xia, R., Tao, R., Qu, Y., Zhang, X., Yu, H., Yuan, C., Zhang, S., & Li, C. 2022. Development and Validation of a Novel and Fast Detection Method for *Cannabis sativa*: A 19-Plex Short Tandem Repeat Typing System. *Frontiers in Plant Science*, 13, 837945.

Zamango, L., Bettin, C., Badocco, D., Di Marco, V., Miolo, G., & Frison, G. 2019. The Role of Time and Storage Conditions on the Composition of Hashish and Marijuana Samples; A Four-Year Study. *Forensic Science International*, 298, 131–137.

5 Other Drug Plants

5.1 Introduction

The focus of this chapter is the forensic examination and identification of drug plants other than cannabis, which was considered in Chapter 4. Some of these plants, including poppies and coca, are illegal, whilst khat is either regulated or illegal in some countries.

5.2 Opium Poppy

5.2.1 Introduction, Background and Legal Status

This section deals mainly with one species, the 'opium' poppy, *Papaver somniferum*. As we shall see in the next section, however, there are perhaps as many as 100 species in this genus and a small number where it is important that a forensic botanist can differentiate them from *P. somniferum*.

The origins of the opium poppy are probably the Eastern Mediterranean, but due to its widespread cultivation and introduction to many countries, there is no wild progenitor. For a detailed history of the opium poppy, the reader is referred to Kapoor (1997).

In addition to its use for opium production, the opium poppy has a number of uses of economic importance. Its seeds are edible and are used as a foodstuff, extracted for oil, and the remnant oil cake is used as cattle feed or as fertiliser. The seeds are also used in pharmacy for making emulsions. Furthermore, *P. somniferum* has escaped from cultivation and has become naturalised in almost all countries with suitable climates, and because of the attractive flowers, the opium poppy is used as an ornamental or garden plant. In many parts of Australia, opium poppy is considered an environmental weed (Anon, 2016). Reflecting its main uses, varieties have been bred with a high opium yield for the medical industry (Yadav et al., 2006) and low alkaloid content for the food industry (Dittbrenner et al., 2009).

All of the above make the legal status of the opium poppy complex. Opium poppy is defined in Article 1 of the UN Convention on Narcotic Drugs 1961 (Anon, 1961) as 'the plant of the species *P. somniferum L*' and poppy straw as 'all parts (except the seeds) of the opium poppy, after mowing'. Schedule 1, Article 25 includes 'concentrate of poppy straw'.

DOI: 10.4324/9781003308508-5

Countries are allowed to authorise the growing of opium poppies for the production of poppy straw and the subsequent extraction of alkaloids for medicinal uses. For example, the State of Tasmania in Australia produces about 25% of the world's production of opium and codeine production and 85% of the world's supply of thebaine. A fact sheet produced by the Tasmanian Department of Natural Resources, however, recognises that a number of ornamental and weed species of Papaver are legal whilst placing the responsibility on the grower to ensure that they are legal species and can be grown without a licence (Anon, 2023).

In many countries around the world, opium poppies can be grown for seeds. However, there have been cases reported of poisoning caused by poppy seeds (DEA, 2019). According to the Drug Enforcement Administration, there have been at least 12 deaths associated with the use of unwashed poppy seed tea. Poppy seed tea is where the user intentionally collects seeds by making cuts in opium pods before they ripen, allowing the latex to seep onto the seed coats, which are then boiled to produce a tea. The seeds themselves do not contain alkaloids. According to Haber et al. (2019), unwashed poppy seeds can be purchased in bulk through private websites and sellers such as eBay and Amazon.

5.2.2 Classification, Nomenclature and Morphology

P. somniferum is a member of the family Papaveraceae, subfamily Papaveroideae within the order Ranunculales (Simpson, 2019). There is one accepted subspecies, *P. somniferum subsp. setigerum,* sometimes referred to as the small-flowered opium poppy. In Australia, this subspecies is more common than the opium poppy (*P. somniferum subsp. somniferum*). The two subspecies can be differentiated by the following differences (Anon, 2016):

- Opium poppy has very few, if any, bristly hairs on its stems and leaves and has a relatively large (about 5 cm across) and almost rounded in shape (globose) fruit.
- Small-flowered opium poppy has stems and leaves that are covered in bristly hairs and a relatively small (about 2 cm across) and slightly elongated (cylindrical or subglobose) fruit.

In addition to the two subspecies, there are as many as 52 botanical varieties or cultivars (Dittbrenner et al., 2009). For example, *P. somniferum var. paeoniflorum* has numerous petals in showy double flowers. There are also numerous ornamental species of poppy, such as *Papaver rhoeas,* the field or 'Flanders' poppy.

According to the United Nations, only *P. somniferum* and *setigerum* (either considered as separate species or subspecies) are of importance for the purposes of narcotics control (Anon, 1961). From a practical viewpoint, they conclude that it is impossible from a botanical viewpoint to distinguish any precise varieties and, hence, also impossible to give a description of the plant based on its external characteristics. A recent paper from Ngernsaengsaruay et al. (2023) on the morphology and anatomy of opium poppies grown in Northern Thailand further emphasised the level of variation seen in actual crops from descriptions in the literature.

Notwithstanding, and cognisant of the variation seen in opium poppies, the following is a general description of the botanical features of *P. somniferum*, (DEA, 1991; Kapoor, 1997).

The opium poppy is an annual herbaceous plant with a thick tapering root. Grown from seed, the entire growth cycle is in the order of 120 days, with plants flowering after about 90 days under long-day (>14 hours daylight) photoperiod conditions. Within two months, plants can reach a height of 0.5–1 m with a primary, long, cylindrical and relatively smooth stem. At maturity, plants can reach a height of 1–5 m. The upper part of the stem without leaves is called a *peduncle*. There may be one or two secondary stems or tillers that grow from the main stem. The leaves are numerous, simple or pinnately lobed, lanceolate to lanceolate-ovate, sessile and alternate, clasping the stem by a cordate base. Leaf margins are toothed and lobed and vary from 4 to 15 inches in diameter at maturity (see Figure 5.1). Leaves are silver to grey-green in colour, glabrous and glaucous (a layer of wax) on both sides.

As plants grow, the main stem and tillers terminate in a solitary, terminal flower bud. During the development of the bud, the peduncle portion of the stem elongates and forms a distinctive hook, which causes the bud to be turned upside down. As the flower develops, the peduncle straightens, and buds become erect and point upwards. Each flower has two smooth, broad green sepals. The sepals fall away, or disarticulate, as the flower opens up to reveal four large petals. The petals may be single or double and can be white, pink, reddish purple, crimson red or variegated and sometimes with a dark marking at the base. The outer two petals are wider and overlap the inner two petals. The petals last for only two to four days before they drop.

Stamens are numerous, hypogynous (attached below a superior ovary) and inserted in two to three rows on the under surface of a flattened receptacle. Filaments are long, flat, ribbon-shaped and flattened at the top. Anthers are linear and attached by a narrow base to the filaments. Anthers are creamy coloured, wavy and twisted after dehiscence. Pollen grains are spheroidal in shape, sometimes flattened at the poles. The pollen aperture is described as tricolpate, and the exine sculpturing is microechinate (Ngernsaengsaruay et al., 2023). Pollen nomenclature will be discussed in more detail in Chapter 9, dealing with forensic palynology. Opium poppies are, in general, insect-pollinated.

The ovary is superior, large, globular and contracted below to a neck (gynophore), which dilates to form the receptacle, narrowing down below into the pedicel. The ovary is unilocular, containing a large spongy placenta bearing numerous ovules and bulging out near the centre to give the capsule its characteristic shape. The stigma is sessile, peltate and spreading over the top of the ovary with eight to 15 short, robust, obtuse oblong rays forming a disk or crown.

Figure 5.1 *Papaver somniferum*. (a) Showing flowering plant parts.

(Continued)

Figure 5.1 (Continued) (b) Showing exuding latex from an unripe fruit.

The fruit is a capsule, more or less globular, supported on a neck (as in the ovary) and crowned by a persistent stigma. The fruit wall (pericarp) consists of outer, middle and inner layers. After fertilisation, and after the petals have dropped, the ovary is ready within 2 weeks for 'lancing' to recover the latex. Latex (opium) is produced within specialist *laticifers* associated with the vascular bundles in stems and leaves, which in turn drains into the middle pericarp layer through the vascular system within the pod. The cells of the middle layer are reported to secrete more than 95% of the opium when the pod is scored and harvested. When ripe, the pericarp wall is hard, dry, brittle and brownish yellow in colour. There is one central cavity with dry placental plates reaching out to about halfway into the centre.

Seeds are small, numerous (up to 5,000 seeds per capsule), reniform (kidney-shaped), variable in colour and are released through pores beneath the disk when the capsule is shaken. The surface of seeds is minutely pitted with small depressions.

5.2.3 Anatomy and Microscopic Features

Ngernsaengsaruay et al. (2023) discuss the anatomy of the stem of opium poppy, describing the presence of two rings of discontinuous and widely spaced vascular bundles with laticifers associated with the phloem tissues of the vascular bundles. These authors also describe the anatomy of the leaf lamina, showing it to be dorsiventral with one to three layers of palisade cells and several layers of spongy mesophyll. There are a number of vascular bundles, with the midrib having the largest bundle. Laticifers are again associated with the phloem tissue.

Wallis (1960) also describes the internal anatomy of the leaf as dorsiventral. The upper epidermis has large polygonal-shaped cells, with the lower epidermis having wavy-walled cells with anomocytic stomata. Wallis (1960) also describes the cells typically seen in fragments of the capsule wall. These include polygonal-shaped cells of the outer epidermis, and much larger and beaded cells of the inner epidermis.

The most comprehensive description of the microscopic detail of the opium poppy is to be found in Jackson and Snowdon (1968), where their focus is on the debris that might be expected in crude samples of opium, these being fragments of the outer capsule wall and pollen grains. Included in their descriptions are features associated with powdered coca leaves, which have been used to adjust the strength of opium in pharmaceutical products.

Jackson and Snowdon (1968) describe the outer epidermis of the capsule as having *tubular* cells, which, in surface view, are polygonal and show considerable variation in size. The anticlinal walls are unevenly thickened and may be beaded. Occasional anomocytic stomata may be present. The cells of the inner epidermis of the capsule are longitudinally elongated in surface view and have lignified walls with conspicuous thickening and beading. There are large, underdeveloped stomata. The epidermis of the placenta is composed of cells with lignified and thickened walls containing slit-shaped pits. In surface view, the cell walls are beaded. The upper and lower epidermises of the leaf are composed of large polygonal cells with thin, slightly sinuous walls. Anomocytic stomata are found on the lower epidermis. Pollen grains are spherical with three pores and a very faintly pitted exine.

5.2.4 Role of DNA Analysis

There have been a limited number of studies of opium poppy DNA for forensic purposes. As with cannabis, early studies focussed on the application of techniques such as RAPD (random amplified polymorphic DNA) and AFLP (amplified fragment length polymorphism) analysis.

Shoyama et al. (1998) did not examine *P. somniferum* but used RAPD to investigate three *Papaver* species, *Papaver pseudo-orientale, Papaver orientale, Papaver bracteatum* and F1 hybrids of *P. bracteatum* and *P. pseudo-orientale. P. bracteatum* is illegal in Japan as it produces thebaine. *P. bracteatum* is also a prohibited plant in Tasmania, Australia (Anon, 2023). Differentiating these three species in their vegetative growth stage is difficult based on morphology alone. Shoyama et al. (1998) demonstrated that RAPD could be used to confirm the parenthood of F1 hybrids and distinguish between these three species and *P. somniferum* and *P. setigerum*.

Saunders et al. (2001) reported on the use of AFLP to analyse accessions of *P. somniferum, P. bracteatum* and *P. setigerum*. The use of a single primer was sufficient to distinguish between different poppy species, and using more than one primer enabled the resolution of genetic relationships within a single poppy species.

For the reason already discussed for cannabis, RAPD and AFLP techniques have fallen from favour and have been replaced by the use of short tandem repeat (STR) and/or single-nucleotide polymorphism (SNP) analysis.

As with cannabis, internal transcribed spacer 2 sequences have been used to distinguish *P. somniferum* from five closely related species of the genus *Papaver* (Cheng et al., 2021).

Liu et al. (2020) reported on the complete chloroplast (cp) genome of *P. setigerum*, identifying 13 cpDNA markers with relatively high levels of variation within *Papaver*, which could be useful for species identification in this genus. Amongst these 13 markers, *psb*A-*trn*H was found to be the best to distinguish *P. setigerum* and *P. somniferum*.

The values of STR, SSR and SNP markers have also been the subject of recent studies.

Young et al. (2020) reported on an evaluation of STR markers for the individualisation of *P. somniferum*, selecting a six-locus STR multiplex. This multiplex was evaluated according to International Society for Forensic Genetics/Scientific Working Group on DNA Analysis Methods guidelines, including the use of a sequenced allelic ladder. The multiplex had a low discriminatory power, with more than two-thirds of the samples tested having just two different genotypes, and, hence, it was not deemed suitable for individualisation. The authors concluded that a search of the poppy genome for new STR markers would be necessary to create a genetic tool capable of individualisation of *P. somniferum*.

Finally, Zhang et al. (2022) reported on the development of simple sequence repeats (SSR) and SNP markers for identifying opium poppy. Eleven SSR markers and two SNP markers were studied to distinguish *P. somniferum* from six other *Papaver* species. A simple multiplex based on three SSR loci was also proposed as a low-cost and rapid approach to identify poppy species. This study did not include *P. setigerum*.

In conclusion, there are a number of approaches available where DNA can assist in identifying poppy species and possibly distinguishing within *P. somniferum* from different seizures. To the best of my knowledge, at the time of writing, there are no commercial kits or laboratories offering a forensic-focussed service to analyse and interpret DNA from poppy species.

5.3 Coca Plant

5.3.1 Introduction, Background and Legal Status

Coca has an ancient history of use in South America where the leaves are considered sacred by some indigenous cultures and chewing or drinking of coca 'tea' is common until today.

The coca plant belongs to the genus *Erythroxylon* in the Family *Erythroxylaceae*. This is a large genus with 259 species listed in Kew's Plants of the World Online (Anon, 2021). Some 17 species of Erythroxylon contain the alkaloid cocaine, but only two species are currently grown in South America for cocaine. These two species are *Erythroxylon coca* and *Erythroxylon novogranatense*. *E. coca* has two recognised varieties, var. *coca* and var. *ipadu*. *E. novogranatense* also has two varieties, var. *novogranatense* and var. *truxillense* (DEA, 1991).

Today, the large-scale growing of coca species is restricted to countries in South America. *E. coca* var. *coca* is the most widely cultivated variety and is cultivated on the eastern slopes of the Andes from Bolivia in the south to central Ecuador in the north. *E. coca* var. *ipadu* is found in southern Columbia, northeast Peru and western Brazil in the Amazon basin. It has a lower level of cocaine alkaloids than the other three varieties and is primarily cultivated for consumption by local communities. *E. novogranatense* var. *novogranatense* is grown in the drier regions of Columbia, and to a lesser extent in Venezuela with *E. novogranatense* var. *truxillense* growing at higher elevations of up to 4,950 feet (DEA, 1991).

Due to its superior flavour and shipping properties, *E. novogranatense* var. *truxillense*, named for the city of Trujillo in northern Peru from which it is exported, is used in the production of extracts to manufacture Coca Cola (Plowman, 1980).

One could be forgiven for thinking that the coca plant has always been restricted to South America, but this is far from the truth. The coca plant has been grown commercially in Nigeria, Sri Lanka, Malaysia, Indonesia, Taiwan and Iwo Jima (Karch, 1998). Karch (1998) provides an informative and highly readable account of the history of coca cultivation from the arrival of Europeans in the New World through to the end of the Second World War and, essentially, the end of commercial cultivation in countries in Asia. The following brief history of cocaine draws on Karch (1998).

Focussing only on the 'botany', and not the geopolitics of coca/cocaine, coca plants were only one of many New World plants introduced to Europe and then to many parts of the world. Kew Gardens in the United Kingdom played a pivotal role, with intrepid botanical explorers sending back newly discovered plants that looked to have economic value. Samples of coca had arrived at Kew by the 1830s as the famous botanist Sir Willian Hooker produced a detailed illustration in 1835. Plants grown from seed at Kew from 1869 had the same appearance as earlier samples that had been collected in Colombia. Colombia at that time was called New Granada, from which the name *E. coca* var *novogranatense* derives. Seeds were sent to the British colonies in Africa, Ceylon (Sri Lanka), Malaya and Jamaica, but importantly not Java (Indonesia), which at that time was a Dutch colony. The exact source of the plants from which seeds were sent to Java is unknown, but it is likely to have been hundreds of miles from the Kew plants. The important difference is that the Java plants, when properly processed, could yield more than twice the cocaine content of the Kew variety.

Cocaine production, however, remained at a very low level for a number of reasons until the key discovery in 1884 that cocaine was effective as a local anaesthetic when applied to the human eye, allowing eye surgery for the first time. Karl Koller is credited with this discovery, although there had been many previous studies that had hinted at the same outcome. The impact on the demand for cocaine was immediate and dramatic. Within months of the results of Koller's work becoming public, Merk alone increased production from three quarters of a pound in 1883 to 3,179 pounds in 1884 to 158,352 pounds in 1886.

The demand for large quantities of leaves was huge, and these had to be transported from remote mountainous areas of South America. Packed leaves were also subject to wastage, as if they became damp; they fermented in transit, whereas dry coca leaves often lost their potency. Enter one Henry H Rusby! Rusby was typical of his day in that he had a medical degree and a botany degree. In 1885, he was sent to South America by Parke, Davis to obtain large quantities of coca leaf. His game-changer was, however, to solve the problem of transport of leaves by inventing a method for making semi-refined cocaine in the country, a method still used by clandestine chemists in South America (DEA, 1991). This transformed cocaine from a perishable commodity to a stable commodity and one that was easier to transport and occupied less cargo space.

Coca leaf from Java continued to be transported in bulk as the alkaloids present needed to be refined to produce cocaine, as this required a more sophisticated chemistry. From 1898 until 1903, this process could only be carried out in Germany as a German company owned the patent. Thereafter, a number of chemical companies began producing cocaine using the same process. Leaf exports peaked in 1920 at 1,650 tons. During the first half of the 20th century, Japan became a major producer of cocaine, firstly using Indonesian-grown leaf and later their own plantations in Japan and then Iwo Jima, Okinawa and Taiwan. The Japanese cocaine industry effectively ended coca as a commercial crop in Indonesia, and by 1937, Java's export of leaf was merely 41 tons. The story of Japanese cocaine is beyond the scope of this chapter. For the reader interested in this era of history, Karsh devotes a chapter which he calls 'Japan's' adventures in the cocaine trade' (Karch, 1998). Changing international attitudes to the legal status of cocaine and limited markets meant the demise of coca growing in Southeast Asia.

The international legal status of coca is covered by the 1961 Single Convention on Narcotic Drugs (Anon, 1961). Article 1 defines the coca bush as 'the plant of any species of the genus Erythroxylon' and coca leaf as 'the leaf of the coca plant except the leaf from which all ecgonine, cocaine and any other ecgonine alkaloids have been removed'. Article 26 commits signatories to the Single Convention to 'so far as possible enforce the uprooting of all coca bushes which grow in the wild. They shall destroy the coca bushes if illegally cultivated'. Furthermore, 'Coca leaf chewing must be abolished within twenty-five years from coming in to force of this Convention' (Article 49, 2.e). Clearly, the objects of the Single Convention with regard to the cultivation of coca have been an abject failure!

5.3.2 Botanical Description of the Coca Plant

5.3.2.1 Morphology

Bentley and Trimen (1875) produced one of the earliest and most comprehensive descriptions of *Erythroxylon coca* in their major work on medicinal plants. The specimen examined was likely to have come from the neighbourhood of Chinchero in Peru (Karch, 1998). It was described as 'a small shrub of 2–4 feet or more, bushy and much branched, with a purplish-brown wrinkled bark, the young twigs smooth' (see Figure 5.2).

Plowman (1980) describes the genus as small shrubs or small trees usually growing in the understory of primary and secondary forests, or in open thickets. The leaves are closely placed, alternate, short-stalked, 1–2 inches long, lanceolate or oval in shape, bright green above and paler and glaucous below and with a prominent glabrous midrib with numerous lateral veins, which are freely anastomosing (rejoining after branching). Leaves were subtended by small stipules with a short stem, combined along their inner edge to form a single, triangular, acute organ with edges with small teeth-like indentations (denticulate).

Flowers are small, relatively inconspicuous, carried on slender, drooping, glabrous stalks, three to four together in the axils of the leaves or on persistent stipules with several small bracts at the base. The calyx is very deeply cut into five triangular-ovate, acute, glabrous segments. There are five petals alternating with the calyx-lobes, imbricate, with a broad claw and broadly oval-oblong in shape. There are ten stamens, hypogynous, equal, longer than the petals, erect and united at the base into a short tube surrounding the ovary. The filaments are white and smooth, and the anthers, oblong, yellow and attached by

Figure 5.2 Bentley and Trimen's historical figure of *Erythroxylon coca*.

the base. The ovary is superior, ovoid, smooth, normally three-celled but normally I-celled with the other two aborted. The resulting ovule is solitary with three styles, which are erect, cylindrical, green and with capitate stigmas. On maturity, the fruit is a small indehiscent drupe, oblong ovoid in shape, pointed and surrounded at its base by a persistent calyx and staminal tube. The surface is smooth, furrowed when dry, and red in colour with a thin mesocarp and endocarp – see Figure 5.3.

According to Bentley and Trimen (1875), 'it is scarcely possible to mistake the leaves of coca for those of any other plant', with the two longitudinal arched lines on the under surface being characteristic. These are often described as veins but are rather folds produced by the way in which the leaves are packed in the bud. This feature is, however, found in other members of the genus *Erythroxylon*.

Readily distinguishing between different varieties of the coca plant is not an easy task (Plowman, 1980; DEA, 1991) and depends on the fine detail of the leaf morphology. Both varieties of *E. coca* (var. *coca* and var. *ipadu*) have broadly elliptic-shaped leaves. The leaf of var. *ipadu* has a rounded apex. The apex of var. *coca* is more pointed and is described as large, thick and dark green in colour. The leaf of *E. novogranatense var. novogranatense* is pale green with a rounded apex and is somewhat narrower and thinner than the leaf of var. *coca* and var. *ipadu*. *E. novogranatense var. truxillense* is very similar to *var. novogranatense* except that it does not have the lines parallel to the central vein of the leaf that are characteristic of many varieties of the coca plant (DEA, 1991).

It should be recognised that there is considerable variation in morphological features due to local climatic, environmental and growing conditions, reflecting a long history of coca cultivation in a wide variety of micro habitats and partial isolation in a large number of Andean valleys (Plowman, 1980).

Figure 5.3 *Erythroxylon* coca plants. (a) Coca foliage. (b) Coca fruit.

5.3.2.2 Anatomy of Coca Leaf

Descriptions of the anatomy of coca leaves, whole and fragmented, are to be found in a number of classic pharmacognosy textbooks, such as Greenish and Collin (1904), Wallis (1960) and Jackson and Snowdon (1968), reflecting the importance of coca leaf in medical botany before cocaine was largely replaced by synthetic substitutes such as procaine.

Wallis (1960) shows the morphology and anatomy of leaves from Bolivian and Peruvian 'coca' – *E. coca* is probably *E. coca* var *coca* and *E. truxillense*, *E. novogranatense* var *truxillense* – highlighting the difference in leaf shape between these two species. The leaf of *E. coca* var. *coca* has a prominent ridge filled with collenchyma, above the midrib, a marked development of sclerenchyma above and below the side veins, and usually shows a line of idioblasts (a cell different greatly from those around it) most abundantly near the lines of the under surface. According to Wallis (1960), Peruvian coca (*E. novogranatense* var *truxillense*) never possesses lignified idioblasts and has only a very slight prominence over the mid vein. The leaf is dorsiventral with a single layer of palisade mesophyll cells and multiple layers of spongy mesophyll cells.

Jackson and Snowdon (1968) describe the anatomy of fragmented leaves of *E. coca* or Bolivian coca. The upper epidermis is composed of fairly large, polygonal cells with moderately thickened walls, which show irregular beading and pitting. Stomata are absent. Occasional prisms of calcium oxalate are seen as single crystals and often in two adjacent cells. The cells of the lower epidermis are smaller than those of the upper epidermis with slightly thinner walls, but they do show irregular beading. The outer walls are papillose (short, rounded nipple like bumps or projections), giving the surface view the appearance of a circle in each cell. Numerous paracytic stomata are present, with subsidiary cells not being papillose.

Prisms of calcium oxalate of varying size and shapes are found associated with groups of fibres, often forming a sheath surrounding the fibre bundles. Fibres are lignified, thick-walled, with few pits and have a narrow lumen.

The lignified idioblasts are found scattered in the mesophyll, often associated with veins. They are somewhat irregular in shape with moderately thickened walls and numerous pits.

Jackson and Snowdon (1968) agree with Wallis (1960) in pointing out that the leaves of *E. truxillense* or Peruvian coca may be distinguished by the absence of lignified idioblasts. The features of fragmented coca leaf are shown in Figure 5.4, taken from Greenish and Collin (1904). These authors identify the diagnostic features of coca leaves as:

- the polygonal papillose cells of the lower epidermis
- the paracytic stomata

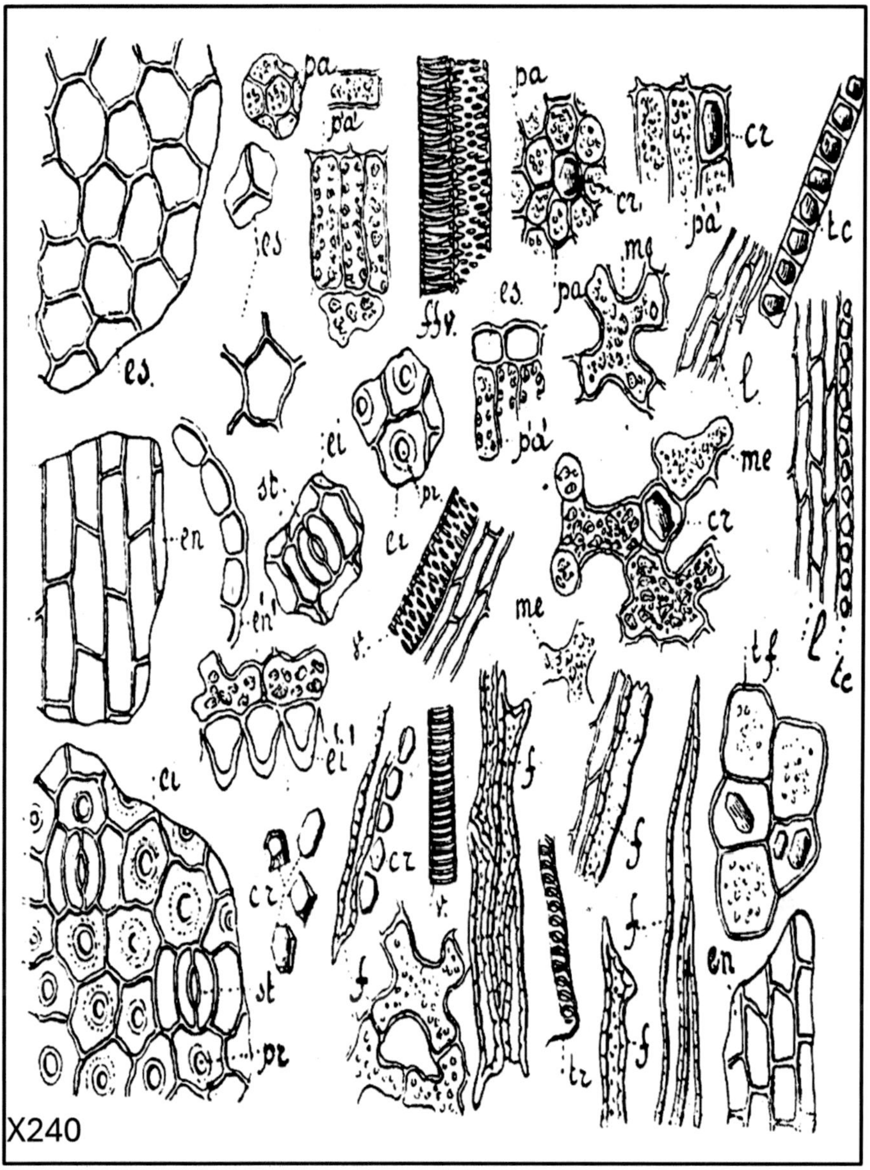

Figure 5.4 *Erythroxylon coca*: microscopic detail of powdered coca leaf.

(From Greenish and Collin (1904), page 65, figure 35.)

(*Continued*)

cr, Prismatic crystals of calcium oxalate.
ei, Lower epidermis with surface view of papillose cells (*pr*).
e'i', Lower epidermis in section.
en, Epidermis over veins, surface view.
e'n', Epidermis over veins in section.
f, Pericyclic fibres.
ffv, Fragments of vessels from midrib.
l, Bast.
me, Spongy parenchyma.
pa, Palisade cells with crystals, surface view.
p'a', The same in section.
st, Stomata, with two cells parallel to the ostiole.
tc, Crystal cells.
tf, Cortical tissue of midrib.
tr, Small vessels. *No hairs either simple or glandular.*

Figure 5.4 (Continued)

- the presence of pericyclic fibres
- the absence of trichomes or hairs
- the prismatic crystals of calcium oxalate

5.3.3 Role of DNA Analysis

As we have seen with cannabis and opium poppies, AFLP analysis played an important role in early applications of DNA analysis. This is also true for coca studies with Johnson et al. (2003) reporting on the identification of *Erythroxylon* species by AFLP DNA analysis. A combination of five AFLP primer pairs proved optimal in differentiating the four taxa, *E. coca* var. *coca*, *E. coca* var. *ipadu*, *E. novogranatense* var. *novogranatense* and *E. novogranatense* var. *truxillense*, from a non-cocaine species, *E. aerolatum*. Based on an analysis of 132 coca leaf accessions, the authors also demonstrated that geographical profiling within Colombian production areas was feasible. The same authors (Johnson et al., 2005) further explored the evolutionary relationship of the four cultivated cocaine producing *Erythroxylon* species, showing that *E. coca* and *E. novogranatense* are indeed two independent species.

A later study by Emche et al. (2011), focussed on phylogeny, also used AFLP DNA analysis to investigate the relationship amongst the cocaine-producing varieties and their non-cocaine/cultivated relatives. Their results showed that the cultivated species were more closely related to each other than any other species of *Erythroxylon*.

The complete genome sequences of *E. coca* and *E. novogranatense* have recently been reported (White et al., 2022). White et al. (2020) have previously shown that the progenitor of cultivated coca was *Erythroxylon gracilipes*.

A project report for a USDA (United States Department of Agriculture) Agricultural Research Service–sponsored research project on DNA fingerprinting of coca leaves (Meinhardt, 2012) described work being conducted to explore the potential for short sequence repeats (SSRs) and SNP markers. SNPs are reported as providing more information than SSRs at lower costs, and this information was used to understand the genetic diversity within the coca plant species. This study was marked as 'terminated', and no publications were recorded. As the author of this report, Meinhardt, is also an author in the previously mentioned 2011 paper by Emche et al., it would appear that the focus of this group turned to AFLP DNA analysis as the most useful approach.

Unlike cannabis and opium poppy DNA studies, I have found no published studies of other DNA approaches.

5.4 Khat Plant

5.4.1 Introduction, Background and Legal Status

Khat (*Catha edulis*) belongs to the family *Celastraceae*, which contains 60–70 genera and 850–900 species. Khat, (qat, chat, jaad or miraa) is the Arabic name for *C. edulis*, a flowering shrub or small tree indigenous to Ethiopia and Yemen, which has spread to other countries in eastern and southern Africa, including Somalia and Sudan. The epithet '*edulis*' signifies that the plant is edible. Although a single species, there are numerous types based on locality and the ecology of where it is grown. For example, 44 different types are recognised in Yemen alone (Dhaifalah & Santavy, 2004).

Similar to the use of coca leaves in South America, khat has an ancient history of its leaves being chewed as a social and cultural custom by indigenous communities.

In 2009, it was estimated that as many as five to ten million people globally used khat on a daily basis (Balint et al., 2009). Khat is also used as a cash crop in many countries, including Ethiopia and Yemen.

Khat contains the alkaloid *cathinone*. Cathinone is not very stable and breaks down to produce *cathine* and *norephedrine* (Getasetegn, 2016). Cathinone has the same sympathomimetic effects as amphetamine and is sometimes referred to as *natural amphetamine* or *herbal ecstasy* (Pendl et al., 2021). After harvest, as khat leaves dry out, the levels of cathinone decrease, hence the preference for fresh leaf material. Chappell and Lee (2010), however, have shown that when khat plant material is dried over a three-day period, there is no appreciable loss of cathinone and that, once dried, cathinone persists for a time frame of at least ten years. These authors believe that it is a common misconception that cathinone is highly unstable once the plant is harvested.

The dried leaves of khat are also called *Abyssinian tea* or *Bushman's tea*. Wallis (1960) refers to khat as an 'other drug described as teas', and Lewis and Elvin-Lewis (1977) includes 'Chat' as a stimulant in their book on Medical Botany. The 'desired' effects of Khat include euphoria, increased alertness, general excitement and hyperactivity, accompanied by relief from hunger and fatigue (Lewis & Elvin-Lewis, 1977). The 'less desired' effects, as reviewed by Getasetegn (2016), include potentially serious effects on the cardiovascular, respiratory and gastrointestinal systems. Amongst other less serious adverse effects is constipation!

The legal status of khat varies greatly from country to country. In the countries where its use is traditional, including Yemen, Djibouti, Kenya, Uganda, Ethiopia, Somalia and Sudan, the production, sale and consumption of khat are legal, and as previously stated, it is cultivated as a cash crop.

Neither is khat included in the 1961 Single Convention on Narcotic Drugs. Cathinone and cathine are included in Schedule 1 and Schedule 3 of the 1971 Convention on Psychotropic Substances. The World Health Organisation (WHO) Expert Committee on Dependence-producing Drugs, Fourteenth Report, confirmed the view that the abuse of this substance (khat) was a regional problem and best controlled at that level (WHO, 1965). In 1980, the WHO classified the plant as a drug of abuse that can produce mild to moderate psychological dependence but did not consider khat to be seriously additive.

Nonetheless, in the last 10–20 years, an increasing number of Western countries have introduced legislation to make khat illegal. The drivers for this change in attitude can largely be put down to two factors. The first has been the increased diaspora in the 'west' of people from those countries where khat has been traditionally used, caused to a large extent by regional conflicts and people seeking refugee status. Khat use has followed immigrant communities, but there is little evidence that the habit has spread to the community at large. The second reason that has driven the policy change has been emerging health concerns. It could be argued that banning the personal use of khat targets specific communities and that prohibition simply drives use into the hands of organised crime (Gebissa, 2010). There is increasing evidence of the illegal importation of khat leaves into several Western countries across North America, Europe and Australia.

Furthermore, the plant can be found growing in many countries in botanic gardens and elsewhere. For example, a major report into law enforcement and khat in Australia found that several plant nurseries around Australia offered khat plants for sale and that khat was growing in gardens in Perth and in the Royal Botanic Gardens, Sydney (Douglas et al., 2012). Regulation of khat in Australia is through a complex regime of overlapping and differing Commonwealth, State and Territory laws, resulting in law enforcement being confused. Douglas et al. (2012) proposed a regulated environment that allows for import and agriculture to be managed, and that might avoid the association between khat and organised crime. These authors also suggested that there was a need for more information for potential user groups on the health effects of using khat. The fact that khat is being grown in Australia is borne out by a recent seizure of khat plants in Queensland (see Figure 5.5).

From a forensic perspective, it is likely that in the future there will be an increase in seizures of khat by border and biosecurity authorities. History tells us that no matter how hard such authorities try, they are never successful in stopping illicit substances at the border. Furthermore, khat plants are in the community. There is a need for border/biosecurity and law enforcement officers to be able to identify or know where to turn, to have alleged khat material identified. The aim of the remainder of this section is to provide the botanical information that would be required to identify khat plants and plant parts.

Figure 5.5 Seizure of khat (*Catha edulis*) plants in Queensland, Australia.

5.4.2 Botanical Description of Khat Plants

5.4.2.1 Morphology

As commented on in the previous section, although accepted as a single species, there is considerable variation in the general appearance of samples of khat from different regions and grown in different conditions.

The following description is drawn from Van der Vossen (2002).

A mature wild khat tree is an evergreen perennial, glabrous, reaching a height of up to 25 m with a single stem branching into a number of vertical stems higher up and with a small, pointed crown. In cultivation, however, khat is a small shrub raised from cuttings, which are multistemmed from the start. Depending on the management of the plant's growth, it can be as short as 0.5 m but typically ranges from 2.5 to 6 m in height.

The bark is thin, smooth and pale grey-green in cultivated plants, and rough on large trees. Branches are terete (round in cross-section), pale to brownish grey in colour. Young twigs are usually flattened and dull green to brownish red in colour. Leaves are alternate on vertical branches and opposite on flowering branches.

The leaves are simple, subtended by triangularly shaped stipules, which are pale green and fall off, leaving a rim-like scar. The petiole is round, 3–11 mm long and pale to dark green. The leaf blade is oblong to elliptical or obovate, tapering at the base to a narrow point and also tapering to a sharp point, forming concave sides along the tip. The leaf margin is crenate to serrate. The upper surface is distinctly pinnate, which is even more marked on the lower surface with a prominent midrib. Near the margin of a leaf, the lateral veins anastomose to form a reticulate venation (Shadan & Shellard, 1962). Mature leaves are leathery and glossy in appearance.

Inflorescences are axillary and dichasial-cymose (a cymose inflorescence in which each axis produces two opposite lateral axes), with many flowers with persistent triangular bracts. Flowers are bisexual, regular, 5-merous, 2–4 mm in diameter with a short pedicel and 1–2.5 mm long. Sepals are fused at the base, broadly ovate to suborbital, 0.5–1 mm long and with fimbriate (fringed) margins. Petals are free, elliptical-oblong, 1–1.5 mm long and white or pale yellow, with finely serrulate to fimbriate margins. The fruit, technically a *samara*, is an oblong, three-valved pendulus capsule, 6–12 mm in length and containing one to three seeds. Seeds are obovoid in shape and have a wrinkled surface formed by small bumps or projections (rugose-papillose).

It is outside the scope of this chapter to describe the management of khat as a crop. What is relevant, however, is how khat is harvested, as it is the harvested material that is most likely to be seen in a forensic context. Typically, young shoots are picked two to three times per week during the growing season. Young shoots are broken off and trimmed to a length of about 40 cm, sometimes 80–100 cm, before tying in to bundles of a size sufficient for

Figure 5.6 *Catha edulis* plants in a market in southeastern Ethiopia.

2 hours of chewing. These bundles are sprinkled with water to keep them fresh, wrapped in banana or other leaves and packed in plastic bags for transport to market (Van der Vossen, 2002) – see Figure 5.6.

5.4.2.2 Anatomy of Khat Leaves

The earliest description of khat leaves goes back to 1893 (Collin, 1893). The most comprehensive description of khat leaf anatomy is the study by Shadan and Shellard (1962), which can be considered as the definitive description of khat leaf anatomy. The following description is largely based on this study.

The leaf in cross-section is dorsiventral with two layers of palisade cells below the upper epidermis and a loosely packed spongy mesophyll with regularly spaced large air spaces above the lower epidermis. Numerous cluster crystals of calcium oxalate are present in the spongy mesophyll.

On surface view, the upper epidermis has a single layer of polygonal cells covered with a fairly thick, smooth cuticle. The anticlinal walls are somewhat sinuous but not thickened to any great extent. No trichomes or stomata are present. The appearance of epidermal cells over the veins is similar to that of the remainder of the upper epidermal cells.

The lower epidermis consists of a single layer of polygonal cells, the walls of which are thickened in a similar manner to those of the upper epidermis. The anticlinal walls are only slightly sinuous. The epidermis is covered in a fairly thick, smooth cuticle. Some of the cells contain one or two small crystals of calcium oxalate, which are either clusters or conglomerates of a few components. No trichomes are present. There are a large number of anomocytic stomata with typically three to four, occasionally five, cells. The appearance of the lower epidermis over the veins differs from the remainder of the epidermis, as there are no stomata, no cells containing calcium oxalate and the cells in the surface view are irregularly rectangular in shape.

Shadan and Shellard (1962) provide a detailed description of the internal anatomy of the midrib and vascular tissues. Of particular note is the presence of a crescent of pericyclic fibres subtending the phloem in the medullary rays. The extent of the development of this fibrous tissue depends on the age of the leaf, with older leaves containing more fibres. Many of the fibres are branched tripartly, and the ends are frequently bifurcated. The fibres are slightly lignified and have lumens of variable width.

The xylem consists of radial rows of vessels, tracheids and fibre tracheids with well-defined medullary rays. In fragmented material, fibre tracheids are lignified and have elongated oval bordered pits. Tracheids may have scalariform and reticulate thickening, whilst others may have elongated bordered pits.

5.4.3 Role of DNA Analysis

I have been unable to find any DNA studies of khat that are specifically aimed at the identification of khat or other forensic applications. The few studies that have been published usually have a phylogenetic focus, although the outcomes can also be applied to determine possible biogeographic origin.

Al-Thobhani et al. (2008) report on the use of RAPD markers on 40 genotypes of khat obtained from locations in Yemen. In all, 152 amplified fragments were obtained using 10 random primers with 78.3% polymorphism. The results indicate that populations from different geographic regions and growth habits in Yemen showed a clear genetic differentiation.

Using 17 SSR markers and 1561 wild and cultivated individuals across historical areas of khat cultivation, Tembrock et al. (2017) showed that Ethiopian cultivars had diverged from their wild relatives, whereas Kenyan khat had diverged to a lesser extent.

Curto et al. (2013) also used SSR markers with next-generation sequencing (NGS) to study the phylogeny of khat. Using 27 SSR markers and 23 samples, geographic differences indicated that this type of analysis can be used to determine the source of cultivated and wild samples.

These two SSR-based studies show the potential for DNA analysis to determine geographic origin, which would be useful to determine the origin of khat seized by border or law enforcement authorities.

Finally, using NGS technology Gu et al. (2018) have reported the complete chloroplast genome of *C. edulis*.

References

Al-Thobhani, M.A.H., Sathyanarayana, B.N., Simon, L., & Sondur, S.N. 2008. First Comparative Genotype Study on Khat (Catha *edulis* Forsk.) Genotypes from Yemen. Mid. *East. & Russ. Journal of Plant Science and Biotechnology*, 2, 1–8.

Anon. 1961. *Single Convention on Narcotic Drugs.* https://www.unodc.org/pdf/convention1961en.pdf

Anon. 2016. *Weeds of Australia – Fact Sheet. Papaver somniferum.* Identic Pty Ltd. https://keyserver.lucidcentral.org/weeds/data/media/Html/papver_somniferum.htm

Anon. 2021. *Kew Plants of the World Online.* https://powo.science.kew.org

Anon. 2023. *Know Your Poppies.* Department of Natural Resources and Environment Tasmania. https://nre.tas.gov.au/Documents?Know%Your%20%Poppies%20Fact%20Sheet.pdf

Balint, E.E., Falkay, G., & Balint, G.A. 2009. Khat: A Controversial Plant. *Wiener Klinische Wochenschrift*, 121, 604–614.

Bentley, R., & Trimen, H. 1875. *Medicinal Plants.* London: J&A Churchill.

Chappell, J.S., & Lee, M.M. 2010. Cathinone Preservation in Khat Evidence Via Drying. *Forensic Science International*, 195, 108–120.

Cheng, L., Lin, Z., Xin, C., Sun, H., & Li, X. 2021. Molecular Identification and Phylogenetic Analysis of Papaver Based on ITS2 Barcoding. *Journal of Forensic Sciences*, 67, 712–719.

Collin, E. 1893. Le *Catha edulis. Journal of Chemical and Pharmaceutical Research* 5, 28, 337–342.

Curto, M.A., Tembrock, L.R., Puppo, P., Nogueira, M., Simmons, M.P., & Meimberg, H. 2013. Evaluation of Microsatellites of *Catha edulis* (qat; Celastraceae) Identified Using Pyrosequencing. *Biochemical Systematics and Ecology*, 49, 1–9.

DEA. 1991. *Coca Cultivation and Cocaine Processing; An Overview.* Washington, DC: US Department of Justice. Drug Enforcement Administration. Office of Intelligence.

DEA. 2019. *Unwashed Poppy Seed.* Drug Control Division. Drug and Evaluation Section. Washington DC.

Dhaifalah, I., & Santavy, J. 2004. Khat Habit and its Health Effect. A Natural Amphetamine. *Biomedical Papers*, 148, 11–15.

Dittbrenner, A., Mock, H.P., Borner, A., & Lohwasser, U. 2009. Variability of Alkaloid Content in *Papaver sominferum* L. *Journal of Applied Botany and Food Quality*, 82, 103–107.

Douglas, H., Pedder, M., & Lintzeris, N. 2012. *Law Enforcement and Khat: An Analysis of Current Issues.* Australia: National Drug Law Enforcement Research Fund.

Emche, S.D., Zhang, D., Islam, M.B., Bailey, B.A., & Meinhardt, L.W. 2011. AFLP Phylogeny of 36 Erythroxylon Species-Genetic Relationships Among Erythroxylum Species Inferred by AFLP Analysis. *Tropical Plant Biology*, 4, 126–133.

Gebissa, E. 2010. Khat in the Horn of Africa: Historical Perspectives and Current Trends. *Journal of Ethnopharmacology*, 132, 607–614.

Getasetegn, M. 2016. Chemical Composition of *Catha edulis* (Khat); A Review. *Phytochemistry Reviews*, 15, 907–920.

Greenish, H.G., & Collin, E. 1904. *An Anatomical Atlas of Vegetable Powders.* London: J&A Churchill.

Gu, C., Tembrock, L.R., Zheng, S., & Wu, Z. 2018. The Complete Genome Sequence of *Catha edulis*: A Comparative Analysis of Genome Features with Related Species. *International Journal of Molecular Sciences*, 19, 525. https://doi.org/10.3390/ijms19020525

Haber, I., Pergolizzi, J., & LeQuang, J.A. 2019. Poppy Seed Tea: A Short Review and Case Study. *Pain and Therapy*, 8, 151–155.

Jackson, B.P., & Snowdon, D.W. 1968. *Powdered Vegetable Drugs. An Atlas of Microscopy for Use in the Identification and Authentication of Some Plant Materials Employed as Medicinal Agents*. London: J&A Churchill Ltd.

Johnson, E.L., Saunders, J.A., Mischke, S., Helling, C.S., & Emche, S.D. 2003. Identification of *Erythroxylum* Taxa by AFLP DNA Analysis. *Phytochemistry*, 64, 187–197.

Johnson, E.L., Zhang, D., & Emche, S.D. 2005. Inter- and Intra-specific Variation Among Five *Erythroxylum* Taxa Assessed by AFLP. *Annals of Botany*, 95, 601–608.

Kapoor, L.D. 1997. *Opium Poppy. Botany, Chemistry and Pharmacology*. Boca Raton: CRC Press.

Karch, S.B. 1998. *A Brief History of Cocaine*. Boca Raton: CRC Press.

Lewis, W.H., & Elvin-Lewis, M.P.F. 1977. *Medicinal Botany. Plants Affecting Man's Health*. New York: John Wiley & Sons.

Liu, L., Du, Y., Shen, C., Li, R., Lee, J., & Li, P. 2020. The Complete Chloroplast Genome of *Papaver setigerum* and Comparative Analysis in Papaveraceous. *Genetics and Molecular Biology*, 43, e20190272.

Meinhardt, L.W. 2012. Report on DNA Fingerprinting of Coca Leaves to Establish Coca Genotypes. Unpublished study.

Ngernsaengsaruay, C., Leksungnoen, N., Chanton, P., Andriyas, T., Thaweekun, P., Rueansri, S., Tuntianupong, R., & Hauyluek, W. 2023. Morphology, Taxonomy, Anatomy, and Palynology of the Opium Poppy (*Papaver somniferum L.)* Cultivation in Northern Thailand *Plants*, 12, 2105. https://doi.org/10.3390/plants12112105

Pendl, E., Pauritsch, U., Kollroser, M., & Schmid, M.G. 2021. Determination of Cathinone and Cathine in Khat Plant material by LC-MS/MS: Fresh vs. Dried Leaves. *Forensic Science International*, 319. https://doi.org/10.1016/j.forsciint.2020.110658

Plowman, T. 1980. *Botanical Perspectives on Coca*. In Jeri, F.R. (ed.), *Proceedings of the Interamerican Seminar on Coca and Cocaine*, Lima, Peru.

Saunders, J.A., Pedroni, M.J., Penrose, L.D.J., & Fist, A.J. 2001. AFLP Analysis of opium Poppy. *Crop Science*, 41, 1596–1601.

Shadan, P. & Shellard, E.J. 1962. An Anatomical Study of Ethiopian Khat (Leaf of *Catha edulis* Forsk). *Journal of Pharmacy and Pharmacology*, 14, 110–118.

Shoyama, Y., Kawachi, F., Tanaka, H., Nakai, R., Shibata, T., & Nishi, K. 1998. Genetic and Alkaloid Analysis of *Papaver* species and Their F1 Hybrid by RAPD, HPLC and ELISA. *Forensic Science International*, 91, 207–217.

Simpson, M.G. 2019. *Plant Systemics*, 3rd ed. Amsterdam: Elsevier.

Tembrock, L.R., Simmons, M.P., Richards, C.M., Reeves, P.A., Reilley, A., Curto, M.A., Meimberg, H., Ngugi, G., Demissew, S., Al Khulaidi, A.W., Al-Thobhani, M., Simpson, S., & Varisco, D.M. 2017. Phylogeography of the Wild and Cultivated Stimulant Plant Qat (Catha *edulis*, Celetraceae) in Areas of Historical Cultivation. *American Journal of Botany*, 104, 538–549.

Van der Vossen, H.A.M. 2002. *Catha edulis* (Vahl) Foressk.ex Endl (Internet) Record from PROTA4U. In Oyen, L.P.A. & Lemmens R.H.M.J. (eds.), *PROTA (Plant Resources of Tropical Africa/Resources vegetales de l'Afrique tropicale)*, Wageningen, Netherlands. https://uses.plantnet.project.org/en/Catha edulis(PROTA).

Wallis, T.E. 1960. *Textbook of Pharmacognosy*. London: J&A Churchill.

White, D., Meinhardt, L., Bailey B., & Pirro, S. 2022. The Complete Genome Sequence of *Erythroxylum coca* and *Erythroxylum novogranatense. Biodivers Genomes.* https://doi.org/10.56179/001c.39776

White, D.M., Huang, J., Jara-Munoz, O.A., Madrinan, S., Ree, R.H., & Mason-Gamer, R.J. 2020. The Origins of Coca: Museum Genomics Reveals Multiple Independent Domestications from Progenitor *Erythroxylum gracilipes. Systematic Biology*, 70, 1–13.

World Health Organisation. 1965. *Expert Committee on Dependence-Producing Drugs*. 14th Report. United Nations Office of Drugs and Crime.

Yadav, H.K., Shukla, S., & Singh, S. 2006. Genetic Variability and Interrelationship Among Opium and Its Alkaloids in Opium Poppy (*Papaver somniferum* L.). *Euphy*, 150, 207–214.

Young, B., Roman, M.G., LaRue, B., Gangitano, D., & Houston, R. 2020. Evaluation of 19 Short Tandem Repeat Markers for Individualization of *Papaver somniferum. Science & Justice*, 60, 253–262.

Zhang, Y., Wang, J., Yang, L., Niu, J., Huang, R., Yuan, F., & Liang, Q. 2022. Development of SSR and SNP Markers for Identifying Opium Poppy. *International Journal of Legal Medicine*. https://doi.org/10.1007/s00414-022-02810-4

6 Plant Fibres

6.1 Introduction

The examination of fibres in a forensic laboratory is usually carried out by fibre examiners, often as members of a chemical criminalistics team. Fibre examiners may have a variety of primary academic qualifications, with most having a background in chemistry. Typically, the forensic examination of fibres is heavily oriented to microscopy and more advanced analytical chemistry techniques (Robertson & Roux, 2018), which reflects the fact that most fibres encountered in forensic examinations will be man-made. Polyester fibres make up more than 50% of worldwide fibre production (Fergusson & Hemmings, 2018). With the exception of cotton and wool fibres, fibres from animal and plant sources form a very small percentage of world production. In particular, in a forensic context, non-cotton fibres of plant origin are quite uncommon today. This partially explains why the skill set to examine plant fibres has all but disappeared in forensic laboratories. Throughout the remainder of this chapter, I will refer to fibres of plant origin as simply plant fibres. These are also commonly called *vegetable fibres*, but as they do not strictly come from vegetables, I prefer plant fibres.

The fact remains that the use of plant fibres remains strong in some countries, especially in the Indian subcontinent and China. Plant fibres also remain important in many societies where they are still in common use. Furthermore, hemp as a fibre crop has made somewhat of a comeback, and it is not uncommon to see garments made of hemp fibre. Of course, prior to the emergence of man-made fibres, products that today are dominated by man-made fibres would all have comprised plant fibres. In aspects such as forensic archaeology, textile conservation, art history and/or fraud, a knowledge of plant fibres is essential.

With increased concerns about plastics, there is likely to be a stronger future interest in replacing some products with natural alternatives. The Food and Agricultural Organisation (FAO) have listed jute, abaca, coir, kenaf and sisal under the combined name of JACKS and claims that JACKS contributes to sustainable development by fostering social, environmental and economic development in the poorest rural areas of the world, helping to offset some of the negative environmental impacts associated with plastic production and disposal.

Taken together, there is still a need for individuals with the skill set to be able to identify plant fibres. At present, this skill set is only likely to be found

DOI: 10.4324/9781003308508-6

in botanical establishments such as national botanical research institutes or gardens.

Hence, the primary aim of this chapter is to provide the reader with sufficient detail to describe and identify the botanical source of a plant fibre from a variety of potential end products, such as textile fabrics, or from cordage and ropes that might typically be seen in a forensic context. It is outside of the scope of this treatment to consider in detail the growing, harvesting, extraction and processing, and textile aspects of plant fibres. Extraction and processing will be covered to the extent that the *technical* plant fibre is determined by these processes.

Bast and Other Plant Fibres (Franck, 2005a) provide this type of information for flax, hemp, jute, ramie, sisal, abaca and coir, as well as some less common plant fibres. *Handbook of Textile Fibres – Natural Fibres* (Cook, 1984) also provides an excellent introduction to plant fibres with a focus on textile science.

With regard to the forensic identification of plant fibres, the single most important source of information is *Identification of Vegetable Fibres* (Catling & Grayson, 1982).

In describing the microscopic appearance of bast fibres, the terms *dislocations* and *cross-markings* are used. Catling and Grayson (1982) discuss the origin and meaning of these terms. The exact cause of dislocations is not entirely clear. One theory is that they are regions in fibre cells where the cell wall has been damaged by compression during growth. Catling and Grayson (1982) define cross-markings as 'the attached wall remains of neighbouring cells or the impressions on the fibre cell wall made by neighbouring cells which have been removed during processing'. These marks may cross the entire cell or may cross only part of the cell, and they may be fine or coarser in appearance. Where the marks are more pronounced, they may be accompanied by distortion of the cell wall. In these instances, Catling and Grayson (1982) suggest that it can be difficult to differentiate between a cross-marking and a dislocation. They conclude that there is some association between dislocations and cross-markings and that it is appropriate to describe them under one heading. Some authors also call dislocations *nodes*. These markings are best seen when fibres are examined with polarising light microscopy – see Figure 6.1.

6.2 Classification of Plant Fibres

Plant fibres can be classified according to the part of the plant from which the 'fibres' are derived. Based on this approach, there are three broad categories of plant fibres:

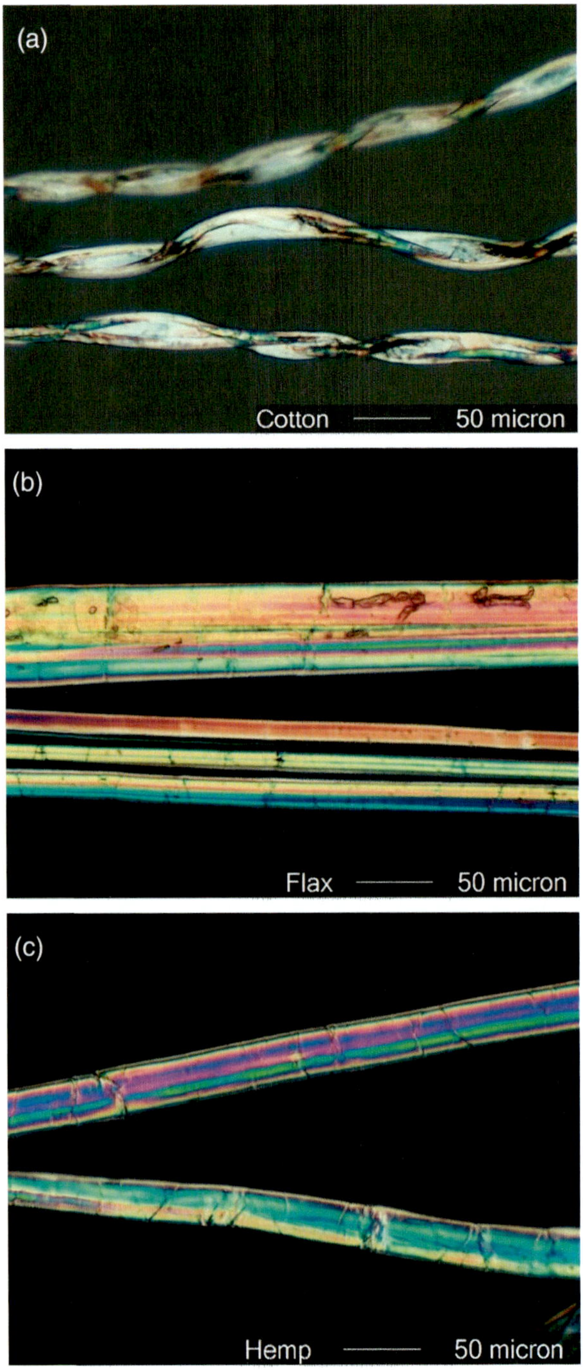

Figure 6.1 Examples of plant fibres as seen under polarising light microscopy: (a) cotton, (b) flax and (c) hemp.

(From MicorlabNM photographic gallery.)

- *Bast or stem fibres* – these come from the fibrous bundles in the inner bark (phloem or bast) of the stems of dicotyledonous plants
- *Leaf fibres* – these run lengthwise through the leaves of monocotyledonous plants
- *Seed and fruit fibres* – included are true seed hairs and the so-called 'flosses'

It is important to be clear at this stage about how the term fibre is used in relation to plant fibres. As described in Chapter 2, in botanical terms, a fibre is an elongated sclerenchyma cell with tapered ends, typically found associated with the cortex, phloem and xylem.

In commercial use, the term *technical fibre* refers to a strand of fibre cells. Thus, the bast fibres of commerce are not individual botanical fibres but rather are groups of botanical fibres cemented together, both end to end and side by side, to form bundles of filaments along the length of the stem (Anon, 1975). Individual botanical cells in commerce are referred to as *ultimates* (Catling & Grayson, 1982).

6.3 Chemical Composition of Plant Fibres

The primary chemical component of vegetable fibres is *cellulose*, with other constituents such as *hemi-cellulose, pectins, lignin* and fats and waxes present in varying quantities depending on the fibre.

Table 6.1 shows the approximate chemical composition of cellulosic fibres.

Cellulose is the primary component of cell walls, comprising 1,4-β-D-glucopyranose units varying in length from hundreds to thousands of monomer units and forming microfibrils. Cellulose microfibrils are embedded in a cross-linked matrix of non-cellulosic molecules. These molecules are the polysaccharides known as *hemicelluloses* and *pectins*, as well as the structural proteins called glycoproteins (Evert, 2006).

Table 6.1 Approximate (%) Chemical Composition of Cellulose Fibres

Fibre	Cellulose	Hemicellulose	Pectin	Lignin	Fat and Wax
Cotton	92–95	5.7	1.2	0	0.6
Flax	62–71	16–18	1.8–2.0	2.0–2.5	1.5
Hemp	67–75	16–18	0.8	2.9–3.3	0.7
Ramie	68–76	13–14	1.9–2.1	0.6–0.7	0.3
Jute	59–71	12–13	0.2–4.4	11.8–12.9	0.5
Sisal	66–73	12–13	0.8	9.9	0.3
Abaca	63–68	19–20	0.5	5.1–5.5	0.2
Coir	36–43	0.2	3–4	41–45	

Hemicelluloses are a heterogeneous group that can include xylans, mannans, glucomannans, galactans and arabans, with xyloglucans often predominant. However, hemicellulose preparations from different tissues show such marked differences in the number and type of compounds present that it has been suggested that the arbitrary nature of hemicellulose questions the usefulness of the term. Xyloglucans consist of linear chains of 1,4-β-D-glucan as in cellulose but with short side chains of xylose, galactose and often terminal fucose (Carpita & McCann, 2000). Xylans are the major non-cellulosic polysaccharides of the secondary walls of all angiosperms and, hence, natural fibres.

Pectins are a diverse group of non-cellulosic polysaccharides. They are characteristic of the primary cell walls of dicots and, to a lesser extent, monocots. The two major constituents of pectin are polygalacturonic acid and rhamnogalacturonan. Pectins are highly hydrophobic and form a gel in which the cellulose–hemicellulose network is embedded (Evert, 2006).

Finally, *lignins* are phenolic polymers that are deposited mainly in cell walls of supporting and conducting tissues. The monomeric composition of lignins in different species varies greatly, but, in general, there are three main monomeric units, *coumaryl*, *coniferyl* and *sinapyl* alcohols. As can be seen in Table 6.1, the amount of lignin in vegetable fibres varies from 0% for cotton to around 12% in jute and over 40% in coir. This is a useful property in differentiating bast fibres, where jute has a much higher level of lignan content than other bast fibres such as flax or hemp. The Swett test (Swett, 1918) can be used to give a qualitative measure of the amount of lignin present. The cell walls of tissues containing lignin give a bright purple-red colour when treated with phloroglucinol in concentrated hydrochloric acid.

The presence of high levels of lignin in fibres can also result in changes in colour when fibres are exposed to light, for example, with jute becoming yellow.

6.4 Bast and Stem Fibres

6.4.1 Introduction

In botanical terms, *bast* is synonymous with phloem. In a commercial sense, the term bast fibre is used to denote fibres obtained from the cortex and pericycle in addition to the phloem. Plant species yielding bast fibres are dicotyledonous, whereas leaf fibres come from monocotyledonous plants.

Bast fibres are referred to as *soft fibres*, with leaf fibres being called *hard fibres*. As previously discussed, the bast fibres of commerce are not single plant fibres or cells but rather bundles of elongated cells (ultimates) cemented together side by side and end to end in long strands. These bundles are usually separated from the remaining plant stem tissues by a process called *retting*, which, simply put, is a process of controlled rotting.

There are four types of retting process (Cook, 1984; Fergusson & Hemmings, 2018):

- *Dam retting* – bundles of stems are immersed in ponds or dams for about 10 days. This method is now only found in Egypt.
- *Dew retting* –stems are spread on the ground and left for several weeks. Wetting by dew and rain promotes fermentation, which rots the tissues comprising parenchyma cells.
- *Tank retting* – stems are mechanically stripped of seed bolls, the 'straw' tied in bundles and packed into a concrete tank filled with water at about 30°C. This process is complete in about 3 days.
- *Chemical retting* – as this method costs more than the alternatives, it is less common. The method uses alkalis such as sodium hydroxide and sodium carbonate with soap.

Tank retting is the preferred method today as it allows greater control over the process.

Following retting, the next stage in fibre production is *breaking*. The fibre bundles and any remaining tissues are passed between fluted rollers to break the woody core into fragments without damaging the fibres. The broken straw then goes through a process called *scutching*, which separates the woody material from the fibre. The fibre is then *hackled*, a process where the fibres are combed by drawing them through sets of pins, each set being finer than the previous one. The coarse bundles are separated into finer bundles and arranged parallel to one another. The long fine fibres are known as *line*, and the shorter fibres as *tow*.

Machinery is used only for the major bast fibres to conduct the above processes. For many bast fibres, following retting, the remaining steps are labour-intensive and carried out by hand.

6.4.2 Flax

Flax fibre comes from the stem of the annual plant *Linum usitatissimum* in the family Linaceae. There are many varieties and cultivars. When used to make textiles, the fabric is called *linen*. The history of linen from ancient times to modern times is in itself fascinating. Cook (1984) records that when Tut-ankh-amen's tomb was opened in 1922, linen curtains, which had been there since about 1250 BC, were still intact. Linen is an unusual fabric in that it becomes stronger when wet and can withstand repeated washings without deterioration. Today, most of the world's production of flax comes from

Russia, China, the 'low countries' (France, Belgium and the Netherlands) and Egypt (Salmon-Minotte & Franck, 2005).

Flax can be grown primarily for either its fibre or its seeds (linseed oil) or for both. Plants grown for fibre are around 90–120 cm in height with a slender stem and a few side branches. When the crop is harvested, it is pulled up by the roots and not cut as with cereal crops. This is to maximise the amount of stem material.

Catling and Grayson (1982) describe the anatomy of the stem in detail. The important component from a fibre perspective is the pericyclic fibres, which lie next to the phloem cells.

The number of fibre bundles in a stem ranges from 15 to 40, and each bundle contains from 12 to 40 ultimates or individual fibre cells (Anon, 1975). The bundles of fibres are held together by pectins.

As discussed above, in order to extract the fibres, the other cellular material has to be removed by *retting*. Some cortical tissue may remain after processing, which gives flax its characteristic colour. Scutching and hackling break up the coarse fibre bundles into finer bundles, but they do not separate the fibre strands into individual botanical fibres. Depending on the quality and degree of processing, an examiner may have samples with fibre bundles and/or individual fibres.

In order to observe the features of individual cells, it is necessary to treat the sample to separate the individual cells. Catling and Grayson (1982) describe a method to macerate samples which is suitable for small samples. This involves placing the sample in a conical flask in an equal mixture of glacial acetic acid and 20-volume hydrogen peroxide in a heated water bath for 7–8 hours. The fibre sample is then transferred to a fresh flask in water and vigorously shaken to separate fibre stands into individual cells prior to the sample being mounted for microscopic examination. An alternative approach uses a 1% solution of sodium hydroxide (Anon, 1975), although this method would appear to be more suited to larger samples rather than for what might be expected in a forensic context.

Assuming that we are dealing with individual fibre cells, at the level of transmitted light microscopy (TLM), the cell walls of flax are typically uniformly thick, with the wall about six times as wide as the lumen. Catling and Grayson (1982), however, comment that it is not unusual to find cells with thin walls and broader lumens. In cross-section, the cells are pentagonal or hexagonal and can be slightly rounded. In a longitudinal view, the cells are typically elongate, cylindrical tubes with an obvious lumen or central canal through the centre. The lumen is narrow but clearly defined and regular in width, narrowing towards the fibre end, which tapers.

The width of the fibre may vary along its length, with frequent dislocations and fewer and faint cross-markings.

Flax has a low level of lignin present. This can be detected by the *phloroglucinol and hydrochloric acid test* (also known as the Swett test). There are various recipes for this test. All involve placing the sample into a solution of phloroglucinol in 95% alcohol, usually at a concentration of 1% or 2%. The sample is allowed to dry and then concentrated hydrochloric acid is added. Owing to the presence of coniferaldehyde groups in lignin, a red colour develops, varying in intensity depending on the degree of lignification. The sample can then be placed in a suitable mountant for microscopic examination.

Finally, if a suitably sized sample is available, a useful test for flax is the *drying-twist test* (Anon, 1975). This test is based on the fact that the walls of fibres have a spiral fibrillar structure with the external fibrillae in flax running in a direction corresponding to an S twist. The method is described in Appendix A.6.

6.4.3 Hemp

As discussed in Chapter 3, hemp fibres come from the stem of plants of *Cannabis sativa* in the family Cannabaceae.

Traditional and historical uses for hemp fibres included rigging of sailing ships, and in sails and ropes, cordage and twines. Today, there is a relatively small-scale use of hemp fibres, either alone or admixed with cotton, in textiles, including garments and headwear.

Hemp is also being used as a composite building material. In the latter use, fibres are extracted using hammer mills. This process results in short fibres, which can also be damaged. If hemp production is to increase, the availability of suitable equipment to produce higher quality long line fibres is a major challenge, as, currently, there are no commercial complete mechanised hemp harvesting equipment (Gregoire et al., 2021). Gregoire et al. (2021) report on a laboratory-scale scutching/hackling device to improve the long fibre scutching yield and minimise the production of tow fibres.

For fibre use, plants are grown to encourage non-branching single stems with an ideal maximum height of 1.8 m, although plants can reach 3 or more metres in height.

Catling and Grayson (1982) describe the anatomy of the stem of Cannabis in detail. In cross-section, stems are nearly orbicular at the foot, more angular or ridged higher up the stem and normally 4–10 mm thick. The primary cortex inside the ridges is narrow and consists of a few rows of thin-walled, tangentially elongated parenchyma cells. In the outer cortex are small groups of pericyclic fibres of larger diameter with thick, weakly lignified walls. The fibres of the secondary phloem are more numerous and occur in tangential bands between layers of sieve tubes with companion cells (Catling & Grayson, 1982). A group of fibre cells can contain between ten and 40 individual cells.

The length of fibre bundles ranges from 1,500 to 2,500 mm, with individual cells ranging from 35 to 40 mm (Sponner et al., 2005).

The phloem also contains laticiferous elements. Cluster crystals of calcium oxalate are present in the pericycle, secondary phloem and rays, often in chambered cells.

In cross-section, hemp fibres show considerable variation, generally being angular with four, five or six sides, whilst some cells are oval or round. Wall thickness and lumen width are variable, but most frequently the lumen occupies between one-third and one-half of the cell (Catling & Grayson, 1982). Fibre cell ends are very variable, with blunt pointed tip ends being most common.

Dislocations are frequent and pronounced, as are cross-markings, which, according to Catling and Grayson (1982), are much more frequent than in flax. Two further features can be used to distinguish flax from hemp. As was discussed, flax has an S twist when subjected to the drying -twist test. Hemp gives a counter-clockwise or Z twist in this test.

A further test that can be applied is to ash the sample and look for the presence of calcium oxalate crystals in the ash. Ashing can be carried out on a piece of mica over a small flame, the ash transferred to a microscope slide in a suitable mountant and examined using polarised light microscopy. As noted above, calcium oxalate crystals have not been observed in flax. A note of caution with the ashing test is that crystals may not be seen in heavily processed hemp, as the parenchyma cells containing them may have been removed during processing. Hence, this test should only be used as a positive test for hemp versus flax when crystals are found.

Finally, the term hemp is used for a number of other fibres such as Sunn hemp and New Zealand hemp. These are not true hemp and once again show how the use of common names can cause confusion.

6.4.4 Jute

Jute belongs to the genus *Corchorus* within the family Tiliaceae. There are many species in the genus *Corchorus*, but only two are grown commercially, *Corchorus capsularis* (white jute) and *Corchorus olitorius* (tossa jute). Three other bast fibre species produce very similar fibres to those of jute and are often included in treatments of jute. These species are kenaf (*Hibiscus cannabinus*), rosella (*Hibiscus sibdoriffa*) and urena (*Urena lobata*). The FAO groups these species along with jute as jute-like fibres.

Jute grows best in hot and humid climatic regions. India and Bangladesh account for over 93% of the world production. Most of the jute grown in India remains in India and is used in domestic production (Krishnan et al., 2005). Somewhere between 80% and 90% of jute in India and Bangladesh is used for manufacturing sacking, carpet backing (due to its non-slip qualities) and hessian cloth.

The jute plant is an herbaceous annual growing up to 2–6 m with a stem diameter of 20 mm – see Figure 6.2. Following retting, in a similar fashion to that described for flax, the strands of fibre can be 2 m long. The stem anatomy is described in detail by Catling and Grayson (1982). As with flax and hemp, the fibre bundles are associated with the phloem, more or less in rectangular groups of between 25 and 50 cells.

Commercial fibre from both species of *Corchorus* is of a golden yellow colour, sometimes with a pink or reddish tinge. Cook, however, describes jute as being 'dirty grey' with a natural silky lustre (Cook, 1984). During retting, some fibre ends become detached from the strands of fibre, giving jute a hairy, rough feel. Individual cells are about 2–6 mm long. Seen in cross-section, cells are polygonal or rounded polygonal with five or six sides. Catling and Grayson (1982) note considerable variation in cell wall thickness and lumen width in jute fibres. Cook (1984) describes the lumen of jute as irregular, becoming very narrow in places and wider towards the end of cells. This contrasts with the even lumen seen in flax fibres.

Figure 6.2 Jute field in Bangladesh.

Cross-marking and dislocations are seen in jute fibres, although dislocations are less numerous than in some other bast fibres. Cross-markings are not always present, and when they are present, they are fine and faint. Cross-marking may extend across the whole width of the cell, or it may only cover part of the cell. In the latter situation, Catling and Grayson (1982) note that it is not unusual to see a second set of marks, offset from the first.

Chains of rhombic-shaped crystals of calcium oxalate are common when the fibre is ashed with an occasional cluster crystal.

As previously mentioned, jute fibres have a high lignin content; the Swett test is a distinguishing test for jute fibres as they show a much deeper red colour than fibres with lower lignin content.

6.4.5 Ramie

Ramie comes from *Boehmeria nivea*, a member of the family Urticaceae. The specific name, and common name, *white ramie*, refers to the appearance of the abaxial leaf surfaces, which are densely covered with unicellular trichomes giving the surface a woolly appearance (Catling & Grayson, 1982). White ramie is also called *China grass*, perhaps reflecting that China has an ancient history of growing ramie and currently produces around 80% of world production. *B. nivea* var. *tenacissima* (sometimes recognised as a separate species, *Boehmeria tenacissima*) is known as *green ramie* and is grown in tropical climates.

Ramie is a perennial sending up numerous stems from an underground rhizome. Stems can reach a height of 2–2.5 m with a diameter of 8–16 mm at the base (Kozlowski et al., 2005).

Catling and Grayson (1982) describe the stem anatomy. As with other bast fibres, the fibre bundles are associated with the phloem. A notable feature is the presence of parenchyma cells embedded within the fibres, with frequent cluster crystals. Ramie is also notable for having mucilage canals or gum-like contents in the lumina of the fibre cells. The latter feature results in ramie requiring special methods to remove the fibres, requiring mechanical stripping of ribbons of fibres from the stems. Some form of degumming is then necessary to produce fibres suitable for spinning (Anon, 1975).

Ramie fibres are exceptionally long with a mean length of 120–150 mm and a mean fibre diameter of 40–60 mm (Kozlowski et al., 2005). Catling and Grayson (1982) report the fibre cells as thick-walled but not being uniform in size and shape, with some being hexagonal or pentagonal and others less distinctly angular. Cross-markings and dislocations are frequent, although cross-markings may be difficult to see due to variations in the thickness of cell walls.

As with cotton, ramie fibres can be mercerised by treatment with caustic soda to remove gum, resulting in fabrics that are smoother.

6.4.6 Sunn Hemp

Sunn hemp, also known as Indian or Bombay hemp, comes from the plant species, *Crotalaria juncea*, a member of the family Leguminosae. As the name suggests, the main country in which this species is grown on a commercial scale is India. Mature plants can reach a height of 3 m with stems reaching 2.5 cm in width. Commercial strands of fibres are removed in a similar fashion to other bast fibres through a retting process with hand processing to remove the fibre bundles. Commercial strands can reach 150 cm in length. Individual cells or ultimates are about 8.5 cm long. In cross-section, Sunn fibre cells are oval with an irregular lumen sometimes filled with a yellow substance (Cook, 1984). The strength of Sunn fibres increases when they are wet. Allied to the fact that Sunn fibres are resistant to microbial attack in seawater means that Sunn fibres are often used to make fishing nets.

6.4.7 Other Bast Fibres

Kenaf (*H. cannabinus*), rosella (*H. sibdoriffa*) and urena (*U. lobata*) are all from the family Malvaceae and are closely related to the jute family, the Tiliaceae. These three species are known under a number of local names. In India, kenaf is known as *mesta* and rosella as *bumli*. All three species are tropical plants and grow to about 3 m. Extraction of fibres and processing are essentially the same as for jute.

U. lobata also belongs to the family Malvaceae. It is grown in Brazil, Madagascar and the Congo region of Africa, where the fibre is known as *Congo jute*.

The stem anatomy of all of the above three species is similar to that of jute, as are the microscopic details and chemical composition. Catling and Grayson (1982) provide detailed descriptions of the two Hibiscus species.

Because of the similarity in the features of the fibres from the three species above with jute fibres, it is safer to identify fibres as jute-like.

6.5 Leaf Fibres

6.5.1 Introduction

As discussed above, leaf fibres are obtained from the leaves or leaf stalks of various monocotyledonous plants, where the fibres are associated with the vascular tissues, seen as veins on leaves. Leaf fibres are extracted by mechanical means, with the result that the removal of parenchyma cells is often less complete than is the case with bast fibres. When examined with TLM, leaf fibres will often have traces of xylem elements present.

The main leaf fibres of commerce are Abaca or Manila hemp (*Musa textilis*), Phormium or New Zealand hemp (*Phormium tenax*) and Sisal (*Agave sisalana*).

6.5.2 Abaca

Abaca, *M. textilis*, also known as Manila hemp, is a species of banana belonging to the family Musaceae. Abaca is primarily grown in the Philippines and Ecuador, with between 70% and 80% from the Philippines.

The abaca plant produces runners or underground stems or *stolons*, and commercial propagation is usually by cutting and transplanting rooted runners. Abaca has a false trunk or stem about 13–38 cm in diameter created by the leaf stalks (petioles), which form sheaths that tightly wrap around to form the *pseudo stem*. When the sheaf-leaves reach an appropriate stage of maturity, they unwrap from the pseudo-stem and take on the shape of banana-type leaves, oblong in shape with a deltoid base. The leaves vary in length, with the outer sheaf-leaves being shorter than the inner ones. A single plant may produce up to 25 stems maturing at different times. When the plant is mature, a flower stalk grows up inside the pseudo-stem. A mature plant can reach a height of 7.5 m. The pseudo-stems are ready for harvesting on flowering. A mature plant can be harvested for on average 15 years, with some plants being productive for up to 30 years.

Fibre extraction is carried out as soon as possible after the leaves have been cut. The first stage, called *tuxying*, separates the fibre bundles from the rest of the plant tissue and is done manually with a knife. This produces ribbons of fibre bundles called *tuxies*, which are about 5–8 cm in width and the length of the leaf. The second stage involves removing the gums, which bind the fibres together and any residual plant material. This can be done manually, semi- manually or mechanically (Franck, 2005b).

Based on the position of the leaf sheaths, four qualities of fibres are recognised. The outer sheaths (*baba*) are dark brown or light purple and green strips. The sheaths next to the outside (*segunda baba*) are striped, very light green and purple. The middle sheaths (*ubud*) are very light green or light yellow and the inner sheaths almost white. The outer sheaths produce the strongest fibres and the inner sheaths the weakest fibres (Cook, 1984).

Commercial abaca fibre consists of strands containing many fibres held together by natural gums. Strand length varies, with good quality abaca having stands up to 4.5 m long.

Individual fibres are 3–12 mm long, cylindrical, smooth-surfaced and regular in width. In cross-section, they are oval or rounded polygons with thin cell walls. Cell wall ends are tapering with rounded or pointed ends. Catling and Grayson (1982) report seeing no bifurcated ends and comment that the tips are much finer and more pointed than those seen in sisal.

The lumen is large, distinct and, in general, uniform in diameter. Dislocations are frequent, but cross-markings are never very common.

A distinctive feature of abaca fibres is that they can contain granular bodies called *stegmata*. These are deposits of silica and are present in longitudinal files (Anon, 1975).

A test based on the presence of these silica bodies that can be used to differentiate abaca from sisal is the *Billingham test* (Anon, 1975). The test method is as follows:

1. Wash the sample with methylene chloride to remove oil and allow it to dry
2. Boil the sample in 5% nitric acid for 5–10 minutes
3. Wash out the excess acid with water and immerse the sample in cold 0.25 N sodium hypochlorite for 10 minutes
4. Remove the sample and dry

Abaca assumes an orange colour, whereas sisal and other leaf fibres are stained a pale-yellow colour.

6.5.3 Phormium

Phormium is a genus of two plant species in the family Asphodelaceae, subfamily Hemerocallidoidea. Some classifications recognise this subfamily as a family called the Hemerocallidaceae.

The two species are *Phormium tenax* (harakeke or New Zealand flax) and *Phormium cookianum*, also known *as Phormium colensoi* (wharariki or mountain flax). Both species are native to New Zealand and Norfolk Island, with *P. colensoi* native to only New Zealand (Moore & Edgar, 1970). The two species readily interbreed (Wardle, 1979). Phormium has a close connection to Māori culture. Commercially, it is grown in countries of South America and South Africa (Anon, 1975). When the term *Phormium* is used, this generally means fibres from *P. tenax*.

P. tenax is a monocot and is a tall perennial plant reaching a height of up to 5 m. It spreads by stout horizontal stems (rhizomes) producing large, undivided, linear and tapering leaves up to 3 m long. The leaves are usually darkish green with, sometimes, coloured edges and central ribs. Note that there are numerous varieties of this species grown as garden ornamentals, some of which have variegated leaves with a range of colours. At maturity, a rigid flower stalk, projecting above the foliage, produces clumps of curving tube-like flowers, which turn bright red – see Figure 6.3.

Fibres are extracted mechanically, and hand-prepared fibres can be woven into a cloth resembling flax. Its main use, however, is for the manufacture of rope and twine.

Figure 6.3 *Phormium tenax* growing near Lake Wanaka, New Zealand.

Individual fibres are smooth with pointed ends and an average length of about 6 mm. In cross-section, fibres are round or polygonal with thick cell walls and a narrow circular lumen tapering and disappearing towards fibre ends (Cook, 1984).

6.5.4 Sisal

There are in the order of 300 species in the genus *Agave* in the family Agavaceae. Only two species of this genus produce commercial quantities of fibre. Sisal is obtained from the leaves of *Agave sisalana*. Although native to Mexico, Brazil is now the largest producer, followed by Tanzania, Kenya, Madagascar, China and then Mexico.

A second species, *Agave fourcroydes*, or henequen, closely resembles sisal and is sometimes referred to as Yucatan sisal or Cuban sisal (Anon, 1975).

Sisal is a tropical and subtropical plant with a lifespan of up to 20 years and produces up to 400 leaves during its lifetime. The leaves arise from a rosette or bole on a short trunk. Leaves are harvested when the plants are 2–4 years old and at intervals until the plant dies. Sisal is harvested once a year, although in some circumstances, it can be harvested two or three times a year. The leaves grow in circles, and it is normal to harvest four or five circles.

As each circle has about 13 leaves, the total number of leaves per plant at harvest is normally between 52 and 65 leaves (Yu, 2005).

The leaves can grow to a length of 1–3 m with a width of 150 mm. They are triangular in cross-section, becoming thinner and more flattened towards the tip, which generally bears a terminal spine (Catling & Grayson, 1982).

Towards the end of its life, the plant produces a flower spike that grows rapidly to a height of up to 10 m. Flowers do not always produce viable seeds. In the axils of the peduncles, small *bulbils* arise, which develop into small plants that fall to the ground and take root. In commerce, these bulbils are collected and grown in nurseries before being planted out. Plants are also propagated from suckers.

Each leaf may contain up to 1,000 fibres (Cook, 1984). As with other bast and leaf fibres, the fibres occur in bundles. Catling and Grayson (1982) describe the internal anatomy of the leaf in detail. Most vascular bundles are more or less parallel to the long axis of the leaf, with occasional smaller transverse bundles connecting with them. Catling and Grayson (1982) also report that, in the specimens they examined, all bundles had xylem and phloem present, although in some bundles, the xylem was very much reduced.

A very obvious feature of sisal is the presence of large acicular crystals appearing in *idioblasts* or intercellular space, either single or in pairs or small groups. These crystals are said to resemble small black bananas.

Fibre is extracted from the leaves by a process of decortication, where leaves are crushed, beaten and brushed by a rotating wheel set with blunt knives. Yu (2005) describes the early processing of sisal leaves in some detail. The aim of processing is to produce high-quality *long fibre*, from 85 to 95 cm in length, white to milky white in colour. Catling and Grayson (1982), however, comment that sisal is seldom well cleaned. Hence, it is common to find the presence of cellular material such as mesophyll parenchyma and xylem cells adhering to the fibres.

Individual cells or ultimates in macerated samples are quite variable. In cross-section, the fibres are circular, polygonal or oval polygonal. The widest fibre cells are in the large vascular bundles and the narrowest in bundles near the epidermis, especially the adaxial epidermis. Cook (1984) describes a number of different cells. 'Normal cells' are on average 2.5 mm in length, straight and stiff. Fibres can sometimes appear saw-edged or scalloped with tapering ends. The lumen varies in thickness and definition, and the lumen is often packed with tiny granules. In macerated samples, in the largest cells, the lumina are most commonly four or five times the width of the cell wall. When the lumen is narrow, it is sometimes constricted. Catling and Grayson (1982) state that the latter feature is found in every specimen that they examined and is sometimes very common. Sisal also contains broader fibres with a characteristic lattice pattern and small pore markings.

Dislocations are uncommon, and cross-markings are absent.

As previously mentioned, acicular crystals are common, sometimes having the shape of a banana.

As was covered in Section 6.5.2, the *Billingham test* and the presence of acicular crystals are features used to differentiate sisal and abaca fibres.

Sisal has a number of end product uses, but most commonly is found in twines and ropes and as weft, and sometimes warp yarns in carpet backing. A useful feature of sisal is that it has a fast absorption and release of water and moisture and is biodegradable, making it suitable for bathrooms and outdoor carpets or rugs (Yu, 2005).

Cook (1984) provides brief descriptions of a further 14 sources of leaf fibres. In the main, the fibres from these species will be used in the countries in which they are grown and will not have broad commercial significance. A forensic examiner should be aware of any localised products relevant to the identification of the source of a plant fibre.

6.6 Seed and Fruit Fibres

6.6.1 Introduction

Cook (1984) lists a total of 11 seed and fruit fibres, including the most common plant fibre, cotton, along with Akund, Coir and Kapok. Only these four fibres will be considered in alphabetical order, and not in the order of importance, as cotton is by far and away the most important commercial plant fibre.

6.6.2 Akund

Akund, or Calotropis floss, is the seed hair obtained from *Calotropis procera* and *Calotropis giganta* belonging to the family Apocynaceae (formerly in the Asclepiadaceae). These two tree-like shrubs are native to southern Asia and Africa and were introduced to South America and the islands of the Caribbean (Anon, 2013). The two species differ in size, with *C. procera* generally being under 2 m and *C. gigantea* reaching heights of 4–5 m (Nelson et al., 2007).

The two types of plant are also known as the *apple of Sodom*, *desert apple*, *giant milkweed*, *rubber bush*, *swallowwort* and *Dead Sea Apple*. As these names suggest, these plants have an historical and biblical connection. The names apple of Sodom and Dead Sea Apple are said to have originated from ancient authors Josephus and Tacitus, who described the plant growing in the area of biblical Sodom (Alder & Kohler, 2023).

The two species are also toxic with all parts of the plant containing a cardiovascular steroid and calcium oxalate crystals (Nelson et al., 2007).

The leaves of both species are ovate or elliptic, thick, glaucous, rubbery and arranged opposite. The flowers appear in clusters along the branches.

They have a prominent crown with recurved petals. The fruit is green coloured, globe-like, resembles an apple in shape and is hollow with the flesh containing a milky sap – see Figure 6.4.

As the fruits mature and dry, they turn a light-yellow colour. The fruits contain numerous seeds (350–500 per fruit), which are light, flat, brown and attached by white silky hairs (pappus) at one end. The seeds with attached hairs are released when the fruit dries out and 'explodes with a puff, like a bladder or puff ball' (Alder & Kohler, 2023).

The individual hairs or fibres have very thin walls and are 30-40 mm in length with an average diameter of 20 μm (Anon, 1975). The fibres are rather featureless, similar to the fibres of kapok from which they can generally be differentiated by the absence of net-like thickenings seen at the base of kapok fibres.

Akund is rarely used as a textile fibre and is found as stuffing in upholstery and pillows.

6.6.3 Cotton

Cotton is the most important plant fibre accounting for about one-third of all textile fibres, and second only to polyester.

It is outside the scope of this chapter to consider the ancient history and modern evolution of cotton as a commercial crop. The interested reader will find an excellent coverage of this topic and a more detailed treatment of the botany of cotton in Cobley and Steele (1976).

In brief, cotton belongs to the genus *Gossypium* in the family Malvaceae. There are numerous wild and cultivated species, depending on the source

Figure 6.4 *Calotropis procera*. (a) Leaves, flowers and seed pods. (b) Seed hairs.

between 20 and 34, which can be separated on the basis of the number of chromosomes, with 'Old World' species being diploid with ($2n$ = 26) and 'New World' tetraploid ($2n$ = 52). Cobley and Steele (1976) detail the evolution of tetraploid species and their relatively recent world domination.

The three New World tetraploid species are *Gossypium hirsutum, Gossypium barbadense* and *Gossypium tomentosum. G. hirsutum* is called 'upland cotton' and accounts for about 80% of the world's cotton production, with 20% coming from 'sea island cotton', *G. barbadense* (Elzebroek & Wind, 2008). *G. barbadense* is also called Egyptian cotton.

Cotton is a perennial shrub; however, due to selection, it is now grown as an annual, much-branched, shrub-like plant covered with glandular hairs up to 3 m high (Elzebroek & Wind, 2008). As cotton fibres come from the seeds, the development of flowers and seeds is the focus of the remainder of this section. Cotton usually has a single stem with dimorphic branches, meaning both vegetative and fertile fruiting branches. The fertile branches are said to be *sympodial* with a succession of short lateral branches. Each fertile branch may have from six to eight solitary flowers. Each flower is subtended by three bracteoles forming the epicalyx. Flowers are large and showy, creamy white to yellow in colour, sometimes with a red spot at the base.

The fruit usually contains 36 seeds, each about 1 cm long, ovoid and dark brown. After about 50 days, the flowers open and split along the carpel edges, exposing the *seed cotton* as soft wads of fine fibres – see Figure 6.5. Each cotton seed may produce many thousands of fibres on its surface, and a single ball can contain 150,000 fibres or more (Cook, 1984).

Two kinds of fibres or hairs arise from the epidermis of the seeds, long hairs, up to 5 cm, called *lint* and very short hairs, 2–7 mm, called *fuzz.*

Figure 6.5 Cotton plants. (a) Maturing plants. (b) Cotton boll opening up as the plant matures.

Cobley and Steele (1976) describe the development of these hairs in detail. Critical to the structure of cotton fibres is the thickening of the cell wall after about 25 days, when cellulose is deposited in the inner cell wall. This cellulose is laid down in spiral bands, and the direction of the spiral may be reversed at any point. At this reversal point, the cell walls of lint fibres are thin, and it is here that the mature lint fibres twist and collapse into a ribbon when it is exposed to the air. The mature lint fibre thus consists of a collapsed tubular cell with a very small lumen and thick walls composed of superimposed spiral bands of cellulose, but having thinner places where the direction of the cellulose spirals is reversed. These twists are termed *convolutions*, and typically an individual fibre may have up to 50 twists per centimetre.

Most commercial crops of cotton are now mechanically harvested. The harvested bolls are then processed at a *ginnery* where the lint is removed from the seed. Fuzz is much more difficult to remove, and it cannot be easily spun as the fibres do not cling together. Fuzz fibres are mainly used in felt, stuffing in upholstery and surgical cotton wool.

Following the gin process, cotton is sent to a mill to be spun. Cook (1984) describes the stages of carding, drawing and combing before the cotton fibres are suitable for spinning. Chemical and finishing treatments can have a marked effect on the appearance and properties of the fibre.

Lint quality is assessed according to its effective staple length, the percentage of short hairs, mean hair weight per centimetre, standard hair weight and tensile strength (Vickery & Vickery, 1979).

A useful feature of cotton fibres is their ability to absorb water, resulting in cotton fibres becoming more plastic and softened, a feature used to advantage in the manufacture of cotton fabrics. This feature is the basis for a process called *mercerisation*, where a strong solution of caustic soda is used to bring about intensive swelling of the fibres so that cotton can be deformed (Cook, 1984). Shrinkage associated with water uptake can be prevented by holding the fibres under tension. The overall process results in increased lustre. During mercerisation, cotton fibres regain their original circular cross-section and tend to lose their convolutions. Higher quality cotton fabrics will often be labelled as double mercerised. In a forensic context, an examiner should be aware that cotton fibres will not always show the convolutions commonly associated with cotton fibres.

6.6.4 Coir

Coir is the fibre extracted from the fibrous mesocarp of the fruit of the coconut palm, *Cocos nucifera*. The coconut is a monocotyledon with the genus *Cocos* being monotypic in the subfamily Cocoideae of the family Palmae. The name *nucifera* comes from the Latin words *nux* and *ferre*, meaning, respectively, *nut* and *to bear* or nut-bearing. As we shall see, the coconut in botanical terms is a *drupe*, a fleshy fruit with a hard stone and not a true nut.

The coconut palm has been called the tree of life. In Sanskrit, it is called *kalpa vriksha*, meaning 'the tree that provides the necessities of life' (Elzebroek & Wind, 2008). Every part of the coconut palm is used by humans. The broader botany of the coconut palm is outside the scope of this chapter. For the interested reader, excellent descriptions of the botany and cultivation of coconut palms can be found in Cobley and Steele (1976) and Elzebroek and Wind (2008). Here, our interest lies in the coconut itself as the source of the fibres called coir.

The inflorescences that ultimately produce fruit are produced singly in the axils of leaves. Cobley and Steele (1976) describe the development of inflorescences and the male and female flowers. The female flower has three carpels, but two of these abort, leaving one carpel to develop into a mature fruit. Around six fruits develop on each inflorescence, with a typical yield being in the order of 60–70 fruits per year per plant. Mature fruits are ovoid to ellipsoidal in shape, 30–45 cm in length and 15–20 cm in diameter. Each fruit has a smooth, shiny green skin or exocarp, which becomes yellow as the fruit ripens. The mesocarp, or husk, is the thick fibrous layer from which the coir of commerce is extracted. Embedded in the mesocarp is the nut, comprising a very hard endocarp enclosing a single seed.

There are two varieties of coir, *brown coir* and *white coir*. Brown coir is harvested from fully ripened coconuts, with white coir being harvested before the coconut is ripe. Brown coir is typically used in mats, brushes and sacking. White coir is finer and weaker than brown coir and is used to make finer brushes, string, rope and fishing nets. India produces 60% of the world's production of white coir and Sri Lanka 36% of the total brown fibre output. Much of the world's production is used in the country of origin, with Sri Lanka being the largest exporter of coir and coir fibre-based products (Figure 6.6).

Mathai (2005) provides a detailed description of fibre production and processing. The number of ultimates in fibre bundles will depend on the degree of processing. The width of a typical coir bundle varies from 10 to 45 cm (Mathai, 2005). Coir has a very high level of lignan, typically around 45%.

Individual fibres or ultimates are oval, less than 1 mm in length, thick-walled and with an irregular lumen. The edges of some fibres have a distinct wavy outline (Anon, 1975; Mathai, 2005). Round stegmata are found when the fibre is ashed.

6.6.5 Kapok

Kapok, or Java kapok, is the cotton-like fluff obtained from the seed pods of *Ceiba pentandra* in the order Malvales, family Malvaceae. This species has

Figure 6.6 Coir market in Salem.

also been called *Eriodendron anfractuosum*. A similar fibre is also obtained from *Bombax malabaricum*, called Indian kapok. Indeed, the species yielding Java and Indian kapok were previously included in the same family, the Bombaceae. The remainder of this section on kapok will deal only with Java kapok.

C. pentandra is a very large deciduous tropical tree that can grow to a height of over 70 m with a trunk girth of up to 3 m with extensive buttress roots. The commercial tree is grown in rainforests of Asia, notably Java (Indonesia), the Philippines, Malaysia and in China as well as South America. There are usually four to six major branches that form an almost horizontal crown of foliage as much as 60 m in width. Individual leaves are palmate with five to nine leaflets. Flowering occurs when the tree is leafless. Kapok does not produce flowers every year, but when the tree does bloom, it is prolific with up to 4,000 fruits measuring up to 15 cm long.

Flowers have five white petals that open at night over a period of two to three weeks, exposing the pale kapok fibres to the wind for dispersal – see Figure 6.7. Kapok is harvested either when pods are cut down or when they fall to the ground. Fibres are removed by hand from bolls.

Figure 6.7 *Ceiba pentandra,* kapok plant.

Individual fibres are 0.8–3.2 cm long, averaging 1.8 cm with diameters of 30–36 μm, oval or circular in cross-section, with very thin walls and a wide lumen. Towards the base of the fibre, the fibre is slightly bulbous with annular or reticulate markings (Anon, 1975). The fibres are essentially hollow cylindrical tubes.

Kapok is seldom used on its own as a textile fibre due to the fibre being somewhat brittle. It is generally used in the form of loose fibres for stuffing upholstery, cushions and life jackets. The latter use is due to the large amount of air trapped in the lumen of fibres, which gives kapok an apparent density of one-tenth to one-twentieth of water and, hence, imparts buoyancy. Kapok will support more than 30 times its own weight and does not waterlog (Anon, 1975).

6.7 Forensic Examinations and Protocols

6.7.1 Introduction

Plant fibres may be classified in two ways: one based on morphology, according to the part of the plant from which they were obtained, or, practically, according to the uses to which they are put, which in turn depends on their properties. Our treatment of plant fibres thus far has been based on a morphological approach. From a forensic perspective, it is useful to consider the second approach of practical uses as this will, to some extent, define the circumstances in which fibres of plant origin may be encountered. Kirby (1963) classifies the following broad groups.

- *Textile fibres* – includes cotton, ramie, jute, flax and similar fibres used for making clothing, bags and fabrics
- *Cordage fibres* – fibres used for making ropes and twines. Most of the leaf fibres fall into this group, including abaca, sisal and henequen. Some stem fibres, such as hemp and flax, can also be used for ropes and twines
- *Brushes and mat fibres* – includes coir, known in the trade as bristle fibres or 'coco' fibre. Coir can be spun into yarn and used in matting
- *Stuffing and upholstery* – this includes fibres from a number of plant species, including several palms and kapok. One palm species, *Chamaerops humilis*, produces fibres known as 'vegetable horsehair', reflecting its use as a substitute for horsehair in upholstery. Kapok and other floss fibres are also included in this group and are used as stuffing. These fibres are more likely to be found in older furniture. etc., as they have been largely replaced by synthetic man-made fibres
- *Paper-making materials* – the most important paper-making materials are the fibres from softwood species – these are considered in Chapter 7. Some other plant fibres can be found in speciality papers, in older documents and in art fraud cases
- *Miscellaneous* – included fibres such as raffia used for tying purposes, and fibres used in making baskets and hats. These types of fibres and uses are common in many societies and cultures

With the move to renewable resources, the use of plant fibres is predicted to increase, especially to replace plastics, as plant fibres are biodegradable. Plant fibres are also finding new applications in the manufacture of composite materials used in a variety of end products in building and in the automobile industry. For example, from 1996 to 2003, there was an almost linear increase in total fibre consumption in the German automotive industry, averaging 10%–20% per year and increasing from 2,000 tonnes to 9,400 tonnes (Mussig et al., 2005). Plant fibre composites in automotives were used in door panels, seat backs, parcel trays, boot linings and as insulation. Hence, forensic examiners need to be aware of the potential to encounter vegetable fibres in situations they may not have previously considered or encountered.

6.7.2 Forensic Protocols

In Chapter 3, quality assurance standards were considered that apply to all forensic protocols and procedures. Plant fibres are no exception. For plant fibres, used in textiles, the normal approach to textile fibres should be followed (Robertson & Roux, 2018). Cotton fibres are amongst the most common fibres encountered in forensic fibre examinations, and there is an extensive literature dealing with forensic aspects of transfer and persistence and their microscopic examination.

Where plant fibres present some unique and different challenges is where they are used in non-textile applications, as there are no studies of transfer and persistence of non-textile plant fibres. Hence, the starting point for any of these practical uses must involve a detailed examination of the product, considering the availability of fibres for transfer. For example, Wiggins (2018) deals with the forensic examination of ropes and cordage, including a small section on the examination of plant fibres with a checklist that includes fibre characteristics (Wiggins, 1995).

A procedure for the identification of bast and leaf fibres is presented by Palenik (2018). As he points out, what is usually encountered is the *technical fibre* or fibre bundles consisting of numerous botanical fibres or ultimates. The examination of bast and leaf fibres involves the examination of the technical fibre and, usually, the examination of fibre ultimates after a process of maceration. As previously described, there are a number of methods used to macerate technical fibres. Most involve treating the technical fibre with a mixture of hydrogen peroxide and glacial acetic acid in a heated water bath until the fibres turn white. This usually takes between 4 and 8 hours. The fibres can then be removed, teased apart, stained with phloroglucinol for lignin and placed in a suitable mountant for microscopic examination.

Procedure for the identification of bast and leaf fibres:

1. Examination of the technical fibre
    - *Preliminary examination* – dependent on the specimen, select a representative sample and examine using low-power (stereo) microscopy, observing colour, texture and stiffness
    - *Fibre scrapings* – scrape the length of the fibre gently with a safety razor. Collect the scrapings over a glass slide and mount in a drop of glycerine–alcohol (1:1) and examine with TLM set up for polarised light microscopy. Features present may include epidermal tissue and other cells, such as parenchyma and vascular tissue. Crystals of calcium oxalate may also be observed
    - *Cross-section* – make a cross-section of the fibre bundle and observe the overall shape and the shape of individual fibre cells
    - *Test for degree of lignification* – take a small sample of fibre and stain with phloroglucinol reagent and observe the colour, which will vary from almost colourless to deep red

- *Clear the fibre* – place the fibre sample on a glass slide under a coverslip in a mixture of 1:1 alcohol and glycerine. Gently heat using a Bunsen burner until the alcohol evaporates. The glycerine will penetrate the fibre, displacing air, allowing the interior of the fibre to be examined for the presence of crystals and other characteristics

2. Examination of fibre ultimates
   - *Length* – measure a number of *ultimates* and determine the average length
   - *Dislocations/nodes and cross-markings* – note the presence, shape and frequency along the length of the cells of dislocations/nodes and cross-markings
   - *Lumen and cell diameter* – note the thickness of the central canal or lumen and how this *varies* along the length of ultimates
   - *Other tissues and cells* – note the presence of any other cell types, such as *parenchyma* and vascular tissue
   - *Fibre cell ends* – note the shape and variation in fibre cell endings
   - *Z or S twist* – assess fibre and/or ultimates for the direction of twist

(Modified from Palenik, 2018.)

The *Hertzog test* can be used to determine whether an ultimate fibre has an S twist or a Z twist. The single mounted fibre is aligned parallel to the vertical cross-hairs of a polarising microscope in what would be an extinction position for a synthetic fibre. A first-order red plate is then inserted, and the colour of the cell wall is noted. Yellow indicates the cellulose helix has an s-twist and blue, a z-twist (Palenik, 2018). This test is useful in distinguishing flax (s-twist) from hemp (z-twist). De Wael (2021) provides a list of s- and z-twists for most of the common bast and leaf fibres.

For a technical fibre, the *drying twist test* method is described in Appendix A.6. (Anon, 1975).

Palenik (2018) recommends that *fibre scrapings* are preferred to the process of ashing fibres to examine for the presence of crystals (Anon, 1975) as the microscopist can observe the distinctive crystals directly and not, as he puts it, as 'pseudomorphs left behind after the ashing process'.

The *Swett test* (Swett, 1918) was devised to distinguish between Abaca and Sisal. A modified version of this test is called the *Billinghame test*. Both tests are based on colour differences between these two fibre types. The Billinghame test, said to produce greater colour differences with more stable colours (Anon, 1975), involves washing the sample (technical fibre) with methylene chloride to remove any oil and allowing it to dry. The sample is then boiled in 5% nitric acid for 5–10 minutes, washed with water to remove excess acid, and immersed in cold 0.25 N sodium hypochlorite for 10 minutes before being removed and allowed to dry. Abaca (manila) assumes an orange colour, whereas sisal and other leaf fibres are a pale-yellow colour.

Cross-sections are usually made from technical fibres. Sections can be made using simple handheld microtomes through to more sophisticated microtomes; however, with the latter, some form of embedding is required. It is more usual to make sections using a handheld *plate method*. This method consists of drawing one or more technical fibres in a suitable medium, if necessary (such as a man-made fibre), through a small hole in a metal or plastic plate until the fibres are firmly held. They are then cut with a razor blade flush with the two sides of the plate. Plastic plates with multiple holes are inexpensive and can be purchased through the McCrone Group. De Wael (2021) describes the use of a handheld microtome, called a MIECO QCSSMS – model M2020; however, this is a relatively expensive option.

A simple scheme for the analysis of plant fibres is presented by Robertson and Roux (2018) – see Figure 6.8 – and a summary of features of the most common bast and leaf fibres is presented in Table 6.2.

The identification of plant fibres may seem to be quite straightforward. This is not the case, and forensic examiners need to take a conservative approach based on looking for differences and a process of elimination. It is unlikely that identifications will meet the botanical standard of being able to be directly linked back to a type specimen from a herbarium.

It is essential that any laboratory conducting this type of examination establishes a comprehensive collection of reference samples, including commercial technical fibres and representative end products.

Appendix A.6: Drying-Twist Test

Principle – if a wet flax fibre is held at one end and allowed to dry, the free end, directed towards the observer, will be seen to move in a clockwise direction. Hemp fibre under similar conditions follows an anticlockwise movement. The test may be applied to material at any stage of manufacture and is successful even with fibres that have been mercerised.

Apparatus – a pair of forceps with fine points meeting accurately. A hot plate or any source of low-temperature heat with a dark background.

Conditions of test – when examining the yarn or cloth, in the latter case, after separating the warp and weft, take cuttings about 25 mm long from various places in the material to obtain a fair sampling. Soak cuttings for a few minutes in water, and with the aid of forceps, withdraw fibres and hold over the hot plate. Observe the direction of twist as the fibre dries.

Precautions – as far as possible, single fibres should be examined. Small bundles of fibres can be examined, provided sufficient time is given over the hot plate to ensure drying has set in.

MACROSCOPIC EXAMINATION

Visual examination and use of low-magnification stereo microscope

SAMPLE

MICROSCOPIC EXAMINATION

Fibres present in bundles

Single fibres

Fibre	Appearance and confirmatory tests
COTTON	Convoluted fibres of uneven diameter: variable lumen
KAPOK	Very thin, smooth-walled fibre with tapered blunt end
FLAX (cottonized)	Similar to cotton but with X nodes along fibre: twist test

Vascular tissue absent

FLAX	As for cottonized flax but in groups
JUTE	Phloroglucinol test – heavily lignified: cross-section: ashing
HEMP	Low lignin content: cross-section: ashing: twist test
OTHER BAST FIBRES	
COIR	Very coarse fibre with short (<1mm) ultimates

Vascular tissue present

ABACA	Cross-section: ashing: Billinghame's test
SISAL	Cross-section: ashing: Billinghame's test

Figure 6.8 Scheme for the analysis of plant fibres. (Reproduced with permission from Robertson & Roux, 2018.)

Table 6.2 Summary of Features of Common Bast and Leaf Fibres

Botanical Origin	Species	Common Name	Length of Ultimates or Fibre Cells (mm)	Extraneous Material	Cross-Markings	Lumen and Cell Wall	Pits	Crystals, Silica Ash
Bast Fibres	*Linum usitatissimum*	Flax	1.6–24.0	Epidermis with paracytic stomata. Parenchyma. Xylem elements.	Rare. When seen, often very regular along the whole length of the fibre cell.	Cell wall occasionally striated. Cell wall thick.	Very fine. Not obvious. Can be seen by using polarised light. Lumen narrow, regular.	No crystals reported or seen.
	Corchorus capsularis and *olitorius*	Jute	0.6–5.3	Few. Mostly parenchyma, sometimes with cubic or cluster crystals. Very occasional vessels.	Few, faint. Occasional marks from chambered cells. Scalloped edges to fibre cells.	Lumen of varying width, often varying regularly along the whole length of the cell.	Bordered, funnel-shaped in side view.	Cubic crystals in chains, sometimes mixed with occasional cluster crystals. Single cluster crystals.
	Cannabis sativa	Hemp	1.0–34.0	Laticiferous elements in unmacerated fibre. Parenchyma of various types. Cluster crystals free or in cells Hairs. Epidermis. More rarely, blocks of tissue and xylem elements.	Variable. Some cells with fine, regularly spaced marks in every specimen. Marks from chambered cells. Several series on one fibre cell. More frequent than in flax. Occasional remains of cells attached to marks.	Cell wall striated. Lumen most commonly three to five times the width of the cell wall.	Slit-like, parallel to the long axis of cell, sometimes coalescing.	Cluster crystal in short chains, often three or four together. Single cluster crystals. *Very* occasional cubic or rhombic crystal in some specimens.
	Bochmeria nivea	Ramie	13.0–82.7	Parenchyma. Cluster crystals free or in parenchyma. Cluster crystals in chambered cells. Few hairs and vessel elements.	Common, fine, nearly always with attached remains of parenchyma cells. Several series on one fibre cell. Occasional marks of chambered cells.	Lumen difficult to see because (a) it varies, (b) cell wall is striated and (c) fibre cells tangle. Lumen commonly two to three times the width of the cell wall.	Elongated, slit-like, parallel to the long axis of the cell, sometimes coalescing.	Cluster crystals in chains. Single cluster crystals.

Leaf fibres	*Agave sisalana*	Sisal	3.0–6.7	No epidermal cells or stomata seen. Mesophyll. Blocks of tissue. Tracheary elements with annular thickening or opposite to scalariform pitting. Spirals, 'butt' fibre cells.	No cross-marks seen. Scalloped edges to some cells.	Lumen has maximum width, five to six times the thickness of the cell wall, most commonly four to five times.	Obvious slit-like, at approximately 40° to the long axis of the cell.	Large acicular crystals, sometimes with organic matter annealed to the surface.
	Musa textilis	Abaca	2.0–9.3	Parenchyma. Stegmata. Epidermis with stomata and sub-epidermal sclerotic layer. Occasional blocks of tissue. Tracheary elements with spiral or annular thickening.	Variable, never very common, occasionally regularly spaced.	Lumen has width two to several times the thickness of the cell wall, most commonly three to five times. Wide variation within and between specimens.	Slit-like, at slight angles to the long axis of the cell.	Silica bodies, very occasional small acicular crystals in some specimens.

Source: **Adapted from Catling, D.I. and Grayson, J.,** *Identification of Vegetable Fibres*, **Chapman & Hall, London, 1982.**

Other fibres – cotton fibres under this test usually twist alternatively in opposite directions in an irregular manner. Ramie and nettle fibre show a drying twist in the same direction as flax, with most other fibres twisting in the same direction as hemp.

(Modified from Anon, 1975)

References

Alder, C., & Kohler, K. 2023. Apple of Sodom. *Jewish Encyclopaedia 2002-2021*. https://jewishencyclopedia.com/articles/1167-apple-of-sodom#:text=by%20Cyrus%Alder%2C520Kaufman%20Kohler,rind%20and%for%20fibers. Accessed 27 April 2026.

Anon. 1975. *Identification of Textile Materials*, 7th ed. Manchester: The Textile Institute.

Anon. 2013. *Akund Floss*. https://www.britannica.com/topic/akund-floss. Accessed 27 April 2026.

Carpita, N., & McCann, M. 2000. The Plant Cell Wall. In Buchanan, B., Gruissem, W., & Jones, R. (eds.), *Biochemistry and Molecular Biology of Plants* (pp. 52–108). Rockville: American Society of Plant Physiologists.

Catling, D.M., & Grayson, J.E. 1982. *Identification of Vegetable Fibres*. London: Chapman and Hall Pub.

Cobley, L.S., & Steele, W.M. 1976. *An Introduction to the Botany of Tropical Crops*, 2nd ed. London: Longman Group.

Cook, J.G. 1984. *Handbook of Textile Fibres 1. Natural Fibres*. Shildon: Merrow Pub.

De Wael, K. 2021. *Microscopy in Forensic Fibre Examinations. A Practical Photo Atlas and Training Tool*. Belgium: Cobalt Blue Coaching.

Elzebroek, A.T.G., & Wind, K. 2008. *Guide to Cultivated Plants*. Wallingford: CAN International.

Evert, R.F. 2006. *Esau's Plant Anatomy. Meristems, Cells, and Tissues of the Plant Body: Their Structure, Function, and development*, 3rd ed. Hoboken: John Wiley and Sons.

Fergusson, S.M., & Hemmings, J. 2018. Fibres, Yarns and Fabrics: An Introduction to Production, Structure and Properties. In Robertson, J., Roux, C., & Wiggins, K.G. (eds.), *Forensic Examination of Fibres* (3rd ed., Chapter 1, pp. 1–60). Boca Raton: CRC Press.

Franck, R.R. 2005a. Overview. In Franck, R.R. (ed.), *Bast and Other Plant Fibres* (Chapter 1, pp. 1–23). Cambridge: Woodhead Pub.

Franck, R.R. 2005b. Abaca. In Franck, R.R. (ed.), *Bast and Other Plant Fibres* (Chapter 8, pp. 315–321). Cambridge: Woodhead Pub.

Gregoire, M., Bar, M., De Luycker, E., Musio, S., Amaducci, S., Gabrion, Z., Placet, V., & Ouagne, S. 2021. Comparing Flax and Hemp Fibres Yield and Mechanical Properties After Scutching/hackling Processing. *Industrial Crops and Products*, 172, 114045.

Kirby, R.H. 1963. *Vegetable Fibres, Botany, Cultivation and Utilization*. New York: Interscience Pub.

Kozlowski, R., Rawluk, M., & Barriga-Bedoya, J. 2005. Ramie. In Frank, R.R. (ed.), *Bast and Other Plant Fibres* (Chapter 5, pp. 207–227). Cambridge: Woodhead Pub.

Krishnan, K.B., Doraiswamy, I., & Sitra, K.P.C. 2005. Jute. In Franck, R.R. (ed.), *Bast and Other Plant Fibres* (Chapter 2, pp. 24–93). Cambridge: Woodhead Pub.

Mathai, P.M. 2005. Coir. In Franck, R.R. (ed.), *Bast and Other Plant Fibres* (Chapter 7, pp. 274–314). Cambridge: Woodhead Pub.

Moore, L.B., & Edgar, E. 1970. *Flora of New Zealand, Vol 11, Indigenous Tracheophyta: Monocotyledones Except Gramineae.* Wellington, New Zealand: Botany Division DSIR.

Mussig, J., Karus, M., & Franck, R.R. 2005. Bast and Leaf Fibre Composite Materials. In Frank, R.R. (ed.), *Bast and Other Plant Fibres* (Chapter 10, pp. 345–376). Cambridge: Woodhead Pub.

Nelson, L.S., Shih, R.D., & Balick, M.J. 2007. *Handbook of Poisonous and Injurious Plants*, 2nd ed. New York: Springer.

Palenik, S. 2018. Microscopic examination of Fibres. In Robertson, J., Roux, C., & Wiggins, K.G. (eds.), *Forensic Examination of Fibres* (Chapter 5, pp. 145–178). Boca Raton: CRC Press.

Robertson, J., & Roux, C. 2018. From Crime Scene to Laboratory. In Robertson, J., Roux, C., & Wiggins, K.G. (eds.), *Forensic Examination of Fibres* (Chapter 4, pp. 99–144). Boca Raton: CRC Press.

Salmon-Minotte, J., & Franck, R.R. 2005. Flax. In Franck, R.R. (ed.), *Bast and Other Plant Fibres* (Chapter 3, pp. 94–175). Cambridge: Woodhead Pub.

Sponner, J., Toth, L., Cziger, S., & Franck, R.R. 2005. Hemp. In Frank, R.R. (ed.), *Bast and Other Plant Fibres* (Chapter 4, pp. 176–206). Cambridge; Woodhead Pub.

Swett, C.E. 1918. Distinguishing Manila from all Other Hard Rope Fibers. *Industrial Engineering Chemistry*, 10, 227.

Vickery, M.L., & Vickery, B. 1979. *Plant Products of Tropical Africa.* London: Macmillan Press.

Wardle, P. 1979. Variation in *Phormium cookianum* (Agavaceae). *New Zealand Journal of Botany*, 17, 189–196.

Wiggins, K.G. 1995. Recognition, Identification and Comparison of Role and Twine. *Science and Justice*, 35, 53–58.

Wiggins, K.G. 2018. Ropes and Cordage. In Robertson, J., Roux, C., & Wiggins, K.G. (eds.), *Forensic Examination of Fibres* (Chapter 3, pp. 89–98). Boca Raton: CRC Press.

Yu, C. 2005. Sisal. In Franck, R.R. (ed.), *Bast and Other Plant Fibres* (Chapter 6, pp. 228–273). Cambridge: Woodhead Pub.

Wood Identification 7

7.1 Introduction and Scope

A recent report estimated that there are about 73,000 tree species globally (Gatti et al., 2022) belonging to one of two groups of plants, Gymnosperms (plants bearing 'naked' seeds) and Angiosperms (seed-bearing plants), with the latter having by far the larger number of species.

Identification of trees can rely on features of bark, leaves, flowers, fruit, buds and seeds. As the focus of this chapter is on one aspect of trees, wood, tree identification in the broader sense is outside of scope. There are numerous texts, however, and increasingly online resources, available for both the amateur and professional with an interest or need to identify trees. As an example, Eucalypts are a common group of plants native to Australia, Papua New Guinea, Indonesia and the Philippines and also as an introduced species in many other countries. There are 934 native Eucalypt species. An identification tool called EUCLID, developed by the CSIRO (Commonwealth Scientific and Industrial Research Organisation) in Australia, is available as a smartphone app or available free online (CSIRO, 2023).

There are many aspects to the use of trees, but the focus of this chapter is where trees are used to produce timber/wood, which then ends up in a multitude of end products such as construction materials, tools, weapons, furniture, packaging, artworks and paper.

As we have already seen in Chapter 2, wood is formed by a process of secondary growth resulting from cell division in the vascular cambium and subsequent expansion and differentiation of these new cells, which then become thickened and lignified.

The examination of wood in a forensic context can be categorised into two broad aspects, applications that relate to the criminalistics value of wood and wood in the regulatory space, related to illegal logging and trading in timber.

In criminalistics applications, wood may take many forms, including timber or timber products used as weapons, wood from built structures where there has been forced entry and wood from sawdust/wood shavings. The analysis of growth rings (*dendrochronology*) may allow the estimation of the age of a plant or enable the estimation of the dating of events.

In the regulatory space, illegal logging and the illegal trade in timber are major problems in many countries and, at a global international level, threaten

DOI: 10.4324/9781003308508-7

entire ecosystems through forest degradation and deforestation. Illegal logging is estimated to be the third largest transnational crime after counterfeiting and drugs (Anon, 2017). The Convention on International Trade in Endangered Species of Wild Fauna and Flora (CITES) lists over 600 tree species, with over 400 used for their timber (UNODC, 2016).

Several major consumer nations, including Australia, Canada, countries of the European Union, Japan and the United States (through amendments to the Lacey Act), have introduced laws intended to curb the global illegal wood trade and to encourage the development and enforcement of laws regulating timber harvest and trade in producer countries (Grant & Chen, 2021).

Regulation and enforcement rely on the accurate identification of timber, as the true taxonomy or origin of timber can be readily misrepresented through fraudulent information in import documentation. In some cases, an importer may end up using wood from illegal raw materials unknowingly, and in good faith, due to incorrect information. For example, DNA testing showed that 40% of timber was incorrectly labelled in Australia (Anon, 2020). In a study of the extent of fraud and misrepresentation in retail forest products in the United States, 62% (45 out of 73) of tested products had one or more fraudulent or misrepresented claims (Wiedenhoeft et al., 2019).

In the same study, the researchers conducted a proficiency-testing programme of experts regarding their forensic wood anatomy capacity and identification accuracy. Total capacity was estimated at only 830 wood specimens per year, and identification accuracy was alarmingly poor, ranging from 6% to 92%. Whilst the UNODC (United Nations Office of Drugs and Crime) have published a best practice guide for forensic timber identification (UNODC, 2016), this can only be meaningful if the requisite skills and capacity exist in the relevant community.

7.2 Forensic Applications and Case Studies

7.2.1 The Lindbergh Kidnapping

One of the most written about case histories involving forensic botany, and specifically wood, is the Lindberg kidnapping of the 1930s. Its historical significance is that it may have been the first time that a witness was qualified by a judge to be an expert. That witness was Arthur Koehler, a wood technologist from the Forest Products Laboratory of the United States Forest Service.

Lindberg was famous for being the first person to fly non-stop across the Atlantic in 1927, later marrying Anne Morrow, the second daughter of the US Ambassador to Mexico, Dwight Morrow. They were the celebrity couple of their day and attracted intense media attention. They lived in a large two-storey house in Hopewell, New Jersey. On the evening of March 1 1932, their infant son was abducted. On April 2, after contact had been made by the kidnapper, or his representative, a ransom of US $50,000 was paid in return for instructions on where to find the baby. Sadly, on May 12, the remains of the child were located only a few miles away from the Lindberg estate (Graham, 1997).

Entry to the second floor of the Lindberg home had been gained by using a roughly constructed ladder, which was left at the scene.

Koehler's first contribution to the investigation was to identify the four types of wood from which the ladder had been constructed. One was a Pinus species, commonly called yellow pine or North Carolina pine, two others were ponderosa pine (*Pinus ponderosa*) and Douglas fir (*Pseudotsuga menziesii*), and the fourth species was birch (probably *Betula alba*). At this stage, Koehler had not examined the actual ladder and had only received 'slivers of wood' (Graham, 1997).

It was not until 1933 that serious attention was placed on the ladder. Undoubtably, the accurate identification of the wood species was an essential requirement. In that era, this would have been a relatively easy task for a person with the experience of Koehler. The importance of this case is that it is a prime example of the application of criminalistics as his further work involved deconstructing the ladder into its component parts, comparative wood anatomy involving comparisons of annual growth rings and other growth patterns, such as knots, physical evidence in the form of marks left on the wood by tools and ultimately physical fits of ladder components back to floorboards in the home of the suspect.

Full accounts of the meticulous, detailed forensic examinations conducted by Koehler are found in Palenik (1983) and Graham (1997). Only a short summary is given in this chapter.

The ladder is described as a 'slender, homemade affair weighting only 38 pounds … built in three sections each about 6ft 8 ins long and held together by birch dowels to form a sort of extension ladder' (Graham, 1997). Being in three parts, the ladder could have been carried in a vehicle to the scene of the abduction. Koehler deconstructed the ladder, numbering the vertical rails and the *cleats* or cross-members.

Cleats 1–10 were ponderosa pine, with cleat 11 being Douglas fir. Side rails 12 and 13 were North Carolina pine, 14 and 15 Douglas fir, 16 North Carolina pine and 17 Douglas fir.

Koehler catalogued a variety of natural and man-made marks on the component parts of the ladder. Some of these marks could be linked to the pattern formed as a board is fed into a planning mill. Using his knowledge of timber manufacture, Koehler was eventually able to trace the potential source of the wood from rails 12 and 13 to the Bronx National Lumber and Millwork Company, narrowing the geographic search area, which, independently, corresponded to where the ransom had been paid. However, the investigators were no closer to identifying a suspect.

The breakthrough finally came in September 1934 when one of the ransom banknotes was used to buy petrol and the license number of the vehicle was recorded by the gas station attendant. The owner of the vehicle was Bruno Richard Hauptmann, a German carpenter who lived in the Bronx, only ten blocks away from the Bronx National Lumber and Millwork Company and only four miles away from where the ransom had been paid (Graham, 1997). Some $14,600 of the Lindberg ransom was found hidden in the garage of Hauptmann.

A search of the attic in the apartment of Hauptmann found that one of the floorboards was about 8 ft shorter than the others. Koehler on visiting the attic was able to show that four square nail holes in rail 16 lined up perfectly with the square holes in the joists. Detailed observations by Koehler, included grain and annual rings, and unique patterns on cleats and rail 16 that could be linked back to a hand plane with a dull, nicked blade found in the garage of Hauptmann. Essentially, Koehler was able to establish that rail 16 had at one time been part of the missing floorboard in the Hauptmann attic.

7.2.2 Other Cases Involving Wood

7.2.2.1 Physical Fits

Forensic scientists should always be aware of the potential for physical fit examinations when examining wood materials.

For example, Townshend (1981) reported on a case study of the theft of a large walnut tree. A section of the stump and a wedge piece of wood from the scene were compared to the end of the tree in possession of the suspects. Based on an examination of grain, growth rings and fracture pattern, the examiners concluded that the wedge piece found at the scene aligned with the end of the tree stump. A cast of the tree stump and a cast of the suspected tree also showed an alignment in microscopic detail.

In another case report, involving a homicide (Hathaway, 1994), four fragments of a broken pool cue physically fit together, showing that they had been part of the same item. The defence position was that multiple cue sticks had been used.

Christophe and Daniels (2008), investigating a hit and run case, in which the suspect had a wooden pallet in the rear of his truck, used Photoshop techniques to demonstrate a physical fit of a questioned wood chip to establish that it had originated from the damaged wooden pallet.

Chips or fragments of wood may be fitted back to forced windows or doors or to tools damaged in gaining entry.

Where rifle and shotgun stocks are sawn off to aid concealment, the parts may be matched by the pattern of the growth rings (Fiddian, 2005). Fiddian also discussed the role wood analysis played in an arson case where a short length of split timber board, used to start a fire, was found to physically fit a similar-sized piece of timber located at the suspect's house. Both pieces of wood were identified as Oregon (*P. menziesii*), but, as Fiddian points out, the nexus between the suspect and the crime scene was much more powerful than if it had only been shown that the same type of timber was present at both locations.

7.2.2.2 Growth Ring Interpretation

Horrocks and Wallace (2001) report on the distortion caused by twine to growth rings of Teatree stems used to support cannabis plants, allowing an estimation of the minimum number of years that twine had been attached.

Yaman and Akkemik (2009) have used growth ring interpretation to date the illegal logging of trees in Turkey.

Estimating postmortem interval (PMI) in cases where the body may have been buried for decades is problematic. Several studies have reported on the use of *dendrochronology*, the study of growth rings, in both roots and stems to provide a minimum PMI for buried human remains (Denne, 1977; Jozsa, 1985; Courtin & Fairgrieve, 2004; Pokines, 2018).

Denne (1977) reports a case in which the length of time a skeleton had been hanging from a birch tree was estimated by dating the disruption to normal growth of the branch by the rope. Denne (1977) also reported a case where damage to roots caused by the digging of a grave near an Elder tree was used to date the time of burial. Damage was noted in the form of wound tissue or callus to the roots.

Jozsa (1985) reported on a series of case studies where a suspect was linked to a crime scene using species identification and physical matching of tree ring patterns. In this case, dating the crime to a calendar year or growing season assisted the investigation. In another case, the dating of fire damage caused to a tree helped substantiate an account of a multiple homicide.

Courtin and Fairgrieve (2004) reported on a case where dating trauma to aerial parts assisted in determining the potential time of death. In this case, a fully clothed, highly decomposed body was found in a forest setting. The body had fallen on a young black spruce (*Picea mariana*) terminal shoot that subsequently had grown up around the body in an attempt to regain its normal vertical growth. This resulted in asymmetrical growth of the terminal shoot and asymmetrical growth rings. Analysis of these growth rings allowed an estimate of the timing of the displacement and, hence, the time the deceased body had fallen on the tree.

It is important to understand that the accurate interpretation of growth rings is not a trivial matter, with growth rings being subject to modifications caused by weather, climate and extreme events. False rings, or incomplete rings, may be formed, which can complicate the clear interpretation of growth rings (Schweingruber et al., 2006). Care needs to be taken when examining trees grown in warm climates where seasonal changes can be more subtle and tree rings less distinguishable (Balouet et al., 2007). Jones and Daniels (2012) assessed estimates of year of death of trees using a number of statistical models, concluding that precision and accuracy errors resulted from many factors, including loss of bark, wood decay, lack of ring formation prior to tree death and human error.

Growth rings in trees may also be used to date environmental information (*dendroecology*) to estimate the age of contamination and characterise the elemental composition of contamination, where trees are natural proxy recorders of historical events. Balouet et al. (2007) describe six case studies related to petroleum products and chlorinated solvent spills where this approach has been successfully applied in a forensic context.

7.2.2.3 *Wood Fragments and Shavings*

Fragments of wood may arise in a number of ways, including splinters that may be created when entry is gained to a premises through a wooden structure, such as a window frame or door. However, not so many years ago, the main source of wood fragments in a forensic context resulted from *safe ballast*.

This was in an era when wages were paid to employees in cash, generally on a Friday. Wages would be delivered to employers on a Thursday and placed in a safe overnight. As an 18-year-old in 1968, I can recall getting my first pay in a brown envelope – such excitement!

The late Jack Fish reported on the identification of wood fragments in 1966 (Fish, 1966), in which he referred to 'highly organised gangs of safe breakers'. Another Fish, Frank Fish, also now sadly deceased, in writing about the formation of the master's degree in forensic science at Strathclyde University, recalled how he was approached in 1954 by the local police to examine safe-packing material to see if wood fragments matched those found associated with a sum of money stolen from a coal merchant's office. As Frank stated, 'this put us in business' not only for safe packings but also with a variety of other botanically based materials. In 1973, one-quarter of the science content of the Strathclyde degree involved analytical microscopy with 'special reference to the identification of leaves, woods, foodstuffs, papers, animal tissues, etc: hairs and fibres' (Fish, 1973). This was the world I first entered in 1976 when I started my journey in forensic science as an academic at Strathclyde. Before taking up my position, I had visited the then Metropolitan Police Forensic Laboratory, where I was told that they were still doing about a case a week involving wood identification. Sadly, the skill base that existed at that time has all but disappeared from core forensic organisations today.

Returning to safes, the main purpose of a safe is to protect its contents, not just against theft, but against fire. A safe consists of two boxes with the space between the boxes being filled with *ballast*. In a modern safe, the ballast is a solid, concrete type of material. Older safes dating back to before 1900, and typical of those found in smaller premises, were packed with heat-insulating materials usually composed of any type of sawdust or wood shavings that was available, along with crystals of potash, ammonium or magnesium sulphate. Diatomaceous earth, iron filings, metal turnings, sand and other workshop sweepings could also be present (Fish, 1966). When the sides or back of these older safes were broken open, ballast spilt out in large quantities and would typically be found on the clothing of perpetrators or their vehicles – see Figure 7.1.

Figure 7.1 Example of an older fire-resistant safe.

Fish (1966) records one particular case involving the theft of a safe from a butcher's shop. The safe was later found in a ditch with the back torn off. A sample of sweepings from a suspect's vehicle revealed the presence of five types of wood along with crystals of potash alum. The same mix of species was found in the control sample of ballast from the abandoned safe, importantly, in the same proportions and particle sizes. The mix of woods present was quite unusual, containing an Australian maple, Burma tulip wood, ebony, oak and boxwood. As Fish points out, not all cases have multiple woods, and there may only be one common wood present. However, even in the latter type of case, there may be other materials present that enhance the potential evidential value.

Today, it is highly unlikely that a safe ballast will be seen in a forensic laboratory, but the possibility of wood fragments from other sources should not be ignored.

Appendix A.7 provides some guidance on the preparation of wood samples for microscopic examination.

7.3 Wood Structure and Anatomy

7.3.1 Introduction

As previously stated, wood derives from either Gymnosperms or Angiosperms. From a timber point of view, the only important order of the Gymnosperms is the *Coniferales*, the cone-bearing plants that provide *softwoods*. The Angiosperms have two main orders, the *Monocotyledons* and the *Dicotyledons*. Included in the monocots are tree-like plants such as bamboo and palms, but these do not produce true wood. Dicot woods are referred to as *hardwoods*. The terms soft and hard wood are quite misleading, as some hardwoods are softer than some softwoods. For example, balsa wood used in crafting is a very soft wood, but it is actually a hardwood!

Depending on the classification system followed, the Coniferales have six families, 43 genera and about 110 species, which include all of the commercial timbers. By contrast, commercial hardwoods come from 77 families, 280 genera and 800 species, representing about 380 commercial timbers (Jane, 1970).

In the simplest terms, wood is a porous, heterogeneous material composed of vertical cells, including vessels and fibres in hardwoods and tracheids in softwoods. In addition to the vertical elements, the cambium forms sheets of cells that run in a radial direction known as *medullary rays*.

Each year, another layer of wood is laid down by the cambium under the external bark, producing a growth ring that is usually composed of two zones, the paler early or spring wood and the darker, denser late or autumn wood.

The arrangement and detailed structure of these vertical and horizontal cells show considerable variation, and it is this variation that allows woods to be identified. In order to study this variation, it is necessary to cut three sections in thin planes with respect to the original tree trunk. These are transverse or cross-section (TS) showing the end-on-end view of the vertical elements.

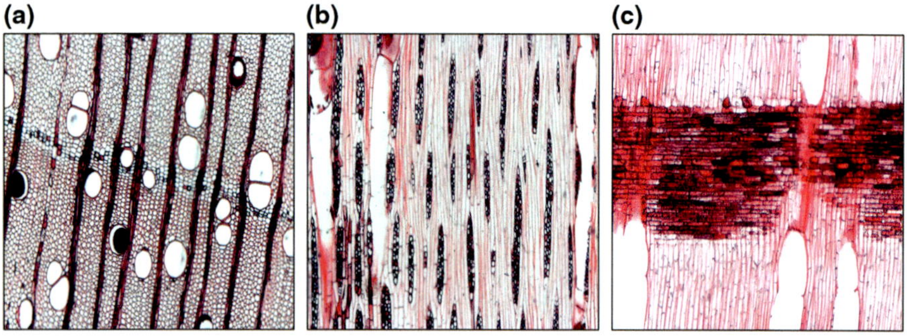

Figure 7.2 Hardwood - Xylocarpus granatum Koen. (a) TS, (b) TLS and (c) RLS.

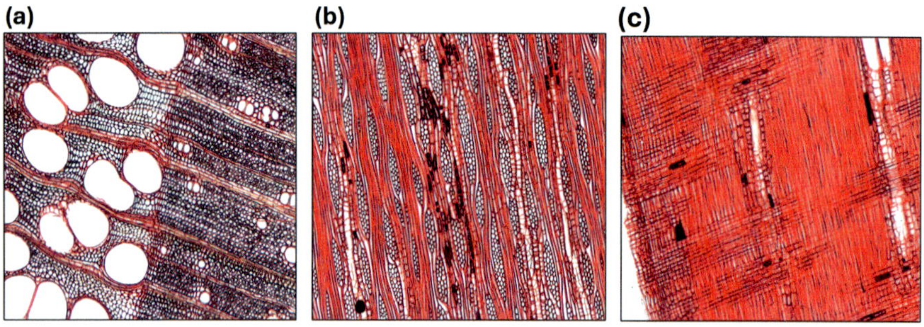

Figure 7.3 Hardwood – Ash – *Fraxinus excelsior*. (a) TS, (b) TLS and (c) RLS.

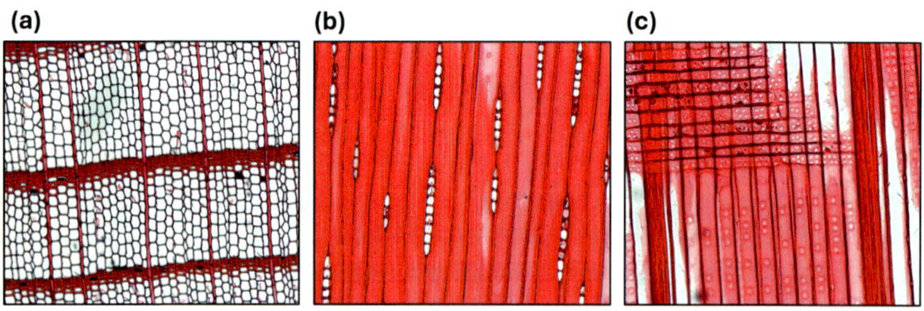

Figure 7.4 Softwood – Redwood – *Sequoia sempervirens*. (a) TS, (b) TLS and (c) RLS.

There are two longitudinal sections. The first is cut in the radial plane (radial longitudinal section (RLS)), which allows a side-on view of the medullary rays. The second horizontal section (tangential longitudinal section (TLS)) is cut at a tangent, enabling the end-on view of the ray cells. By studying the sections in these three planes, it is possible to build up a solid three-dimensional picture of wood (Fish, 1966). Figures 7.2–7.4 show these three sections for a softwood and a hardwood sample.

7.3.2 Brief Revision of Anatomy Fundamentals

Chapter 2 introduced plant anatomy. This section aims to extend and build on that introduction and lay the foundation for a more detailed consideration of those features that are found in softwood and hardwood species that form the basis for identification.

A comprehensive treatment of the structure of wood can be found in Jane (1970), as revised by Wilson and White, who produced a slightly shorter book following the same approach as Jane (Wilson & White, 1986).

Practically, all the cells of which wood is composed are derived from the *vascular cambium*. There are two basic types of cells, *fusiform initials*, spindle-shaped axillary elongated cells, and the *ray initials*, where the cells are about as broad as they are high.

Considering only the cells produced by the vascular cambium that contribute to wood, the cells resulting from the fusiform initials differentiate into one of four cell types, *parenchyma*, *fibres*, *tracheids* or *vessel* elements. Not all cell types will occur in every wood, but parenchyma is invariably present, although in some conifers restricted to the rays.

Parenchyma are brick-shaped cells that may be squatter or more elongated.

Fibres are long, pointed cells in which the pits are simple and slit-like. The walls may be thick or thin. A thick-walled fibre with slit-like simple pits is termed a *libriform fibre*, and its main function is support.

There are intermediates between fibres and tracheids called *fibre–tracheids*. These are long with pointed ends and have bordered pits with slit-like apertures.

Tracheids are elongated cells with pointed or rounded ends and thin walls with perforated pits. There is considerable variation, however, in cell wall thickness and length.

Vessels consist of several cells separated only by a middle lamella and connected end to end, forming long tubes. Where the end wall more or less disappears, there is a *simple perforation*. Where only parts of the wall disappear, the perforated end plate may have multiple perforations.

In some woods, growths may be observed inside the vessel elements. These originate from surrounding parenchyma cells and are often found at the transition from sapwood to heartwood. Ray or axial parenchyma cells adjacent to vessels may resume growth and grow out into the vessel in the form of little bladders, filling the vessel with parenchymatous-like masses termed *tyloses.*

A tylosis is thus an extension of a living cell. After the death of their living contents (protoplasts), the remaining cell walls may significantly affect the penetrability of wood. Similar growths into intercellular spaces are termed *tylosoids.*

Turning now to consider the cells formed by the *ray initials*, these produce sheets of cells extending through the wood in a radial direction, which are termed *rays*. Rays vary enormously in size and shape, from one to many cells broad and deep. They may contain only one cell type or a number of cell types.

A quite common feature found in rays is the presence of *ducts* or cavities, which typically contain deposits of resinous or gum material. Ducts are formed in one of two ways. *Schizogenous ducts* are formed by the splitting apart of contiguous cells and are lined with parenchyma cells. *Lysigenous ducts* are formed by the breakdown of cells. Ducts can also be found in axillary cells. Ducts are common in commercial softwood timber.

7.3.3 Softwood Versus Hardwood Anatomy

Softwoods are simpler in structure than hardwoods. In softwoods, the vertical cells are mainly tracheids. The horizontal cells of the medullary rays are often exclusively parenchyma, but ray tracheids may be present. Often, these form the upper and lower margins of the ray. Vessels and fibres common to hardwoods are not found in softwoods.

In hardwoods, the vertical cells are more varied with vessels, fibres and tracheids. The vessels show major differences in size and distribution. Parenchyma may be abundant and distributed in characteristic ways. The horizontal cells of the rays may be uniseriate and only a few cells high, or at the other extreme, large multicellular structures, 2–3 cm high and many cells wide. Although hardwood rays are wholly parenchymatous, several distinct types of cells may be present.

Thus, in hardwoods, we have a wide range of possible variation at a more macroscopic level. With softwoods, identification relies on attention to fine detail, especially on the appearance of pitting.

As previously discussed, to fully describe the anatomy of a wood, it is necessary to examine the transverse, radial and tangential surfaces.

7.3.4 Softwoods

As has been discussed, the development of new wood is a seasonal affair. At the start of the growth cycle, the cambium produces tracheids, which are large, thin-walled, hexagonal in outline and those of contiguous rows alternate. As the season progresses, tracheids become more rectangular in cross-section, radially compressed and with thicker cell walls. The arrangement of radial rows moves from alternate to opposite. The transition in wall thickness may be sudden or more gradual. Intercellular space between tracheids is uncommon but is found in some species where the tracheids are more rounded and, thus, do not fit closely together.

The pits of axial tracheids are bordered, usually circular in surface view, with a small circular aperture. The pit membrane is not of equal thickness throughout, with a circular thickened area occupying the centre, the *torus*, surrounded by a thinner zone, the *annulus*. In face view, the pit may appear as three concentric rings. In most species, these pits are found only on the radial walls, but a few may be found on the tangential walls.

The distribution of these pits on the radial wall is a useful diagnostic feature. They may occur down the full length of the tracheid or at intervals along its length. Frequently, they occupy the entire width of the tracheid. In some species, the pits are smaller and may form from two to four rows on the radial wall and may be either opposite or alternative.

There are a number of other features associated with axial tracheids, which are of value in wood identification. These include the following:

- *Tori extensions*
- *Scalloped torus* – characteristic of one genus, *Cedrus*; the margin of the torus is indented or scalloped in appearance
- *Crassulae or bars of Sanio* – these are formed by localised thickening of the middle lamella and primary cell walls in the neighbourhood of pits. This is a common feature, but they are not found in two genera, *Agathis* and *Araucaria*
- *Spiral thickening* – found in two well-known timbers, Yew (*Taxus baccata*) and Douglas fir (P*seudotsuga* sp), mainly in early wood tracheids
- *Callitroid thickening* – transverse bars of localised thickening of the secondary cell wall, which overhang the bordered pits. Restricted to one genus, *Callitris*, from which it derives its name
- *Trabeculae* – radial bars of thickening, which pass across the lumina of tracheids, usually lying across several cells in a horizontal row

In addition to vertical tracheids, two other types of tracheids occur in softwoods, which, when present, are of value in identification. These are *strand tracheids* and *ray tracheids*.

Strand tracheids are formed by fusiform initials, but by deposition of cross-walls or septa, several brick-like cells are formed. They resemble parenchyma but can be distinguished by the presence of bordered pits. Strand tracheids are often found associated with vertical resin ducts.

As previously stated, tracheids are not found in the rays of all softwoods, but when present, they are formed from ray initials. Ray tracheids have pits lying across all walls of the cells. When present, they often form the upper and lower marginal cells of rays, and there may be more than one row. In some genera (*Picea* sp), minute peg-like outgrowths of the horizontal and end walls of the ray tracheids project into the lumina. These dentate tracheids are also found in some pine species. Pines as a group may be separated on the basis of the number of leaves or needles on each compound leaf. Thus, we have two to three needle pines and five-leaved pines. None of the five-leaved pines has dentate tracheids, whilst two to three leaf pines have ray tracheids with well-developed dentations.

The only other cell type found in softwoods of diagnostic significance is *parenchyma*.

Vertical parenchyma is never abundant and is easily identified in transverse section due to the lack of dark coloured contents. The end wall may have beads of thickening and is referred to as being nodular. The position of vertical parenchyma may be described as scattered or zonal. Pitting on the walls is always simple.

With those exceptions where ray tracheids are present, the softwood ray usually comprises only parenchyma cells. Typical medullary ray parenchyma cells are radially elongated and stretch across several vertical tracheids seen as *cross-field* areas when viewed in radial section – see Figure 7.2. A cross-field is defined as the area bounded by the intersecting walls of a single longitudinal tracheid and a single ray parenchyma.

The wall between the radial parenchyma cell and the tracheid is pitted, and the type of pitting is one of the most useful features in softwood identification. Six main types of pits are recognised (Phillips, 1948; IAWA, 2004):

- *Window-like* – with usually one to two large, simple pits
- *Pinoid* – one to six pinoids, three or more pits common. Small to fairly large, depending on the number per cross-field, simple or with reduced borders, and often irregular shape as compared with the more or less rectangular window-like pit
- *Piceoid* – small bordered pits with slit-like pores, which extend beyond the border at either end.
- *Cupressoid* –small, bordered pit where the pit is elliptical and never extends beyond the border
- *Taxodioid* – small, bordered pit where the pit pore is larger than the border in most parts of the pit. Thus, the wall slopes outwards from the pit membrane

- *Araucarioid* – individual pits are mainly cupressoid, but they are arranged in alternate rows of usually three or more, with a tendency to crowd. Individual pits often have a polygonal outline. This type of pit is restricted to the Araucariaceae.

Pitting should always be observed in *early wood.* Pit apertures may also be modified in compression wood.

Variation can also be seen in medullary rays when observed in a tangential longitudinal section. Rays are often uniseriate, and biseriate in part or less often biseriate. The number of cells high can be diagnostic, with less than 15 cells (low rays) or more than 30 cells (tall rays). Where horizontal resin ducts are present, the shape of the rays can be altered. Most often, the duct is solitary and occupies the centre of the ray. For example, in Douglas fir, the duct is central, whereas in Larches and Spruce, the duct is eccentric. These features require a sufficient sample for accurate assessment.

Finally, the presence of *resin ducts* in epithelial parenchymatous tissue is a common feature of woods and especially softwood species – see Figure 7.2. A resin duct forms when the middle lamella between cells disappears, and adjoining cells enlarge and separate, leaving a resin canal in their midst. These schizogenous ducts run vertically between the axial tracheids and radially in the rays. In most cases, vertical ducts occur singly. The epithelial cells surrounding the ducts may be thin-walled or thick-walled. The number of cells in a tangential section can also be diagnostic.

7.3.5 Hardwoods

All hardwoods are dicotyledons, and their wood is considerably more complex than that of coniferous wood. In softwoods, the bulk of the woody tissue consists of axial tracheids with narrow, predominantly parenchymatous rays.

In hardwoods, the axial elements are more varied, almost always having four or five types of cell present, *vessels*, *parenchyma*, *libriform fibres*, *tracheids* and *fibre-tracheids.*

As has been discussed previously, *vessels* are formed by the partial or complete disappearance of the end walls in an axial row of cells. There is a trend, moving from advanced to less advanced species, for no elongation of vessel elements, and in some species, the vessels can even be shorter and broader than the fusiform initials.

Tracheids, not universally present in hardwoods, when present, can be one of two types. One of these, the *vasicentric tracheid*, is so called because it occurs in association with the vessels. Compared to the tracheids of softwoods, these cells are not greatly elongated and have rounded ends and bordered pits in all their walls. They are often also distorted in shape. The second type, the *vascular tracheid*, resembles small vessels in that they are formed in axial columns, but their end walls are pitted and not perforated. Vascular tracheids may be thought of as incompletely differentiated vessel elements.

Vessels and tracheids do not have a support function, which falls in hardwoods to *libriform fibres*. These cells are elongated, on average 2–3 mm long, often, but not always, thick-walled and with few, often slit-like pits.

Between the vasicentric tracheid and the libriform fibre, there are all stages of intermediate cells, some performing mainly a transport function and others a support function. In describing hardwoods, little distinction is made between libriform fibres and fibre tracheids. In descriptive work, the focus is on fibres, parenchyma and vessel elements.

As with softwoods, three sections are normally examined, TS, TLS and RLS. The TS is as important and informative as the other two sections when examining hardwoods. Discrimination and identification are achieved by studying the variation of features caused by these basic cell types and how they are arranged and organised.

As has been discussed, most hardwood species produce seasonal growth that results in annual rings.

Based on the distribution of vessels, as seen in TS, hardwoods are classified as either *diffuse-porous* or *ring-porous*. In diffuse-porous woods, the vessels are more or less evenly distributed throughout the growth ring. In ring-porous woods, the vessels are in a distinct ring due to there being larger vessels in the early wood. If the vessels are in a distinct ring but are not markedly larger than the late wood pores, or if the early wood vessels are larger but the ring is not distinct, the wood is said to be *semi-ring porous*.

Vessels may be arranged singly (solitary) or in groups (multiple). When in groups, they can be further classified based on their distribution in radial multiples or in clusters with no other cells between the individual vessels. Chains of vessels may run obliquely or more or less tangentially. Even when solitary, vessels may have a close topographical relationship. Eucalypts have a characteristic arrangement in which the vessels have a flame-like distribution.

In longitudinal surfaces where the vessels are visible, they form grooves, termed *vessel lines*.

Other features of vessels of diagnostic importance include the types of perforation plates between elements, types of thickening, the presence of tylosis and intravascular pitting. Pitting is not as important a feature in hardwoods as it is for softwood, where the type of pitting is one of the most diagnostic features. In some woods, spiral thickening may be present on the vessel walls.

The parenchyma in hardwoods is generally more abundant than is softwoods. When viewed in TS, the distribution of parenchyma is often distinctive and of great value for diagnostic purposes. Three basic distributions are recognised.

- *Boundary* – where the parenchyma is at the beginning or end of a growth ring
- *Paratracheal* – where the parenchyma is associated with vessels or tracheids
- *Apotracheal* – where the parenchyma is not associated with vessels or tracheids

Further subdivision based on these three types is possible, giving 11 subdivisions.

With some exceptions, the medullary rays in hardwoods are almost entirely composed of parenchyma cells. Parenchyma cells in rays are of three types:

- *Procumbent cells* – radially elongated parenchyma
- *Upright or square cells* – parenchyma where the vertical height is equal to or greater than the radial length. Often, this cell type forms the 'tail' of a ray
- *Tile cells* – cells that lose their protoplasm soon after formation and, hence, do not elongate

Rays may also be classified based on the appearance of the ray as seen in RLS and TLS. Rays are first subdivided into *homogeneous* and *heterogeneous*. As the names imply, in a homogeneous ray, only one cell type is found, these being procumbent cells. In a heterogeneous ray, all the above three cell types may be present.

Other features of rays that are of diagnostic value include height, breadth and appearance in the transverse section (curved, straight). Where rays are seen to be broader at the boundaries of growth rings, they are said to be *noded*.

Although pitting is of less diagnostic value for hardwoods than for softwoods, cross-field pitting between the vascular cells and ray parenchyma is present and can be classified into three types: large, rounded pits greater than 10 u in diameter; large gash-like pits, which are elongated and also greater than 10 u and small pits, which are less than 10 u and often much smaller.

Finally, it is possible to come across oil cells and cellular inclusions in wood. Calcium oxalate is found as prisms, acicular or microcrystals. Silica nodules may be present. Resin canals may also be present, although, unlike in softwoods, there is less certainty in the number or position of such canals in hardwoods.

7.4 Wood Identification

7.4.1 Types of Material

As previously discussed, the types of wood-based materials requiring identification vary from logs, large pieces of lumber, timber products and composite materials such as plywood, particle board, splinters, sawdust and wood pulp used to manufacture paper.

Very small fragments may not show a full range of features. Some woods may be altered due to various treatments or have been stained and do not have their original colour. Wood may be decayed, charcoaled or poorly preserved, making it difficult to section and prepare. Cell sizes are altered in charcoaled woods, making quantitative features of limited value (Wheeler & Baas, 1998).

In art applications, it will not usually be possible to take large samples, if samples at all, from frames or panels to establish authenticity and provenance. Hence, identification will be more challenging, and often limited, in the absence of unaltered pieces of wood large enough to make TS, RLS and TLS that are most likely to provide a diagnostic combination of features.

7.4.2 Macroscopic Versus Microscopic Features

Some woods have distinctive combinations of macroscopic features and can be identified with the aid of a hand lens. In experienced hands, macroscopic features can be used to quickly establish whether wood is correctly labelled or to which family it is likely to belong (Wheeler & Baas, 1998).

Using a punch card system, a lens key for the identification of hardwoods described 84 features to identify 450 species (Anon, 1970). A more recent key from the CSIRO uses 50 features (Ilic, 1990). Macroscopic keys typically have fewer features than microscopic keys. The main limitation from a forensic viewpoint is that macroscopic features may not be present due to sample size and condition.

Hence, it is not that microscopic features are necessarily better than macroscopic features, but they may be applied across a wider range of sample types and sizes.

7.4.3 Identification Procedures

Identification by comparison is a frequently used approach in natural history. Numerous wood anatomical atlases covering many geographic areas exist. Wheeler and Baas (1998) list many of these resources, drawing attention to the CSIRO Hardwood Atlas, which contains photographs of TS, RLS and TLS for 1,800 species and was intended to be a portable wood anatomical slide collection (Ilic, 1991). Whilst comparative studies may not result in an identification, they are also often an essential starting point for forensic examinations, as if two woods are different, it may not be necessary to identify the woods.

Having examined a specimen and created a description of its features, a common approach in botany is to use a *dichotomous key*. A dichotomous key presents a series of paired contrasting choices. Rather like the digital world, one of these two options has to be chosen to move on to the next couplet until eventually reaching a name that, for wood, could be at almost any level of taxa, generally only to genus and only rarely at the level of species. Jane (1970) provides examples of dichotomous keys for both softwoods and hardwoods.

Dichotomous keys are useful for an unknown where there is a small number of possible matches and for material without missing features. The limitation for forensic examinations is that features may well be missing due to the sample size or condition and the inherent problem that one mistaken choice of feature can lead to a misidentification.

For these reasons, a *synoptical key* is more commonly used for wood identification. This type of key relies on multiple entries, which list for each diagnostic feature the taxa having that feature. This type of key was first devised in the Forest Products Research Laboratory in 1936 (Clarke, 1938) and was based on the use of a multiple-entry punched card using perforated edges. The card was then clipped where a feature was present.

Later keys for softwoods (Phillips, 1948) and hardwoods (Brazier & Franklin, 1961) followed. The hardwood key used 86 features and the softwood card 53 features. A card would be prepared for each species, the features being recorded by clipping the card opposite the appropriate holes. The user would select a feature and then place a needle through the relevant clipped hole and shake the cards, some 400 for hardwoods. Cards that had the relevant feature present would fall out. This process would be repeated until either the examiner had run out of features or only one card remained.

Today, this manual approach has been replaced by computer-assisted programs. Wheeler and Baas (1998) discuss in detail the evolution of computer-assisted wood identification leading to the development of *Insidewood*, a web resource for hardwood anatomy (Insidewood, 2004; Wheeler, 2011). Feature selection still draws on the original manual systems described above.

Additional features have been added over time as the number of features is no longer limited by the physical limit to the number that could be accommodated in a card system. The IAWA (International Association of Wood Anatomists) has published lists of microscopic features for hardwoods (IAWA, 1989) and softwoods (IAWA, 2004). The hardwood list contains 163 anatomical criteria, and the softwood list has 124 criteria. These two publications effectively replace all earlier publications and are central resources for the identification of woods. Using these two resources, Gasson et al. (2021), in a beautifully illustrated paper, describe the wood anatomy of CITES-listed tree species.

Even these IAWA lists are not intended to be comprehensive and cover all woods, with the focus still on commercial timbers. Wheeler and Baas (1998) list some other resources covering trees from specific geographic areas.

Some features are more useful than others for identification. Regardless, they must be consistent from one sample to the next. Some features are useful because they occur in relatively few taxa, and, indeed, 46 of the 63 anatomical features used by Clarke (1938) occur in less than 25% of the world's hardwoods and 20 occur in less than 10% (Wheeler & Baas, 1998). For example, aggregate rays with tile cells account for a mere 1%. Woods with a small number of these rare features, in combination with more common features, can be quickly identified.

When multiple entry cards were in use, the keys were accompanied by quite detailed notes explaining the features that most users would take time to read before attempting to use the key. The danger with computer entry keys is that users may be more likely not to read explanatory material, which could result in misuse and misinterpretation of features (Wheeler & Baas, 1998).

Wheeler (2011) also cautions that it is important to remember that thousands of species are not included in the Insidewood database and that descriptions are based on a few samples and may not represent the full range of variation of a species. As was the situation with earlier card-based systems, the appropriate use of Insidewood is to produce a short list of possible species, with the expert then using a comparative analysis approach to reach a final conclusion.

Comparative analysis depends on having access to reference samples. Few forensic laboratories will now have their own reference collection or in-house expertise and will rely on having formed a relationship with an external provider. Wood reference collections are called *Xylaria*. Kew Gardens in the United Kingdom keep a record of Xylaria around the world. As of 2010, there were 83 operational wood collections, but disturbingly, some 80 had either closed or been absorbed into other collections, resulting in a loss of professional expertise (Lynch & Gasson, 2010).

Of greater concern is that not all accessions are based on herbarium vouchers, risking that some woods may not have been correctly identified. Many specimens may also lack accurate geo-location, limiting their value for geo-provenancing (Gasson et al., 2021). In an attempt to improve the reliability of wood reference collections, a programme called *WorldForest ID* has been developed by a consortium of government and non-government organisations, with the aim of building a comprehensive collection of internationally traded timber species (Gasson et al., 2021).

Ultimately, however, accurate identification depends on the ability of individuals to correctly recognise features. Dormontt et al. (2015) comment on the 'paucity of appropriate wood anatomical expertise worldwide', a view supported by Gasson et al. (2021), who comment on the reality of the declining population of expert anatomists despite the increased global trade in wood products.

In order to aid or reduce the need for human expertise, machine learning and other neural networks are being investigated with the aim not to replace or eliminate the need for expert wood anatomists but to provide alternative methods. Hermanson and Wiedenhoeft (2011) have reviewed the use of machine vision in the context of automated wood identification systems. Gurau et al. (2013) report on an image processing method, based on a program called *ImageJ analysis*, as a supporting tool for wood species identification. This same software was used to assist in the identification of wood species in the conservation of wooden cultural heritage (Timar et al., 2012). Using transverse sections only, Lens et al. (2020) report on the use of convolutional neural networks achieving 95.6% accuracy in identifying species from a collection of 112 neotropical trees. Machine learning is, however, still predominantly at the research and pilot implementation stage.

In a study on the use of scanning electron microscopy and complex optical microscopes to identify archaeological and historical wood samples, whilst these techniques may offer exciting possibilities, the author cautions that the danger is that the novice may think that advanced technology is a substitute for specialised training in wood anatomy (Cartwright, 2015).

Appendix B.7 lists digital reference databases used for the analysis of wood anatomical features for the purpose of wood identification.

7.4.4 Non-Microscopic Methods for Wood Forensic Examinations Including the Role of DNA Analysis

A detailed consideration of methods used to examine wood is outside the scope of this chapter. The Global Timber Tracking Network (GTTN, 2020) provides a comprehensive overview of different timber tracking methods and an overview of current practices in data analysis for wood identification. There is, however, a substantial gap between the potential and realised application of most of these technologies (Dormontt et al., 2015).

A summary of techniques and their potential capabilities is presented in Table 7.1.

Dormontt et al. (2015) argue that no one scientific method is capable of addressing all diagnostic forensic timber identification questions and that an integrated approach is necessary, whilst stressing that the great strength of wood anatomy is that completely unknown samples can be quickly identified down to genus level in a way that is faster and cheaper than any other method that can identify genus.

From a botany perspective, only DNA analysis will be considered further, with a focus on only recent studies and emerging technologies.

DNA in plants can be derived from nuclear genomes (nDNA) and plastid genomes, chloroplasts (cpDNA) and mitochondria (mtDNA).

Table 7.1 Method Capabilities

Identification Need	Wood Anatomy	Machine Vision	Dendro-Chronology	Mass Spectrometry	Near Infrared Spectroscopy	Stable Isotopes	Radiocarbon	Genetics
Genus	Yes	Yes	No	Yes	Yes	No	No	Yes
Species	Occasionally	Occasionally	No	Yes	Yes	No	No	Yes
Provenance	Occasionally	Unknown	Occasionally	Yes	Yes	Yes	No	Yes
Individuals	No	No	Yes	No	No	No	No	Yes
Age	No	No	Yes – with growth rings	No	No	No	Yes	No

Source: UNODC (2016), Table A8.1, page 176.

Nuclear genomes evolve at varied rates, but fast-evolving loci such as microsatellites and some single-nucleotide polymorphisms (SNPs) are suited to identifying populations. Plastid genomes evolve relatively slowly, making them suited for distinguishing species and genera. Differences in plastid DNA can be detected by either sequencing sections of DNA, the so-called barcodes, by identifying SNPs (cpSNPs) or by studying size differences in microsatellite regions, also called single sequence repeats or short tandem repeats (STRs). Plastid DNA is generally poor at identifying individuals, populations and closely related species, but has a major advantage in that they are present in high copy numbers per cell, increasing its sensitivity in low-quality wood samples (Jiao et al., 2018).

The choice of DNA technology also depends on the scope of the analysis. Discrimination of distantly related species or geographic provenance can usually be performed by analysis of a few chloroplast gene sequences, such as *rbcl* and *matk* barcode regions, or in combination with nuclear loci with highly conserved flanking regions, such as Internal Transcriber Spacer (*ITS*). Hu et al. (2021) found that ITS2 was the most efficient locus for Chinese timber species identification, but a combination of ITS with three other barcode loci (*rbcl*, matK and *trnH-psbaA*) was required to distinguish closely related species involving internationally traded species. Schardosin et al. (2019) also reported on the use of the ITS1 region to identify seized logs in a case study from Brazil. Jiao et al. (2018) showed that a combination of three cpDNA barcode regions, *psbA+trnH+trnK*, gave 100% discrimination of sandalwood (*Santalum album*) specimens. Finally, Ng et al. (2020) report on the use of a combination of four cpDNA and 14 STR markers for a geo traceability system to classify merbau (*Intsia palembanica*) into northern and southern regions of peninsular Malaysia.

A major challenge in applying DNA analysis for wood is that the DNA is difficult to extract and is often highly degraded (Deguilloux et al., 2001). Rachmayanti et al. (2009) concluded that the main factors influencing the success rate of amplification were the size of the amplified fragments and the processing status of the wood. Amplification of DNA extracted from the outer sapwood was more successful than for DNA extracted from the transition wood and heartwood, with, however, the disadvantage that there were more polymerase chain reaction–inhibitory substances in the sapwood extracts. A decrease in the quantity of DNA in heartwood compared to sapwood in both fresh and dried wood is consistent with a number of other studies (Abe et al., 2011; Tang et al., 2011; Jiao et al., 2012). Using ancient DNA techniques, DNA has also been extracted in archaeological studies of ancient timbers, successfully amplifying at least one marker for 56% of samples (Akhmetzyanov et al., 2020).

Due to the highly degraded nature of DNA extracted from wood, the use of SNPs has become the technique most applied in the analysis of DNA from wood. These can be either nuclear (nSNPs), mitochondrial (mtSNPs) or chloroplast (cpSNPs). Although SNPs have lower diversity than, for example, STRs, markers can be developed relatively easily, quickly and at low cost. SNPs have an additional benefit in that they can be studied using automated high-throughput platforms that allow for the simultaneous analysis of large numbers of SNPs. Hence, SNPs have been used for species identification (Chaves et al., 2018: Honorio Coronado et al., 2019: Paredes-Villanueva et al., 2019), provenance determination (Blanc-Jolivet et al., 2018: Tysklind et al., 2019: Hartvig et al., 2020) and individualisation (Dormontt et al., 2020).

In the study of Blanc-Jolivet et al. (2018), 253 SNP loci were used to screen 916 Larix spp. Individuals and eight genetic clusters were identified, showing an East to West pattern in Eurasia. This research group have published several studies (not referenced in this chapter) where nuclear and/or chloroplast SNPs have been developed for a variety of other timber species.

Dormontt et al. (2020), building on an earlier study (Jardine et al., 2015), used 128 SNPs and three insertion/deletion markers to genotype 394 individual samples from bigleaf maple (*Acer macrophyllum*) to produce a reference population database and demonstrated that the resulting DNA assay was reliable, species-specific and effective at low DNA concentrations. This same group (Dunker et al., 2020) have also developed a set of 156 SNP markers for teak (*Tectona grandis*).

Recent progress in developing DNA methods for wood and timber in a forensic context is encouraging. Low et al. (2023), however, in reviewing the status of methods for the identification of timber species and the origin of harvest for the top 322 global priority timber taxa, found that for species identification, data existed for 100% of taxa using wood anatomy, 86% using genetics, 41% for DART TOFMS (Direct Analysis in Real-Time of Flight Mass Spectrometry) and only 6% for NIRS (Near Infrared Spectroscopy). For origin identification, data existed for only 24% of taxa, with genetics accounting for 23%. Importantly, they also point out that no studies have developed forensic-ready tests for global timber taxa. Based on the current rate of research, they estimated that it will take 27 years to generate geographic data for all 322 priority taxa!

To the best of my knowledge, no mainstream forensic laboratory offers DNA analysis for wood samples. Hence, relationships need to be established with external providers whose main interests will lie outside of forensic work and who may have little or no experience of the stringent quality standards required for forensic work. The UNODC have published a best practice guide for forensic timber identification (UNODC, 2016).

Appendix C.7 lists laboratories that provide DNA analysis.

Finally, as has been pointed out by many authors, DNA testing is not a replacement for macroscopic or microscopic examinations. With, however, the declining numbers of specialists with the expertise to conduct microscopic examinations, it is probable that DNA analysis will play an even increasing role in the future. This situation is not likely to improve unless there is a greater emphasis on traditional botany and specialist training available. Kew Gardens in the United Kingdom and the McCrone Institute in the United States are two organisations that run training programs in wood anatomy – see www.kew/org/science/training-and-education/continuing-professional-development/wood-identification and www.mcroneinstitute.org/V/1498/Wood-and-vegetable-Fiber-Microscopy-1540. At the time of writing, there were no dates for courses at Kew.

Appendix A.7: Preparation of Wood for Microscopic Examination

The following is based on a Forest Products Research Laboratory (FPRL) publication from 1951.

Preparation will depend on the size and condition of the sample. The aim is to prepare a sample where TS, RLS and TLS can be made. For fresh green or sapwood, little to no preparation may be necessary as long as the sample is kept moist with water or alcohol. For seasoned timber, some degree of softening is usually required. For timbers that are relatively soft, the sample is immersed in gently boiling water until it becomes waterlogged and sinks. The sample should then be sectioned whilst still hot. An alternative method for fragile samples is to place the sample in water in a vessel connected to a vacuum pump and exhaust air. This is repeated until the sample is waterlogged.

After initial treatment, samples can be softened in a number of ways. After waterlogging, samples can be placed in a mixture of equal parts of glycerine and alcohol for about three days. This treatment is suitable for even fairly hard woods.

Without previous waterlogging, the use of peroxide and acetic acid is recommended. The sample is boiled for at least 1 hour in a mixture of one part of glacial acetic acid and two parts of hydrogen peroxide in a reflux condenser. Where the sample size allows, a larger block should be cut, as the outer 1.5 mm of the block may be too soft for sectioning.

For small samples, or where it is desirable to examine individual cellular elements, the same approach as above can be used, with the fluid being poured off and the chips washed in two or three changes of distilled water. An alternative is to use commercial hypochlorite bleach solution.

The preparation of charcoals or badly decayed wood will normally require some form of embedding. Guidance on appropriate techniques can be found in BRAS (1976).

Sections of softened wood may be cut freehand or using a microtome. Guidance on the use of microtomes can be found in any textbook on

microtechniques. For very small samples, a useful technique is to freeze the sample in ice using a Peltier stage and then make freehand sections.
For sections and/or macerations, these are normally stained before being placed in a suitable temporary or permanent mountant. The use of Safranine will differentiate lignified (red-stained) and unlignified (green-stained) cells. Sections or macerated cells are placed in a 1% aqueous solution of Safranine for 5 minutes, the solution is drained off, and the samples are washed with at least three changes of distilled water and then twice with 97% alcohol. The samples are then covered with 1% Fast Green in clove oil and alcohol (1:3) for 2 minutes, the solution is drained off and the sample is washed with at least three changes of 97% alcohol. Finally, the samples are transferred to clove oil for 5 minutes and then cedarwood oil for 1 minute before being placed into the mountant.

Appendix B.7: Resources for Wood Anatomical Analysis

InsideWood – over 700 descriptions of the microscopic anatomy of hardwoods, including a searchable database of over 45,000 images – freely available on the internet

Richter and Dallwitz (2000) – an interactive identification key with integrated database of the microscopic descriptions and illustrations for 404 commercial timber species – freely available on the internet at delta-intkey.com

CITESwoodID – an integrated identification key with an integrated database of macroscopic descriptions illustrated for 44 CITES-protected species – 2014 onwards at Richter, Gembruch and Koch, at delta-intkey.com

Richter and Dallwitz (2016) – an interactive key with integrated database of microscopic descriptions and illustrations for 53 coniferous taxa, at http://delta-intkey.com

Wood database of the Forestry and Forest Products Research Institute – a database system of Japanese woods based on the IAWA list of hardwood identification, including a multiple-entry key and image database

The Forest Species Database – a database for the automatic classification of forest species, composed of microscopic images for 112 species of hardwoods and softwoods from 85 genera and 30 families

The Forest Species Database – macroscopic – a database for the automatic classification of forest species composed of macroscopic images from 41 forest species of the Brazilian flora

Microscopic Wood Anatomy of Central European Species – web-based identification key of central European species

(Based on GTTN, 2020)

Appendix C.7: Laboratories Offering Anatomy or DNA Timber Testing Services

Wood Anatomy

Royal Botanic Gardens Kew (UK) – https://www.kew.org
Thunen Institute (Germany) – https://www.thuenen.de/en
WoodID Lab (Italy) – https://woodislab.it
Innovation Environmental (UK) – https://innovationpropertyuk.com
Test-Tech (UK) – https://www.test-tech.co.uk
IPS Testing (USA) – https://ipstesting.com
Forest Research Institute Malaysia (FRIM) (Malaysia) – http://www.frim.gov.my
USDA Forest Products Laboratory – https://www.fpl.fs.fed.us

DNA Analysis

Royal Botanic Gardens Kew (UK) – https://www.kew.org
Thunen Institute (Germany) – https://thuenen.de/en
Naturalis Biodiversity Centre (Netherlands) – https://www.naturalis.nl
FERA (UK) – https://fera.co.uk
FRIM (Malaysia) – https://www.frim.gov.my
IBL (Forest Research Institute) (Poland) – https://ibles.pl
USDA Forest Products Laboratory – https://www.fpl.fs.fed.us
Australian Centre for Evolutionary Biology and Biodiversity, University of Adelaide – https://www.adelaide.edu.au/directory/andrew.lowe
DoubleHellix – https://doublehelixtracking.com

(Based on NEPCon, 2017)

References

Abe, H., Watanabe, K., Yoshida, K., Kuroda, K., & Zhang, C. 2011. Changes in Organelle and DNA Quality, Quantity, and Distribution in the Wood of *Cryptomeria japonica* Over a Long-term Storage. *IAWA Journal*, 32, 263–272.

Akhmetzyanov, L., Copini, P., Sass-Klaassen, U., Schroeder, H., Arjen de Groot, G., Laros, I., & Daly, A. 2020. *DNA of Centuries-Old Timber Can Reveal Its Origin*. www.nature.com/scientificreports; https://doi.org/10.1038/s41598-020-77387-2

Anon. 1951. *The Preparation of Wood for Microscopic Examination*. Forest Products Research Laboratory. Norwich, UK: Her Majesty's Stationary Office.

Anon. 1970. *Identification of Hardwoods. A Lens Key*. 2nd ed. London: HMSO.

Anon. 2017. *Transnational Crime and the Developed World*. Global Financial Integrity. https://gfintegrity.org/wp-content/uploads/2017/03/Transnational_Crime-final.pdf. Accessed 27 April 2026.

Anon. 2020. DNA Testing Shows 40% of Imported Timber Incorrectly Labelled. *Timberbiz*. https://www.timberbiz.com.au. Accessed 9 December 2020.

Balouet, J.C., Oudijk, G., Smith, K.T., Petrisor, I., Rudd, H., & Stocklassa, B. 2007. Applied Dendroecology and Environmental Forensics. Characterizing and Age Dating Environmental Releases: Fundamentals and Case Studies. *Environmental Forensics*, 8, 1–17.

Blanc-Jolivet, C., Yanbaev Y., Kersten B., & Degen, B. 2018. A Set of SNP Markers for Timber Tracking of Larix spp. in Europe and Russia. *Forestry*, 91, 614–628.

Brazier, J., & Franklin, G.L. 1961. Identification of Hardwoods: A Microscope Key. *Forest Products Research Bulletin*, No 46.

Cartwright, C.R. 2015. The Principles, Procedures and Pitfalls in Identifying Archaeological and Historical Wood Samples. *Annals of Botany*, 116, 1–13.

Chaves, C., Blanc-Jolivet, C., Sebbenn, A., et al. 2018. Nuclear and Chloroplastic SNP Markers for Genetic Studies of Timber Origin for *Hymenaea* Trees. *Conservation Genetics Resources*, 11, 329–331.

Christophe, D.P., & Daniels, C. 2008. An Unusual Technique for Physical Match Comparison. *AFTE Journal*, 40, 396–398.

Clarke, S.H. 1938. A Multiple-Entry Perforated-Card Key with Special Reference to the Identification of Hardwoods. *The New Phytologist*, 38, 369–374.

Courtin, G.M., & Fairgrieve, S.I. 2004. Estimation of Postmortem Interval (PMI) as Revealed Through the Analysis of Annual Growth in Woody Tissue. *Journal of Forensic Sciences*, 49, 781.

CSIRO. 2023. *Identifying Eucalypts Made Easy*. EUCLID, Eucalypts of Australia. https://www.CSIRO.au/en/research/plants/native/Identifying-eucalypts-made-easy

Deguilloux, M.F., Pemonge, M.H., & Petit, R.J. 2001. Novel Perspectives in Wood Certification and Forensics: Dry Wood as a Source of DNA. *Proceedings of the Royal Society of London. Series B*, 269, 1039–1046.

Denne, M.P. 1977. Dating of Events from Tree Growth and Wood Structure. *Journal of the Forensic Science Society*, 17, 257–264.

Dormontt, E.E., Boner, M., Braun, B., et al. 2015. Forensic Timber Identification: It's Time to Integrate Disciplines to Combat Illegal Logging. *Biological Conservation*, 191, 790–798.

Dormontt, E.E., Jardine, D.I., Van Dijk, K.J., Dunker, B.F., Dixon, R.R.M., Hipkins, V.D., Tobe, S., Linacre, A., & Lowe, A.J. 2020. Forensic Validation of a SNP and INDEL Panel for Individualisation of Timber from Bigleaf Maple (*Acer macrophyllum* Pursch). *Forensic Science International: Genetics*. https://doi.org/10.1016/j.fsigen.2020.102252

Dunker, B., Dormontt, E.E., van Dijk, K.J., Dixon, R.R.M., Jardine, D.I., Kireta, D., Nurtahjaningsh, I.L.G., Win, T.T., Rimbawanto, A., & Lowe, A.J. 2020. A Set of 156 SNP Markers for Teak (*Tectona grandis* Linn.f.). *Conservation Genetics Resources*, 12, 205–207.

Fiddian, S. 2005. Botany in Forensic Science. In Freckelton, I., & Selby, H. (eds.), *Expert Evidence* (Chapter 91). Sydney: Thomson Reuters.
Fish, F. 1973. Academic Forensic Science: Retrospect and Prospect. *Journal of the Forensic Science Society*, 13, 243–247.
Fish, J.L. 1966. The Identification of Wood Fragments. *Journal of the Forensic Science Society*, 6, 67–75.
Gasson, P.E., Lancaster, C.A., Young, R., et al. 2021. WorldForestID: Addressing the Need for Standardized Wood Reference Collections to Support Authentication Analysis Technologies; a Way forward for Checking the Origin and Identity of Traded Timber. *Plants, People, Planet*, 3, 130–141.
Gatti, R.C., Reich, P.B., Gamarra, J.G.P., et al. 2022. The Number of Tree Species on Earth. *Proceedings of the National Academy of Sciences*, 119, e2115329119.
Graham, S.A. 1997. Anatomy of the Lindberg Kidnapping. *Journal of Forensic Sciences*, 42, 368–377.
Grant, J., & Chen, H.K. 2021. *Using Wood Forensic Science to Deter Corruption and Illegality in the Timber Trade*. https://www.worldwildlife.org/publications/using-wood-forensic-science-to-deter-corruption-and-illegality-in-the-timber-trade
GTTN. 2020. *Overview of Current Practices in Data Analysis for Wood Identification. A Guide for the Different Timber Tracking Methods*, Schmitz, N. (ed.). Global Timber Tracking Network. https://doi.org/10.13140/RG.2.2.21518.79689
Gurau, L., Timar, M.C., Porojan, M., & Loras, F. 2013. Image Processing Method as a Supporting Tool for Wood Species Identification. *Wood & Fiber Science*, 45, 1–11.
Hartvig, I., So, S., Changtragoon, S., et al. 2020. Conservation Genetics of the Critically Endangered Siamese Rosewood (*Dalbergia cochinchinensis*); Recommendations for Management and Sustainable Use. *Conservation Genetics*, 21, 677–692.
Hathaway, R.A. 1994. Physical Wood Match of a Broken Pool Cue Stick. *AFTE Journal*, 26, 185–186.
Hermanson, J.C., & Wiedenhoeft, A.C. 2011. A Brief Review of Machine Vision in the Context of Automated Wood Identification Systems. *IAWA Journal*, 32, 23–250.
Honorio Coronado, E.N., Blanc-Jolivet, C., Mader, M., et al. 2019. Development of Nuclear and Plastid SNP Markers for Genetic Studies of *Dipteryx* Tree Species in Amazonia. *Conservation Genetics Resources*, 11, 333–336.
Horrocks, M., & Wallace, R. 2001. Distortion of Teatree Stems by Twine as a Means to Determine the Number of Years that the Stems Have Been Used to Support Cannabis Plants. *Journal of Forensic Sciences*, 46, 854–856.
Hu, J.L., Ci, X.Q., Dormontt, E.E., Conran, J.G., Lowe, A.J., & Li, J. 2021. Assessing Candidate DNA Barcodes for Chinese and Internationally Traded Timber Species. *Molecular Ecology Resources*, 22, 1478–1492.
IAWA Committee. 1989. IAWA List of Microscopic Features for Hardwood Identification. *IAWA Bulletin*, 10, 219–332.
IAWA Committee. 2004. IAWA List of Microscopic Features for Softwood Identification. *IAWA Bulletin*, 25, 1–70.
Ilic, J. 1990. *CSIRO Macro Key for Hardwood Identification*. Australia: CSIRO.
Ilic, J. 1991. *CSIRO Atlas of Hardwoods*. Berlin: Springer-Verlag.
InsideWood. 2004. *Published on the Internet*. http://insidewood.lib.ncsu.edu/search
Jane, F.W. 1970. *The Structure of Wood*, 2nd ed., revised by Wilson, K., & White, D.J.B. London: Adam and Charles Black.

Jardine, D.L., Dormontt, E.E., van Dijk, K.J., Dixon, R.R.M., Dunker, B., & Lowe, A.J. 2015. A Set of 204 SNP and INDEL Markers for Bigleaf maple (*Acer macrophyllum* Pursch). *Conservation Genetics Resources*, 7, 797–801.

Jiao, L., He, T., Dormontt, E.E., Zhang, Y., Lowe, A.J., & Yin, Y. 2018. Applicability of chloroplast DNA barcodes for Wood Identification Between *Santalum album* and its Adulterants. *Holzforchung*. https://doi.org/10.1515/hf-2018-0047

Jiao, L.J., Yin, Y., Xiao, F., Sun, Q., Song, K., & Jiang, X. 2012. Comparative Analysis of Two DNA Extraction Protocols from Fresh and Dried Wood of *Cunninghamia lanceolata* (Taxodiaceae). *IAWA Journal*, 33, 441–456.

Jones, E.L., & Daniels, L.D. 2012. Assessment of Dendrochronological Year-of-Death Estimates Using Permanent Sample Plot Data. *Tree-Ring Research*, 68, 3–16.

Jozsa, L.A. 1985. Contribution of Tree -Ring Dating and Wood Structure Analysis to the Forensic Sciences. *Canadian Society of Forensic Science Journal*, 18, 200–210.

Lens, F., Liang, C., Guo, Y., et al. 2020. Computer-Assisted Timber Identification Based on Features Extracted from Microscopic Wood Sections. *IAWA Journal*, 41, 660–680.

Low, M.C., Schmitz, N., Boeschoten, L.E., et al. 2023. Tracing the World's Timber: The Status of Scientific verification technologies for Species and Origin Identification. *IAWA Journal*, 44, 63–84.

Lynch, A., & Gasson, P. 2010. *Index Xylatorium*, 4th ed. Melbourne: Royal Botanic Garden.

NEPCon. 2017. Timber Testing Techniques, A Guide to laboratory Techniques to Determine Species and Origin of Timber Products. Thematic Article Series No. 1, info@nepcon.org

Ng, C.H., Ng, K.K.S., Lee, S.L., Tnah, L.H., Lee, C.T., & Zakaria, N.F. 2020. A Geographical Traceability System for Merbau (*Intsia palembanica* Miq.), an Important Timber Species from Peninsular Malaysia. *Forensic Science International: Genetics*. https://doi.org/10.1016/j.fsigen.2019.102188

Palenik, S. 1983. Microscopic Trace Evidence – The Overlooked Clue. *The Microscope*, 31, 114.

Paredes-Villanueva, K., Blan-Jolivet, C., Mader, M., et al. 2019. Nuclear and Plastid SNP Markers for Tracing *Cedrela* Timber Species in the Tropics. *Conservation Genetics Resources*, 12, 239–244.

Phillips, E.W.J. 1948. Identification of Softwoods by Their Microscopic Structure. *Forest Products Research Bulletin*, No 22.

Pokines, J.T. 2018. Two Cases of Dendrochronology Used to Corroborate a Forensic Postmortem Interval. *Journal of Forensic Identification*, 68, 457–464.

Rachmayanti, Y., Leinemann, L., Gailing, O., & Finkeldey, R. 2009. DNA from Processed and Unprocessed Wood: Factors Influencing the Isolation Success. *Forensic Science International: Genetics*, 3, 185–192.

Richter, H.G., & Dallwitz, M.J. 2000. *Commercial Timbers: Descriptions, Illustrations, Identification, and Information Retrieval*. https://www.delta-intkey.com

Richter, H.G., Gembruch, K., & Koch, G. 2014. *CITESwoodID: Descriptions, Illustrations, Identification, and Information Retrieval*. https://www.delta-intkey.com

Schardosin, F.Z., Gonzalez de Cademartori, P.H., Morrone, S.R., Bolzon de Munez, G.I., & Nisgoski, S. 2019. The Use of ITS Region as a Tool for Wood Identification in Brazil: A Case Study of a Tree from the Genus *Hovenia* sp. *Brazilian Journal of Wood Science*, 10, 123–134.

Schweingruber, F.H., Borner, A., & Schulze, E.D. 2006. *Atlas of Woody Plant Stems. Evolution, Structure and Environmental Modifications*. Berlin: Springer-Verlag.

Tang, X., Zhao, G., & Ping, L. 2011. Wood Identification with PCR Targeting Noncoding Chloroplast DNA. *Plant Molecular Biology*, 77, 609–617.

Timar, M., Gurau, L., & Porojan, M. 2012. Wood Species Identification, a Challenge of Scientific Conservation. *International Journal of Conservation Science*, 3, 11–22.

Townshend, D.G. 1981. Examination of Tree Stumps. *AFTE Journal*, 13, 32–36.

Tysklind, N., Blanc-Jolivet, C., Mader, M., et al. 2019. Development of Nuclear and Plastid SNP and INDEL markers for Population Genetic Studies and Timber Traceability of *Carapa* Species. *Conservation Genetics Resources*, 11, 337–339.

UNODC. 2016. *Best Practice Guide for Forensic Timber Identification*. New York: United Nations.

Wheeler, E.A. 2011. Insidewood - A Web Resource for Hard wood Anatomy. *IAWA Journal*, 32, 199–211.

Wheeler, E.A., & Baas, P. 1998. Wood Identification – A Review. *IAWA Journal*, 19, 241–264.

Wiedenhoeft, A.C., Simeone, J., Smith, A., Parker-Forney, M., Soares, R., & Fishman, A. 2019. Fraud and Misrepresentation in Retail Forest Products Exceeds U.S. Forest Wood Science Capacity. *PLoS One*, 14, e0219917.

Wilson, K., & White, D.J.B. 1986. *The Anatomy of Wood: Its Diversity and Variability*. London: Stobart & Son.

Yaman, B., & Akkemik, U. 2009. The Use of Dendrochronological Method in Dating of Illegal Tree Cuttings in Turkey: A Case Study. *Baltic Forestry*, 15, 122–125.

Identification of Foodstuffs and Other Fragmented Plant Material

8

8.1 Introduction

The focus of this chapter is the examination of fragmented plant material in the context of the forensic examination of stomach contents, vomit and faeces.

The analysis of stomach contents can be used within limits to determine the time of death or postmortem interval (PMI) and/or provide information as to what a deceased person has recently eaten. This type of information may assist investigators where the deceased person may have eaten at a particular type of restaurant or food outlet (Bock & Norris, 2016; Schneck, 2013).

A common factor in cases involving physical assault, often allied with excessive alcohol consumption, is that a victim may vomit, and traces may transfer to a perpetrator or be left at a scene. A victim may also defaecate during an assault, and a transfer of faecal material may take place to an assailant. As discussed in Chapter 1, my introduction to the world of forensic science was a case where an elderly female had been sexually assaulted and had defaecated. It is also not uncommon for perpetrators to leave a 'present' at the scene of a crime in the form of faeces! The need to examine scats and faeces from animals may arise in cases involving the theft of animals, bestiality and wildlife crime.

The identification of foodstuffs is central to the forensic examination of stomach contents, faeces and vomit, but other non-foodstuffs may also be present. The obvious example will be cases of poisoning, intentional or unintentional. There continue to be relatively frequent examples of ingestion of poisonous mushrooms, mainly where individuals have collected mushrooms in the field and have misidentified the species. Foraging for edible plants is becoming quite common, with the potential for misidentification of plants. The reality is that most plants are, or have the potential to be, poisonous. Over time, plants have been bred to reduce potentially toxic components, are processed to remove toxic parts of the plant, or toxic components are removed during cooking. As quoted by Moffat (1980), it was Paracelsus (1493–1541) who said, 'All substances are poisons: there is none which is not a poison. The right dose differentiates a poison from a remedy'.

Finally, there has been a resurgence of interest in herbal medicine, especially in Chinese herbal medicine. Byard (2010), in reviewing the potential significance of traditional herbal medicine, points out that there are at least 7,000 species of medicinal plants in China. One of the problems with herbal

DOI: 10.4324/9781003308508-8

medicines is that they are not subject to strict regulation or control, as they are often marketed as dietary substances and not as drugs. Of course, many plant species do have a legitimate role to play in a regulated pharmacy setting. Here, it is critical that species are correctly identified and that they are free from adulteration (Upton et al., 2011).

Byard (2010) discusses examples where species have been intentionally or accidentally substituted during the preparation of herbal medicines, where they have been intentionally adulterated with standard drugs and where incomplete processing has led to residual toxins with potentially devastating consequences. The identification of herbal products, whilst outside the scope of this book, requires the same expertise as used in forensic analysis of fragmented or powdered vegetable drugs – this is the world of *pharmacognosy*, a subject no longer to be found in academic courses in western countries but experiencing a resurgence of interest in many other parts of the world (Upton et al., 2011). As a matter of personal interest, I was the last person to teach pharmacognosy during my time at Strathclyde University.

8.2 The Human Digestive System

In order to understand the products of the human gastrointestinal tract (GI), some understanding of the human digestive system is necessary. The GI tract can be thought of as an open-ended tube, 8–9 m in length, extending from the mouth to the anus and comprising the pharynx, oesophagus, stomach, small intestine, large intestine or colon and the rectum – see Figure 8.1.

Accessory organs are the teeth, tongue, salivary glands, liver, gall bladder and pancreas. Together, the digestive tract is responsible for extracting digested products and disposal of waste products.

Some of the earliest documented studies of human gastric emptying come from the early 1820s, when William Beaumont, a US Army surgeon, working at Fort Mackinac, treated a young employee of the American Fur Company, Alexis St. Martin, who had been accidentally shot in the stomach at close range with a shotgun. Not expected to survive from his injuries, he did survive but was left with a hole, or *fistula*, that connected the inside of his stomach to the outside world. St. Martin was hired as a handyman by Beaumont. In the years that followed, Beaumont performed experiments on digestion using the stomach of St. Martin. Beaumont would tie pieces of food to a silk string and insert it through the hole into St. Martin's stomach, removing it at intervals to study how well it had been digested. Beaumont also established that digestion was primarily a chemical process and not a mechanical one. Beaumont published an account of his experiments in 1833 (Beaumont, 1833).

Although the approach of using experimental windows, called *gastric fistulas*, is still used in animal studies, an approach such as this with humans

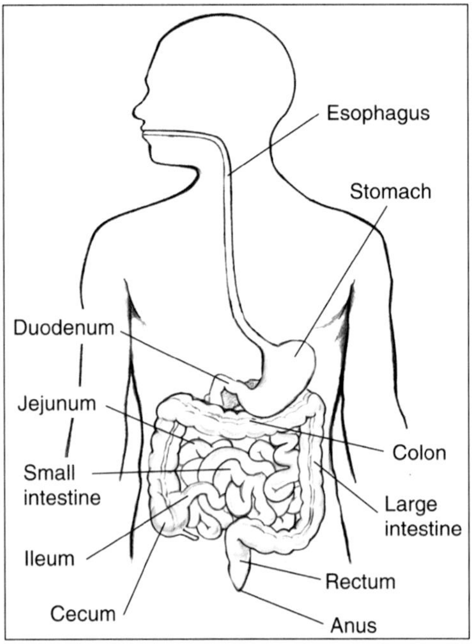

Figure 8.1 Drawing of gastrointestinal (GI) tract.

(With permission from the National Institute of Diabetes and Digestive and Kidney Diseases.)

would simply never meet modern ethical standards. There is, however, an extensive literature dealing with food digestion to develop knowledge to create healthier and more sustainable foods (Mackie et al., 2020). Numerous *in vitro* artificial systems have been developed to replicate the human stomach, such as the *Human Gastric Simulator* (HGS) (Mackie et al., 2020; Sensoy, 2021).

For example, the physical digestion of brown and white rice has been studied using an HGS. It was shown that the bran layer on brown rice limited the physical disintegration of rice, delaying gastric emptying and, as a result, possible starch digestion. This research offers a plausible mechanism for human studies that show that brown rice lowered glycaemic responses compared to white rice (Kong et al., 2011; Panlasigui & Thomson, 2006).

The results of *in vitro* digestion models are often different to those found using *in vivo* or clinical trials because of the difficulties in accurately simulating the highly complex physicochemical and physiological events occurring in the digestive tract. Mackie et al. (2020), however, have commented positively that the *in vitro* models are increasingly being validated with *in vivo* data whilst also pointing to the future use of artificial intelligence methods.

The physical and chemical processes occurring in food digestion in an adult human digestive system are summarised in Table 8.1.

Table 8.1 Summary of the Physical and Chemical Processes Occurring in the Human Digestive System

Section (Adult)	Physical Process	Chemical Process	Conditions
Mouth	**Mastication** Food breakdown Mixing	**Enzymatic hydrolysis** Starch breakdown (α-amylase) Lipid breakdown (lingual lipase)	pH: 5–7 Transit time: 5 s–2 min
Oesophagus	**Peristalsis** Bolus transport		8–10 s (solid) to 2 s (liquid)
Stomach	**Contractions/Peristalsis** Food breakdown Mixing Gastric juice secretion Gastric transport Gastric sieving Gastric sieving	**Enzymatic hydrolysis** Protein breakdown (pepsin) Lipid breakdown (gastric lipase) **Acid hydrolysis** Food softening and dissolution (gastric acid)	pH: 1–3 Transit time: 15 min – 4 h 1-3 L/day 3 cycles/min
Small intestine	**Peristalsis** Chyme transport **Segmentation** Mixing **Absorption** Monosaccharides Amino acids/peptides Monoacylglycerides, glycerol, free fatty acids Phosphates, nitrogenous bases, pentose sugars	**Enzymatic hydrolysis** Starch breakdown Protein breakdown Lipid breakdown Nucleic acid breakdown	pH: 5–6.7 Transit time: 1–5 h Pancreatic juice secretion: ≈1.5 L/day
Large intestine	**Peristalsis** Chyme transport **Segmentation** Mixing **Absorption** Water, ions, minerals, vitamins, fats and organic molecules	**Fermentation** Production of short-chain fatty acids and other by-products	pH: 5–7 Transit time: 12–24 h Microbiota: -10^{11} to 10^{12} (>1,000 different species)

Source: Modified from Sensoy (2021).

Bock and Norris (2016) provide somewhat different estimates of food processing with a total time of between 10 and 23 hours, comprising 2–6 hours in the stomach, 2–8 hours in the small intestine and 6–9 hours in the large intestine.

The reality is that the rate of gastric evacuation is subject to many factors, including the composition, temperature and size of the meal. Consumption of alcohol, drugs and water prior to or during the meal also affects the rate of evacuation, with alcohol consumption typically accelerating gastric evacuation. Human factors such as activity levels after a meal, obesity, gender (males have a faster rate of evacuation than do females) and health conditions can all influence evacuation rates (Bock & Norris, 2016).

In a state of unconsciousness, all gastric functions are delayed, and food can remain in the stomach for 12 hours or more. In a state of excitement, food may pass through more quickly than normal (Anon, 1978).

Given the number of variables and different estimates for gastric evacuation, a conservative approach is recommended when providing any estimate of PMI and a safer approach is to only estimate a minimum time.

8.3 Collection, Sampling and Presumptive Testing for Digestive Tract Materials

8.3.1 Stomach Contents and Vomit

Stomach contents are routinely collected from deceased persons during postmortem examination and should have been placed in formalin equal to 5%–10% by volume. If stomach or intestinal contents are received for examination that are not in formalin, they should immediately be placed in formalin to denature all potentially infective agents. If samples are received frozen, they can be defrosted by immersion in a 10% formalin solution.

Where stomach contents may be required for toxicological and/or drug analysis, discussions should be held with the appropriate toxicologist and agreement reached on how to proceed.

Stomach contents should be poured through a 16-mesh sieve, gently washing the retained material with water. Retained material can then be sorted into fractions based on size, shape, colour and appearance, placing different materials in suitable containers such as small petri dishes.

Depending on the degree of digestion and the nature of foodstuffs present, it may be possible at this stage to identify some foodstuffs based on visual appearance. In many cases, however, a more detailed examination will be required. Depending on the amount and nature of fragments present in stomach contents, it may only be necessary to subsample a representative fraction of the recovered material.

Initial examination will involve using stereo/low magnification microscopy during which material can be further subdivided and, where necessary, teased apart and placed onto microscope slides in a temporary mountant.

It may be necessary to remove plant pigments to make microscopic detail more easily observed. This can be done on a glass slide by adding a small amount of chloral hydrate (50 g of chloral hydrate in 20 ml of water) and gently heating the slide over a Bunsen burner. Where there is sufficient material, an alternative approach is to place the material in a small test tube with chloral hydrate and then heat the test tube in a water bath, teasing the material during heating to a point where it separates but does not disintegrate. The treated material can then be sampled and placed onto microscope slides with a suitable temporary mountant.

The use of a variety of dyes or stains should be considered. As covered in earlier chapters, the presence of lignified cells can be detected with the use of phloroglucinol. Starch granules can be detected using a solution of iodine (1 g iodine + 1 g potassium iodide in 100 ml of water), where the starch is stained a deep blue colour. Proteins may be detected using a ninhydrin test and fats using Sudan IV (5 g Sudan IV in 95 ml 70% alcohol).

Detailed examination requires the use of transmitted light microscopy (TLM). Crystals of calcium oxalate can be a very useful feature. These are easily visualised using polarising light microscopy (PLM).

Apling (1980) suggests a general scheme of examination as follows:

- mount in water or dilute glycerine, and examine directly and between cross-polaroids
- irrigate with iodine in potassium iodide
- prepare a second mount in chloral hydrate and clear
- prepare a third mount in phloroglucinol, allow the alcohol to evaporate, add a drop of concentrated hydrochloric acid and examine

The treatment of *vomit* will depend on the nature of the vomit sample. If the quantity is significant and still moist, then it can be treated in the same way as stomach contents. If the presumptive vomit is in the form of dried stains, then these should be sampled at the first opportunity, paying close attention to ensuring sampling recovers any visible variation. Dried samples can then be rehydrated in dilute formalin and then examined using the same protocol as for stomach contents.

8.3.2 Faecal Material

Faecal material in human adults is usually brown in colour due to the presence of urobilinogen formed from bilirubin by a reduction process in the intestine. Faecal material may, however, be green, black or red depending on diet, drugs and pathological conditions (Anon, 1978). Infant faeces for the first few days after birth are dark green/brown due to the presence of *meconium*. Thereafter, infant faeces are yellow until solid foods are introduced around 6 months of age.

Faeces is associated with having a distinctive odour due to decomposition products, including indole and skatole. The strength of the odour is associated with the amount of meat in the diet and the activity of gut biotics.

A presumptive test for the presence of faecal material is based on the presence of *urobilinogen*. The test relies on the formation of a green, fluorescent zinc–urobilin complex. Urobilinogen is oxidised to urobilin, which is soluble in alcohol. In the presence of neutral alcoholic zinc salts, the green complex is formed (Anon, 1978).

The method is as follows:

- prepare an aqueous extract of the stain or material in a small test tube
- put an equal amount of water into a second test tube as a control
- add three drops of alcoholic mercuric chloride to each tube
- add an equal amount of alcoholic zinc chloride and shake
- examine the liquid under ultraviolet light. A stable apple green fluorescence occurs if urobilinogen is present

Faeces may present as visible turds or as a contact material. Faeces should always be considered as a biohazard. Dry stains should be carefully scraped off or sampled and may be rehydrated using a dilute formalin solution.

As in any forensic sampling protocol, it is important to ensure the samples taken are representative of any visual variation and are sufficient for further testing.

Depending on the sample size, it may be necessary to use a similar approach as used for stomach contents, with sieving of a bulk sample and then separation of recovered fragments into separate small petri dishes for subsequent examination. For scrapings, the sample size may allow the rehydrated material to be examined with or without clearing on microscope slides. Due to digestion processes, only those plant cells that are most resistant to digestion will be present. Faeces often contain only fragments such as single cells, trichomes or crystals. It is often not possible to identify these to a single species, and the examination is a comparison of material found at a scene or of stains recovered from a victim and suspect.

In some cases, faecal material may not be of human origin. For example, in cases of alleged bestiality or where livestock have been stolen. The examination of animal faecal material is similar in respect of the approach that can be applied. However, obviously, the nature of the fragments in animal faeces will reflect their diet. For herbivores, an understanding of grass and other plant fragments will be necessary.

If it is suspected that the material may be of avian origin, the presence of large amounts of uric acid is an indicator of avian origin. The following test can be used to detect uric acid (Thrasher, 1962):

- place a sample of the suspect material in a cavity slide and place it on a hot plate
- add two drops of concentrated HNO_3 to the cavity
- evaporate to dryness and continue to heat for 1–3 minutes
- allow to cool
- a positive reaction is indicated by the development of an orange to deep red colour, depending on the amount of uric acid present
- add 1 drop of 40% sodium hydroxide – in the presence of urea, an intense red colour develops within a few minutes.

The need for a thorough, systematic and detailed recording of stomach contents, vomit and faecal material cannot be overstated. Where possible, this should be completed before any comparison process to meet contemporary standards aimed at minimising the potential for context bias.

Outside of the scope of this chapter is a consideration of non-plant components arising from meat consumption. The forensic scientist examining stomach contents, vomit and faecal material should become familiar with the appearance of meat and meat products (Anon, 1982).

8.4 Resource Material for the Identification of Plant-Derived Foodstuffs

8.4.1 Introduction

The Biology Methods Manual of the Metropolitan Police Forensic Laboratory (MPFSL) (Anon, 1978) contained no less than 51 pages of descriptions of specific foodstuffs, reflecting an era when forensic botany was still a significant aspect of the laboratory's expertise. Given the many thousands of plant species that could be found in foodstuff, it may appear an impossible task to be able to identify all potential sources, especially when one considers that we are not dealing with vouchered herbarium material but, in the context of this chapter, incomplete and fragmentary materials. The reality, however, is not quite as daunting as it may seem. Bock and Norris (2016) estimate that the contemporary diet of modern North American humans is only about 70 plant foods and that many can be dealt with in groups, such as beans.

It is worth remembering that 'you don't need to remember everything. You just need to know where to find them' (Fashionpedia, 2021).

With respect to food microscopy, many of the most informative books date back to the late 1800s and early 1900s and are long out of print. The good news, however, is that some of these older texts are available as facsimile versions, and whilst the figures are not of the same quality as the originals, the text is mostly of good quality.

Hence, a shorter version of the four-volume series by Andrew and Kate Winton, the classic series of books on the microscopy of vegetable foods, can now be obtained (Winton, Moeller & Winton, 1916).

Similarly, a modern reprint can be obtained of the 1909 version of Clayton's Compendium of Food-Microscopy, based on the earlier work of Hassall (Clayton, 1909).

Perhaps surprisingly, an excellent source of botanical descriptions of some plant foods, especially herbs and spices, can be found in older textbooks of pharmacognosy (Greenish & Collin, 1904; Wallis, 1957, 1960). Jackson and Snowdon (1992) include descriptions of 20 herbs and spices in their book on medicinal plants, culinary herbs and spices.

Jolliffe and Jolliffe (1976) developed a computer-aided identification for powdered vegetable drugs and later a similar scheme for food materials based on 30 features with 12 main categories, including trichomes, stomata, vessels/tracheids, fibres, starch and calcium oxalate crystals (Jolliffe & Jolliffe, 1979). Written in BASIC, clearly this is no longer a viable program, but it would not take a great effort to create a contemporary version of this program based on the existing descriptions. A husband-and-wife team, the final contribution of Georgina Jolliffe was as a technical editor for the American Herbal Pharmacopoeia on Botanical Pharmacognosy (Upton et al., 2011).

Whilst a forensic botanist is unlikely to have a background in food processing and preservation, some knowledge of processing, including thermal, chemical and mechanical processing of different types of foods, can be critical in understanding the appearance of processed and/or cooked foods (Duckworth, 1966; Seidemann, 1966; Evers, 1979; Jewell, 1979).

Bock and Norris (2016) have an online resource showing images for over 80 common plants. These authors point out, however, that the best reference collection is the one you prepare yourself from raw and cooked plants relevant to where you live and the circumstances of each case.

The remainder of this chapter will deal with some of the most important food plants and products, starting with starch.

8.4.2 Starch

Starch is stored in plants in seeds, fruits and storage organs such as tubers, roots and modified stems, and is present in many foodstuffs. Starch is also extracted and used in a variety of end products, often as a thickening agent due to its properties when heated. When starch is heated in an aqueous environment, it reaches a temperature where its structure is lost due to *gelatinisation*. It is sometimes incorrectly stated that starch always loses its microscopic structure when cooked, but this is not the case. In baked products, the degree of gelatinisation depends on the amount of water present, and in many baked products, only partial gelatinisation is seen (Evers, 1979).

In a forensic context, starch may be found in trace materials such as seeds and other plant parts, in stomach contents and vomit, and as starch granules in food products. Forensic archaeologists examine starch on ancient tools to determine the diet of ancestral peoples.

Chemically, starch comprises polymers of anhydroglucose ($C_6H_{10}O_5$) units. Two types of polymers are formed depending on how these units are connected. Where the units are connected by 1,4 linkages, they form long linear chains, called *amylose*. Where 1,6 linkages are present, the molecules are branched, called *amylopectin*. Starch from different plants will contain differing amounts of these two components, resulting in a variety of physical and chemical properties (Bowen, 2003) (Figure 8.2).

Starch granules, or grains, vary in size (from 1 to over 150 μm), distribution, grain shape, presence or absence of aggregates, position and conspicuousness of the hilum, shape of the hilum, conspicuousness of growth rings on

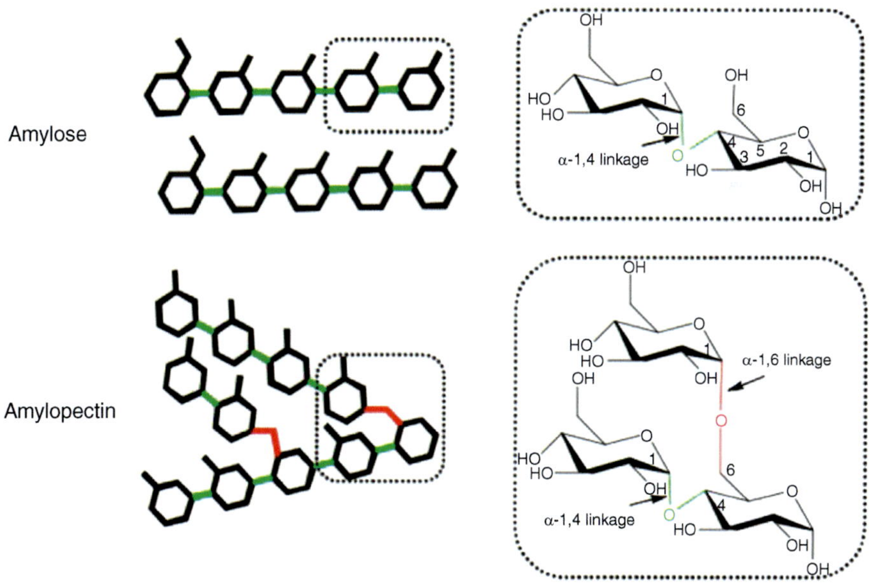

Figure 8.2 Chemical structure of amylose and amylopectin components of starch.

the grains and form (Winton et al., 1916). Winton et al. (1916) describe the various forms as globular, lenticular, ellipsoidal, ovoid or pear-shaped, truncated and polygonal.

The morphological features of starch granules from different plants are sufficiently characteristic that they can be used to identify plant species. An analytical key for the identification of commercial starches has been published in Winton et al. (1916).

Starch is laid down from a point called the *hilum*, the position and visibility of which can be an important feature in starch identification. This highly organised process results in concentric rings and imparts a semi-crystalline property to starch. The regulation of growth ring formation is still not fully understood, with a complex interplay of several factors being proposed (Pilling & Smith, 2003). These concentric rings are quite visible in larger starch granules, such as potato starch, when viewed with TLM.

Starch is formed through starch-forming bodies, *leucoplasts*. If a single grain is formed within the leucoplast, it will grow uniformly on all sides and have a round outline, a central hilum and concentric growth rings. If the grain is formed on the sides of the leucoplast, it will grow only on the side of the attachment and, as a consequence, will be elongated with an eccentric hilum and eccentric rings. If several grains are formed within the leucoplast, the sides in contact will be flattened. Grains on the surface of the aggregate will have both rounded and flattened surfaces, whilst those on the inner part, in contact with grains from all sides, will be polygonal (Winton et al., 1916).

When starch is examined using PLM, the semi-crystalline nature of starch results in exotic shapes and interference colours with a 'Maltese cross' appearance that Bowen (2003) describes as an 'aesthetic experience'. Polarised light also locates the position of the hilum as the dark crosses, which appear on the bright grain, intersect at that point. In grains with a central hilum, the lines will intersect at the centre forming an X-shaped cross. In grains with an eccentric hilum, they often intersect so near one end that they appear V-shaped. In starch grains from legumes, there are two Y-shaped crosses that unite to impart a distinct feature for legumes – see Figure 8.3.

Finally, as Bowen (2003) points out, when single grains are encountered, the task of identification is far more difficult as the range of features for a particular starch may not be displayed. However, possible sources may be eliminated if there are clear differences. For example, there would be little difficulty in distinguishing potato starch from maize starch.

Table 8.2 summarises the features of some of the more common sources of starch. Seidemann (1966) has also published an atlas of starch microscopy with descriptions of important varieties.

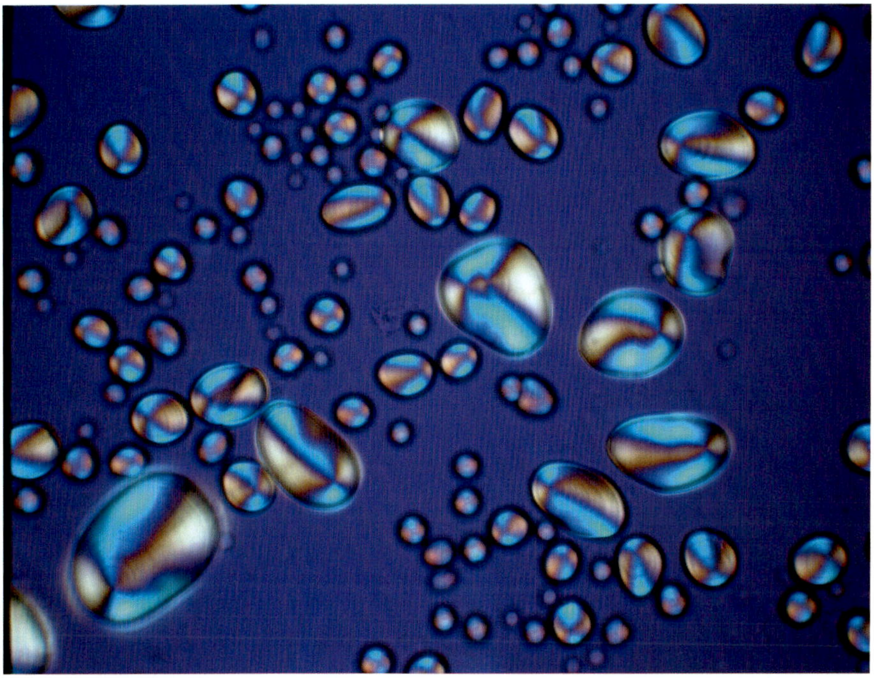

Figure 8.3 Potato starch in polarised light.

8.4.3 Herbs, Spices and Flavourings

In botanical terminology, an herb is a small, non-woody seed-bearing plant in which all the aerial parts die back at the end of each growing season (Allaby, 1998).

In the context of this chapter, herbs used for culinary purposes and food seasonings are generally the leafy green or flowering parts of plants, fresh or dried, whereas spices are usually dried and from other parts of the plant, including seeds, bark, roots and fruits. These definitions become somewhat blurred when considering the use of herbs and spices for medicinal uses.

Many of the commonly used herbs come from the family Labiatae. These include peppermint, spearmint, rosemary, sage and thyme. Being leaf material, identification requires a knowledge of leaf anatomy, including how to describe epidermal cells, stomatal types and, typically, plant hairs or trichomes. Aromatic herbs are characterised by having glandular trichomes.

The term spice conjures up images of the spice trade and the Far East, and, indeed, many of the spices of commercial importance come from plants grown in tropical regions. As well as their use in restaurants and domestic

Table 8.2 Microscopic Features of Selected Starch Grains

Source	Size (μm)	Shape	Hilum	Rings	Polarisation Crosses
WHEAT Several species of *Triticum*	Large: 15–40 Small: 1–10 Grains, simple	Large grains lenticular, more or less regular circular in outline Small grains usually globular, less often polygonal	Central	Indistinct	Poorly defined
BARLEY *Hordeum* species	Large: 10–30 Small: 2–10 Small grains often in groups	Large grains lenticular, kidney-shaped or subangular Small grains spherical or spindle-shaped	Central	Indistinct	Poorly defined
OATS *Avena sativa*	Mix of simple and compound grains consisting of 5–200 grains Simple grains: 2–10 Compound: 35–60	Simple grains, mostly rounded, some spindle or lemon-shaped Compound grains oval or rounded grains from the centre, angular	Central	Indistinct	Poorly defined
RICE *Oryza sativa*	Mix of simple and compound grains Simple: 4–6 Compound: varies in size based on the number of grains – up to 150	Sharply polygonal grains Predominant grains near the surface of aggregates are rounded on some edges	Central, but not always evident	Not evident	Distinct crosses
MAIZE *Zea mays*	Several polygonal or irregularly rounded grains are often united – round or oval aggregates **not** present 2–30, average 10	Grains from the horny endosperm sharply polygonal. Floury endosperm grains are irregularly globular	Central and distinct radiating clefts in many grains	Not evident	Brilliant with distinct crosses

POTATO *Solanum tuberosum*	Mix of small and large grains Small: 10–30 Large: 30–100 Average: 70	Small – rounded Large – irregular, oval and subspherical	Visible at the narrow ends of grain	Distinct concentric rings	Distinct crosses
MARANTA/ ARROWROOT *Maranta arundinacea*	Aggregates rare or absent Grains: 30–50 On occasion, up to 75	Ovoid, pear-shaped or broadly spindle-shaped	Visible at the broad end with fissures Resembles the wings of a bird in flight	Distinct but less prominent than in the potato	Distinct crosses
CASSAVA (Tapioca) *Manihot utilissima*	Aggregates of 2–3. Less often, 4–8 grains – usually broken up in commercial starch Small: <15 Large: 20–35	Usually kettledrum or shoe-heel–shaped Flattened surfaces correspond to surfaces in contact with twin aggregates	Central and distinct	Indistinct	Distinct crosses
SWEET POTATO *Ipomoea batatas*	Twin or triplet aggregates mostly separate in commercial starch Small: 5–15 Large: 25–35 up to 55	Bell-shaped with one or two, less often more, flat or slightly concave surfaces	Distinct with radiating fissures Eccentric	Indistinct	Distinct crosses
SAGO *Metroxylon sagu*	Mix of medium to large grains Medium: <20 Large: 30–50 up to 80	Mostly ovoid – some angular, where other grains have been attached	Distinct, either as a point or with fissures	Distinct	Distinct crosses
BEANS OR LEGUMINOUS SPECIES	Aggregates are rare Mostly 50, but up to 100	Ellipsoidal grains predominate, with reniform, trefoil and other irregular forms less common	Distinct, elongate, often with branching clefts	Distinct	Distinct with a distinctive shape due to the elongated hilum
BANANA (raw) *Musa* species	Variable sizes from 7 to 65, mostly 20–40	Very characteristic with elongate cigar-shaped, elongate or bean-shaped grains	Distinct, usually at the broader end	Distinct	Poorly defined, with one half much larger than the other

Note: These descriptions are based on a number of references, but the primary source is Winton et al. (1916).

cooking, spices and flavourings are of importance in food manufacturing in products such as pickles, chutneys and sauces. For example, curry powder is prepared from a mixture of ground spices in varying proportions and may include pepper, chilli, paprika, coriander, cardamom, turmeric, ginger, fenugreek, cloves, mustard, cumin and allspice.

As most spices are fruits and seeds, an understanding of the structure of fruits and seeds is necessary to describe and attempt to identify them.

As discussed in Chapter 2, a fruit consists of an outer coat (pericarp) enclosing one or more seeds. Fruits have two scars, one marking the former attachment to the plant and the other the remains of the floral parts. Seeds have only one scar, the hilum. The pericarp and seed consist of several layers as described below.

The *pericarp* can be subdivided into the outer *epicarp*, middle *mesocarp* and the inner *endocarp*. The epicarp has the general characteristics of an epidermis with stomata and one or more layers of modified hypodermal layers. The mesocarp is usually composed of parenchyma cells and may contain oil in cells, glands or ducts, and starch. Sclereids are common, either scattered or in a layer. The endocarp or inner layer may be entirely lignified.

Seeds consist of an outer layer or seed coat. This layer is the most characteristic layer of the seed, often diagnostic, with one or more hypodermal layers. The endosperm or perisperm is a storage tissue for oil, starch and aleurone layers. Finally, the seed will contain the embryo.

Many of the spices come from the family Umbelliferae. The fruits of this family consist of two mericarps attached to a forked stalk, or carpophore, and are called *cremocarps*. The position of the fork, visible when the mericarp is separated, is diagnostic, as are longitudinal ducts, or *vitae*, and surface ridges.

It is beyond the scope of this chapter to describe the multitude of herbs and spices, and, in this instance, it really is a case of 'knowing where to find' the relevant detailed information. A primary reference is Winton et al. (1916). As previously mentioned, a facsimile copy of this book is available, print on demand. Part IX of this book includes an extensive section covering almost 150 pages on spices and condiments. The figures in the reproduction have lost some quality and definition, but are still useful and can be understood with some level of expert knowledge.

Other essential reference texts are Greenish and Collin (1904) and Jackson and Snowdon (1992). This later version of an earlier text, which focused on powdered vegetable drugs (Jackson and Snowdon, 1968), includes descriptions of 20 culinary herbs and spices. This book contains verbal descriptions accompanied by excellent line drawings of the diagnostic microscopic features. Figures 8.4 and 8.5 provide examples of the microscopy of the component parts of an herb, sage and a spice, coriander.

The MPFSL Biology Methods Manual (Anon, 1978) also has an extensive section on plant foodstuffs with excellent descriptions and diagrams of diagnostic features. To the best of my knowledge, however, this privately printed manual is no longer in print.

8.4.4 Cereals

Cereals all belong to the grass family, Gramineae or Poaceae. The classification of grasses, their morphology and anatomy will be considered in Chapter 10. The primary focus of this chapter is on grains as foodstuffs.

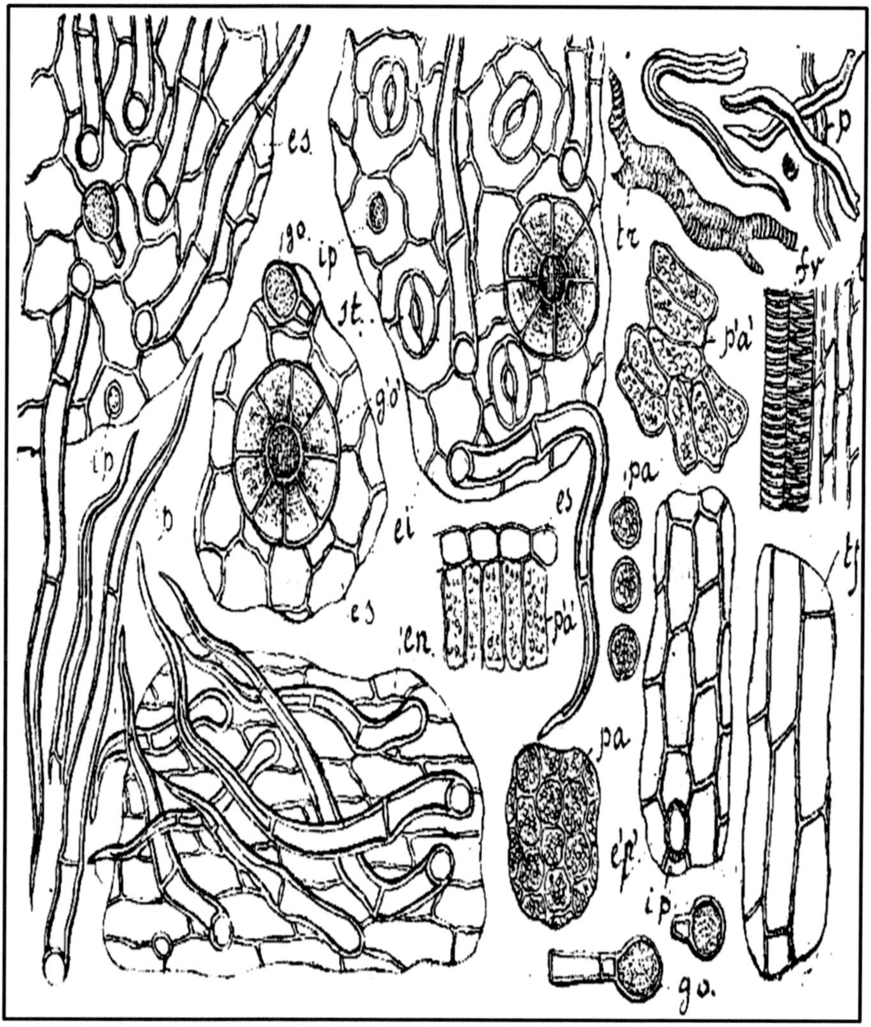

Figure 8.4 Microscopy of powdered sage leaves.

(From Greenish & Collin, 1904, page 81, figure 44.)

(*Continued*)

el, Lower epidermis.

en, Neural epidermis.

e'p', Epidermis of petiole.

es, Upper epidermis.

fv, Vessels from midrib.

go, Unicellular gland.

g'o', Octocellular gland.

ip, Scar of hair.

p, Simple hairs.

pa, p'a', Palisade cells.

st, Stomata with the ostiole at right angles to the walls separating two cells from one another.

tf, Cortical tissue of midrib.

tr, Tracheids.

Figure 8.4 (Continued)

The anatomical structure of all cereal grains is similar, differing from one cereal to another in detail only. Grains of wheat, rye, maize and grain sorghum are *naked caryopses* in that they consist of the fruit coat or pericarp and the seed. The seed is composed of the seed coat, germ and endosperm. Grains of oat, barley and rice have additional structures present outside of the seed coat, the fused glumes (palea and lemma), which constitute the husk. These grains are referred to as *covered* or *coated caryopsis.*

A typical naked caryopsis consists of various layers as follows.

Bran consists of all layers of the fruit to the aleurone of the seed coat. The outer palea and lemma are referred to as the *hull* and amount to about 13% in weight of barley, 20% in brown rice and as high as 25% in oats. In maize, the term hull corresponds to the bran in other cereals (Kent, 1975).

The following describes the anatomical features of some common cereals. These descriptions are based largely on Greenish and Collin (1904) and Winton et al. (1916).

8.4.4.1 Wheat (Triticum aestivum)

Figure 8.6 shows the microscopic features of wheat flour.

The epidermis consists of rectangular, thick-walled cells, pitted to give a beaded appearance, 40–180 µm long and 12–35 µm wide. Towards the apex, they are shorter, and some are polygonal. The walls are slightly and unevenly lignified. There are a few stomata. Covering trichomes in the apical part are unicellular, taper to a fine point and have walls that are thicker than the lumen along most of their length. The trichomes are lignified. The hypodermis consists of thick-walled pitted cells like those of the epidermis. Transverse cells, or *cross-cells*, are thick-walled, pitted and beaded, lignified and with the long axis at right angles to those of the epidermis and hypodermis. These cells are aligned in rows, side by side and not end to end. The cells of the inner epidermis of the pericarp are called *tubular cells*. These small cells, 175–250 μm long by 10–15 μm wide, are more or less detached.

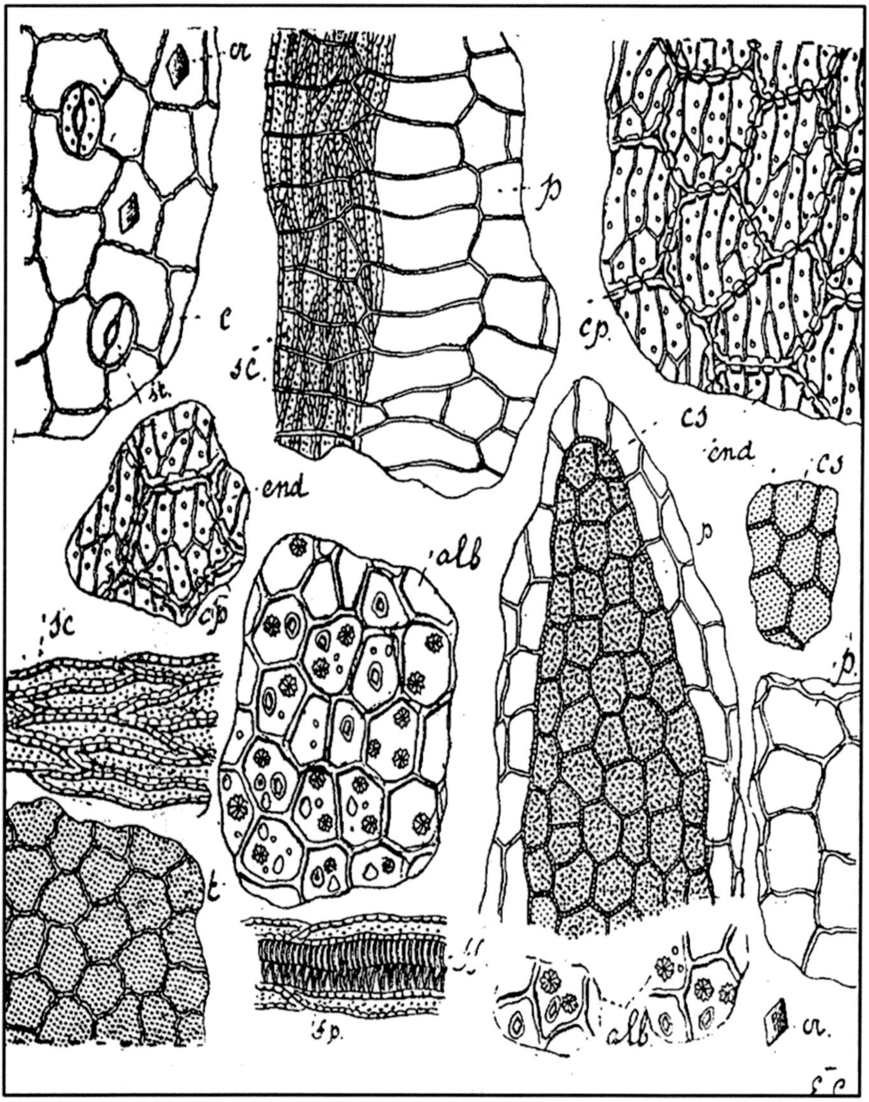

Figure 8.5 Microscopy of powdered coriander fruits.

(From Greenish & Collin 1904, page 151, figure 76.)

alb, Endosperm.
cp, Large sclerenchymatous cells of inner part of pericarp.
cr, Crystals of calcium oxalate from epidermal cells.
cs, Fragment of secretory duct.
e, Outer epidermis of pericarp.
end, Inner epidermis of pericarp.
ffv, Fragments of fibro-vascular bundle.
fp, Sclerenchymatous fibres.
p, Parenchymatous tissue from pericarp.
sc, Fibrous sclerenchymatous layer of pericarp.
st, Stomata.
t, Seed coat. (X240)

Figure 8.5 (Continued)

8.4.4.2 Rye (Secale cereale)

The epidermis of rye pericarp is like that of wheat. The walls are lignified. Trichomes from the apex of the caryopsis are more slenderer and less pointed than those of wheat with the apex being more rounded. Hypodermis cells are not pitted. Transverse cross cells are lignified and are similar in arrangement to those seen in wheat, but less regular in size. The longer side walls are less thick and less distinctly pitted than in wheat; the end walls are rounded and are not pitted, and there are well-marked small intercellular spaces between the ends of the walls. These features are shown in Figure 8.7.

8.4.4.3 Barley (Hordeum vulgare)

Barley differs from wheat and rye because the flowering glumes (palea and lemma) adhere to the pericarp. Glumes are modified leaf structures and therefore display the features typical of grass leaves.

A more detailed consideration of grass leaf anatomy will be considered in Chapter 10. In the current context, four layers from the palea may be observed. The first layer, or outer epidermis, consists of elongated rectangular cells with sinuous walls alternating with short cells or twin-cells arranged in parallel rows. The epidermis is lignified. The next layer of hypodermal cells consists of fibres, and is about three cells thick and is also lignified. The third layer consists of a layer of spongy parenchyma with wavy crenate walls and numerous large oval to circular pits. The inner epidermis of the palea consists of delicate sub-rectangular cells with scattered stomata and sharply conical, unicellular trichomes.

The pericarp of the barley caryopsis has an epidermis of sub-rectangular cells, becoming polygonal at the apex, where there are unicellular conical trichomes, often curved, bearing the base. The transverse cells are in two layers.

Figure 8.8 shows the microscopic features of barley flour.

8.4.4.4 Rice (Oryza sativa)

Although rice is a covered grain and has the outer layers or palea present, these are not found in processed white rice.

The outer epidermis of the palea consists of rectangular, highly sinuous and partially lignified cells. Scattered trichomes are present, especially over the ribs. These are unicellular, sharply pointed, with very thick walls and somewhat lignified at their base. Many of the trichomes are broken off,

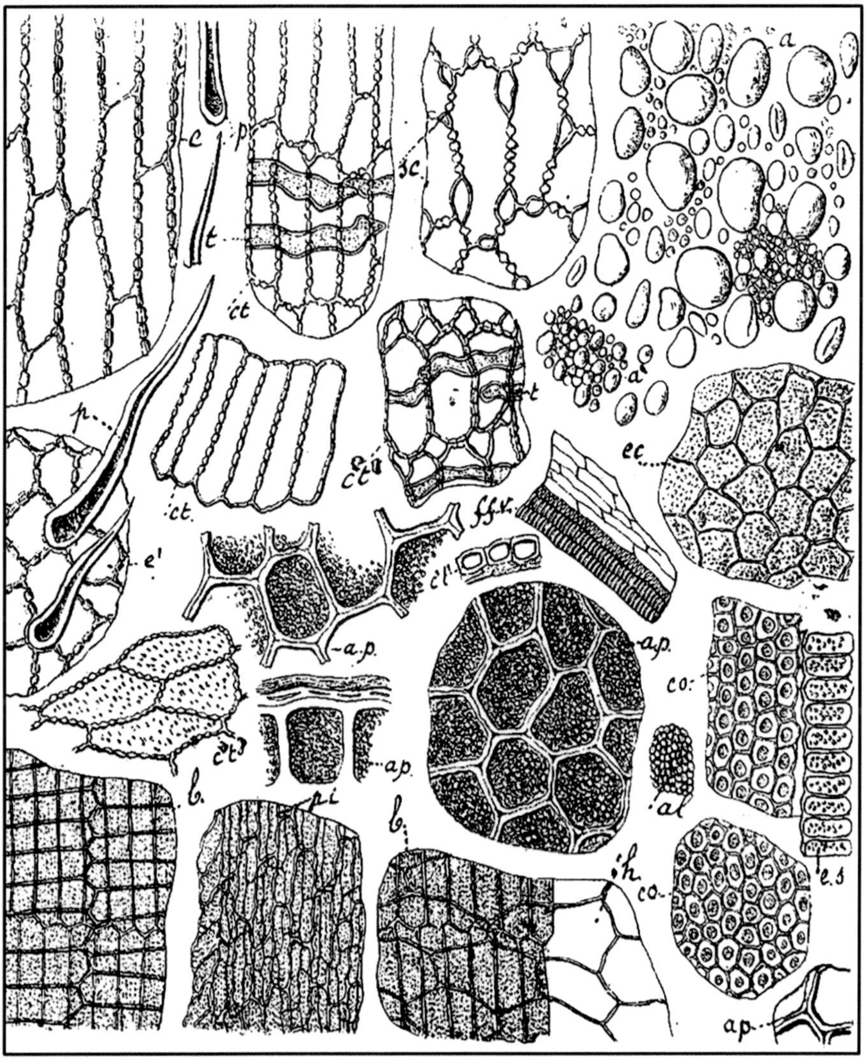

Figure 8.6 Microscopy of wheat flour.

(From Greenish & Collin, 1904, page 27, figure 20.)

(Continued)

a, a′, Starch grains *without evident hilum, larger than barley starch, smaller than rye starch.*
al, Aleurone.
ap, Aleurone layer.
b, Brown layer.
co, Débris of cotyledon.
ct, *Transverse cells with pitted walls,* surface view.
c′ t′, *Transverse cells,* section.
e, e′, Outer epidermis of pericarp with *thickened and pitted cell walls*
ec, Scutellum.
es, Epidermis of scutellum.
ffv, Débris of the fibro-vascular bundle of the raphe.
h, Cells of the hyaline layer.
p, Hairs, entire or broken, *with lumen rather abruptly enlarged at the base.*
pi, Cells with brown pigment.
sc, Lignified cells of the hypoderma with *pitted walls.* (X240)
t, Tubular cells, surface view.

Figure 8.6 (Continued)

leaving a scar, or cicatrix, with a very short piece of trichome attached. The hypodermis consists of several layers of lignified fibres.

8.4.4.5 *Oats* (*Avena sativa*)

The main diagnostic features of oats are:

- the smooth, rounded (not ribbed as in barley) flowering glume, with an epidermis of thick wavy-walled, elongated cells, circular cells and twin-cells and with star-shaped spongy parenchyma (not rectangular as with other cereals)
- the palea has a more delicate structure, having keels barbed with coarse hairs forming a saw-toothed edge
- the epicarp with long, slender trichomes often narrowing at the base
- rounded aggregates of polygonal starch grains and spindle-shaped forms

Oatmeal, rolled oats and similar breakfast cereals contain all the above elements, though in some of these products, the starch grains will have been distorted by cooking.

8.4.4.6 *Maize* (*Zea mays*)

Only the naked fruit of maize will be considered and not the outer covering leaf structures.

The epicarp consists of porous-walled, elongated cells like those seen in wheat, except that the walls are much thicker. A thin cuticle covers the exposed surface. The mesocarp comprises six or more layers of cells, like the epicarp, but with even thicker walls. In the outer layers, these cells are 600 μm or more in length and 20–40 μm broad. In the inner layers, the cells are broader, flatter and thinner-walled, grading into those of the next layer. The cross-cell layer in maize consists of spongy parenchyma, in contrast to the close layer of cross cells as seen in wheat, rye and barley. This spongy parenchyma comprises branching and anastomosing cells with narrow radiating arms and large intercellular spaces. Tube cells and the inner spermoderm and perisperm can only be visualised when the whole kernel is soaked in water and stripped of the outer pericarp. These cells are not highly diagnostic.

Figure 8.9 shows the microscopic features of maize flour.

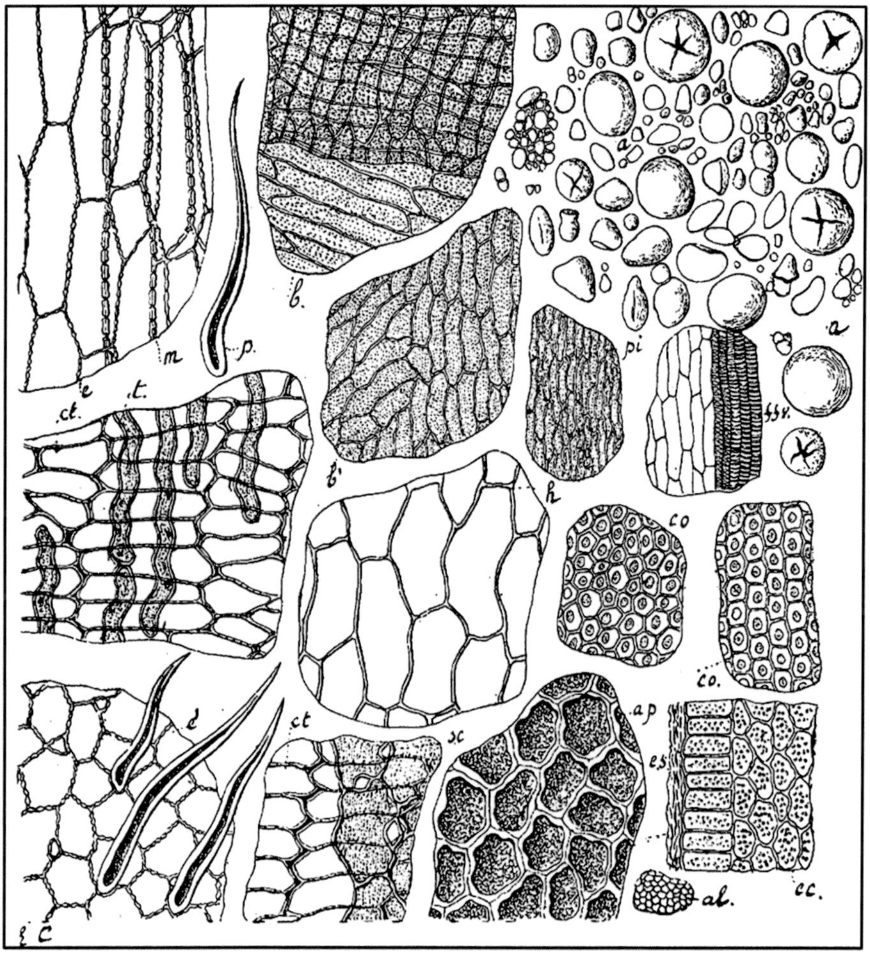

Figure 8.7 Microscopy of rye flour.

(From Greenish & Collin, 1904, page 29, figure 21.)

(Continued)

a, Starch the grains are *larger than those of wheat*, and often show a *stellate hilum.*
al, Aleurone.
ap, Aleurone layer.
b b′, Brown layer.
co, Cotyledon.
ct, Transverse cells; *the cell walls are thinner and less pitted than those o wheat.*
e, e′, Epidermis of the pericarp with *pitted cell walls.*
ec, Scutellum.
es, Epidermis of scutellum.
ffv, Fibrovascular bundle of the raphe.
h, Hyaline layer.
m, Hypoderma.
p, Protective hair.
pi, Pigment cells.
sc, Sclerenchymatous cells, *less abundant than in wheat.* (X240)
t, Tubular cells.

Figure 8.7 (Continued)

Maize kernels can almost be considered an indicator food for the time taken to digest food, as the non-digested remnants are readily recognisable in faecal material.

Winton et al. (1916) describe several other cereals, including sorghum and millet, which are important in many parts of the world as food sources.

8.4.5 Fruits

The focus of this section is on fruits eaten fresh, preserved or processed into jams and other commercial products. The family Rosaceae provides us with fruits, such as apples, pears, plums, cherries, raspberries and strawberries. The family Rutaceous provides us with a variety of citrus fruits. In most cases, apart from the whole fruit, the skin and seeds provide the most useful diagnostic features. Jams, marmalades and sauces may contain histological elements of the source fruits. The following is a summary of the main diagnostic features of some of the more important fruits. More detailed descriptions of most of these are to be found in Winton et al. (1916).

8.4.5.1 Family: Rosaceae

Apple – *Pyrus malus*: fruit a pome, composed of a receptacle and pericarp, hairs on the skin around depressions of the calyx and stem (see Figure 8.10). The bulk of the fruit consists of thin-walled parenchyma cells. Epidermis consists of thick-walled cells divided by thin walls to form two to five quadrilateral daughter cells with the appearance of windows, giving the name to these cells as *window cells*. Epidermal cells are up to 75 μm in diameter. Endocarp sclereids are thick-walled. Crystals occur in rosettes. Apples have five wedge-shaped carpels, which at maturity are open and connected, each with two

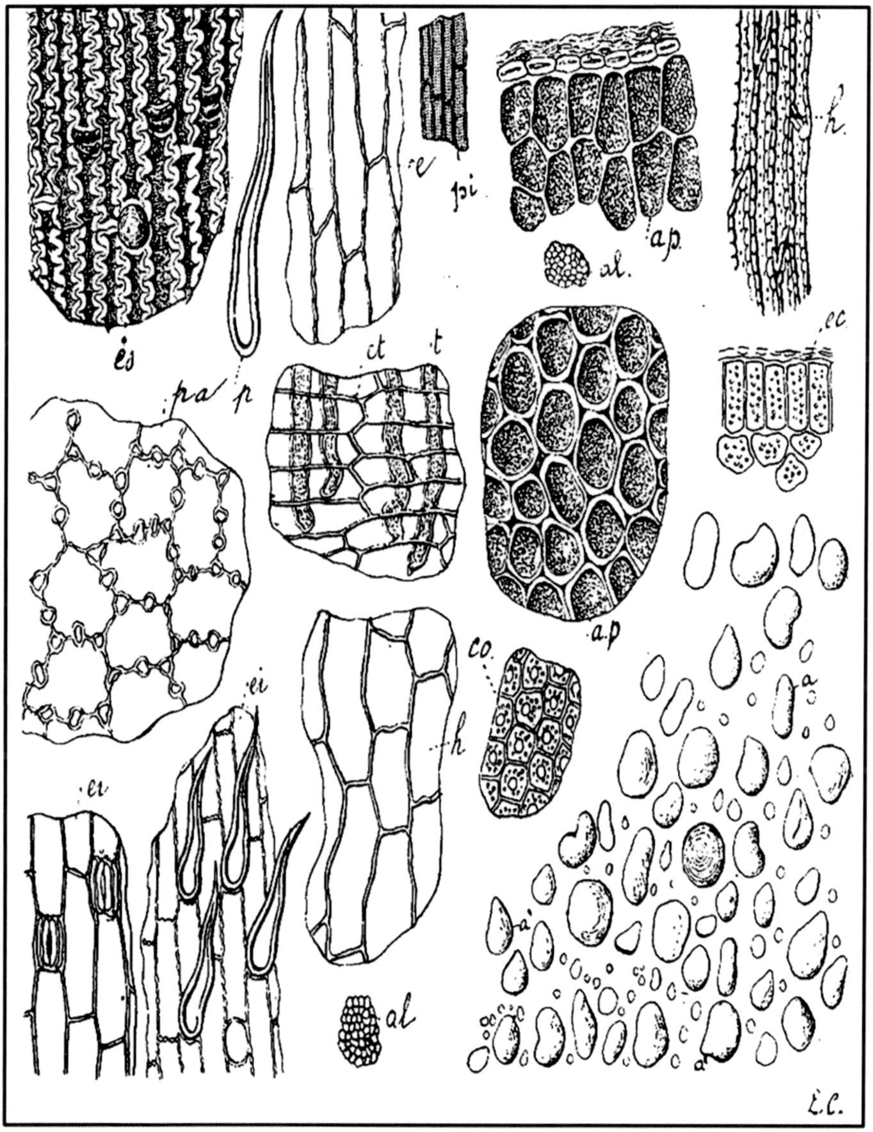

Figure 8.8 Microscopy of barley flour.

(From Greenish & Collin, 1904, page 31, figure 22.)

(*Continued*)

a, a′, Starch grains, *smaller than those of wheat, and often reniform or exhibiting protuberances.*
al, Aleurone.
ap, Aleurone layer, *with two or three rows of cells.*
co, Cotyledon.
ct, Transverse cells.
e, Epidermis of pericarp.
ec, Scutellum.
ei, Inner epidermis of palea, *with little hairs.*
es, Outer epidermis of palea, *with sinuous cell walls.*
ffv, Fibro-vascular bundle of raphe.
h, Fibrous hypoderma of palea (in upper right corner of plate),
h, Hyaline layer.
p, Hairs on the epidermis of pericarp.
pa, Parenchyma of palea, *with intercellular spaces.*
pi, Pigment cells.
t, Tubular cells.

Figure 8.8 (Continued)

brown, flattened ovoid seeds. Seeds are not intentionally eaten and, hence, unlikely to be found as food residues. A full description of the microscopic features of the apple seed can be found in Winton et al. (1916).

Pear – *Pyrus communis*: fruit a pome, obovate, with a variable depression with hairs at the calyx end. Window-like epidermal cells up to 35 μm, groups of stone cells with radiating parenchyma (see Figure 8.11). Endocarp sclereids not strongly thickened and with a broader lumen and thinner walls than those of apple. Clusters of strongly thickened, lignified stone cells in the fruit flesh are characteristic of pears. Crystals monoclinal.

Peach – *Prunus persica*: fruit a stone fruit or drupe with a hairy epicarp, a fleshy mesocarp and a dense furrowed stone or endocarp. The epicarp consists of polygonal cells, stomata and numerous hairs forming a dense velvety coat. The hairs vary in length from short papillae to longer than 1 mm with narrow bases, very thick walls and a narrow lumen. The large sac-like cells of the mesocarp often contain rosette-shaped crystals. The inner layer of the endocarp is a 200–300-μm layer of narrow transversely elongated stone cells.

Apricot – *Prunus armeniaca*: fruit a stone fruit or drupe, like peach, but not as hairy. Stone almost lens-shaped with a distinct keel. Epicarp cells less than 40 μm and hairs broad at the base with walls narrower than the lumen. The cells of the mesocarp have bundles of rosette-shaped crystals like those of peach. Accompanying the bundles are thick-walled fibres.

Plum – *Prunus domestica*: plums never have a hairy epicarp, differentiating them from peach and apricot. The stone is smaller than that of the apricot and more elongated. Epicarp cells are beaded and seldom over 60 μm. Thick-walled fibres in the mesocarp are rare.

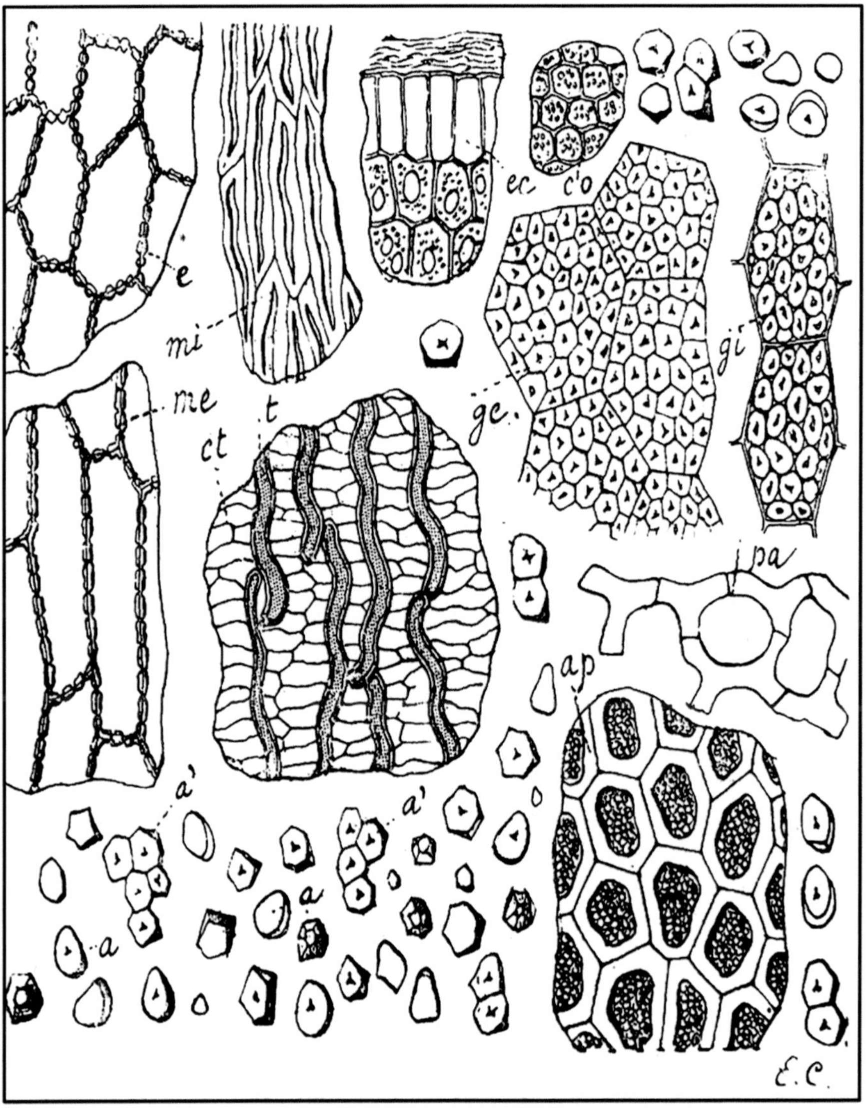

Figure 8.9 Microscopy of maize flour.

(From Greenish & Collin, 1904, page 37, figure 25.)

(Continued)

a, *Rounded grains of starch* from the mealy part.
a', *Angular starch* from the horny part.
ap, Aleurone layer; the cell walls readily swell in water, more readily still in caustic potash.
co, Cotyledon.
ct, Transverse cells, often irregular, separated from one another and resembling the tubular cells which they cross at right angles.
e, Outer epidermis of pericarp.
ec, Scutellum.
ge, Masses of starch from the horny part.
gi, Masses of starch from the mealy part.
me, mi, Layers next to the epidermis.
pa, *Branching cells with lacunæ.*
t, Tubular cells, *very numerous.* (X240)

Figure 8.9 (Continued)

Strawberry – *Fragaria chiloensis*: a strawberry comprises an aggregate of dry one-seeded fruits (achenes) on a fleshy receptacle. The receptacle consists of a fleshy pith and cortex with a narrow zone of fibrovascular bundles from which branches shoot off through the cortex to the achenes. The epidermis is sparingly pubescent. When present, the hairs are often over 1 mm long, gradually tapering from the widest part near the base to the point. The achenes are ovate, pointed and about 1 mm long.

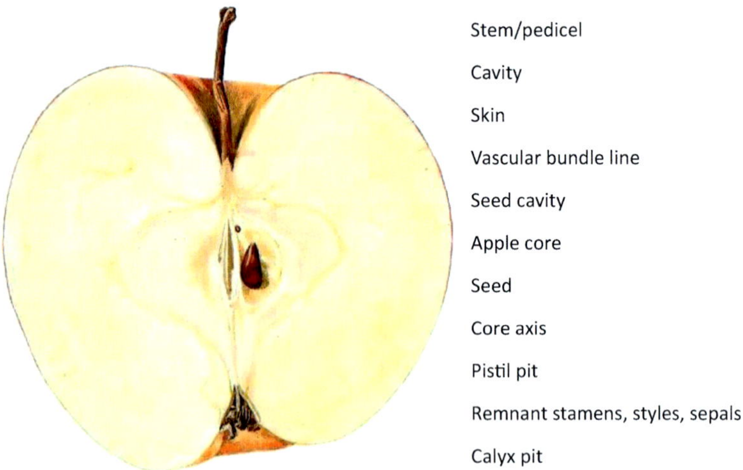

Figure 8.10 Cross-section showing the anatomy of an apple.

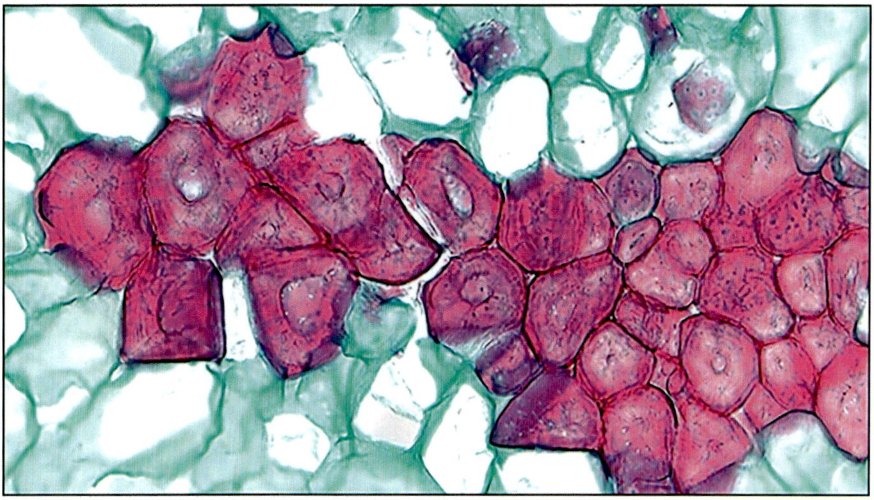

Figure 8.11 Bundles of lignified stone cells or sclereids, which are responsible for the gritty texture in pears.

Each achene is attached to the receptacle a little above its base and contains a single anatropous seed. The style, about 2 mm long, arises from the ventral side, a little above the point of attachment. The style is important in distinguishing strawberries from other edible rosaceous fruits due to its constricted base and large transparent epidermal cells.

Detailed descriptions of all the cell layers in the receptacle and achenes can be found in Winton et al. (1916).

Raspberry – *Rubus idaeus*: fruit of many drupelets, each containing an elongate, furrowed stone, the whole loosely attached to the receptacle. When picking a raspberry, the drupelets separate from the receptacle, clinging together in the form of a cup. The exposed epidermal cells of the pericarp consist of polygonal cells and hairs. The hairs are numerous, with five or six epidermal cells forming a rosette about each hair base. Hairs vary greatly in length and can be as long as 700 μm. Mostly, the cell walls are thin and nearly uniform. The style is highly characteristic, being up to 4 mm long, with a broad base, with hairs and small wrinkled epidermal cells. The stones or pips have a wrinkled surface unlike other genera and are smaller than blackberry

Blackberry – *Rubus fruticosus*: fruit of many drupelets, each containing a furrowed stone, the whole closely attached to the receptacle

In detail, the blackberry is similar in general structure to the raspberry, but can be distinguished based on the following differences. The drupelets are glabrous and are firmly attached to the receptacle by broad bases and do not separate on picking the fruit. Indeed, the receptacle has no distinct epidermis as the surface is almost completely covered by the bases of the drupelets, the epicarp being continuous with that of the adjoining drupelet. The pips resemble those of the raspberry in shape and markings, but are much larger. The styles are about 2 mm long, do not have hairs at the base, do not broaden from the base and are not wrinkled.

8.4.5.2 Family: Rutaceae

Orange – *Citrus aurantium*: a citrus fruit is called a *hesperidium*, a fruit whose fleshy parts are divided into segments, the whole being surrounded by a separate skin (Allaby, 1998). The rind comprises the epicarp and part of the mesocarp with the segments enclosed by a membranous endocarp. Some varieties of orange contain no seeds, but when present, the segments contain two or more seeds and club-shaped vesicles. Embedded in the rind are numerous cavities, often over 1 mm in diameter, in which is secreted essential oils. The epicarp cells are polygonal and thick-walled. Stomata have nearly concentric rings of epidermal cells, giving them a circular appearance. The mesocarp comprises layers of spongy parenchyma with large intercellular spaces.

When oranges are used to make marmalade, the features described above will usually be visible and are diagnostic for orange, noting that all common citrus fruits have similar microscopic features.

8.4.5.3 Miscellaneous fruits

Winton et al. (1916) include descriptions of members of the family Saxifragaceae, including red and black currents (*Ribes rubrum* and *Ribes nigrum*), gooseberries (*Ribes grossularia*) and the family Ericaceae, including cranberries (*Vaccinium macrocarpon*), blueberries (*Vaccinium myrtillus*) and huckleberry (*Gaylussacia resinosa*).

Grape – Order Vitaceae – *Vitis vinifera*: fruits are botanically berries and are typically green or red, but other colours are also possible. The berry has a smooth epicarp, often with a bloom, a pulpy mesocarp, but lacking a conspicuous endocarp. Each of the two locules normally contains two seeds, although there are commercial varieties that are seed-free. Seeds are pear-shaped, 5–8 mm long. Grapes in dried form are called raisins, currants or sultanas. The varieties used, and the process and time to dry the grapes vary in different countries.

As commonly only the skin and fruit flesh are encountered, grapes lack unique diagnostic features, as the polygonal epidermal cells resemble those of many other products, and the pulp cells are not characteristic. Parenchyma cells, which form a subepidermal layer of two to six layers thick, often contain calcium oxalate crystals in the form of raphides.

Banana – Order Musaceae – *Musa sapientum*: bananas are botanically berries. They are elongated, varying in length but typically 10–20 cm long. The outer tough rind is easily separated from the inner pulpy fruit flesh, which shows three indistinct locules with minute brown abortive seeds. As the rind is inedible, only the fleshy mesocarp is of interest. The fruit flesh is a mass of rounded pulp cells, which, in the outer layers, are nearly isodiametric. Inner mesocarp cells form radial chains of cells containing large, irregular starch grains with an eccentric hilum, 20–40 μm, occasionally up to 75 μm. The fusiform starch grains are variably described as cigar-shaped, sickle-cell–shaped (where two grains join end to end) or rod-shaped. They have distinct rings. Starch grains, however, largely disappear as the banana ripens. Chains of large oleoresin cells are associated with fibro-vascular bundles.

Pineapple – Order Bromeliaceae – *Ananas comosus*: the fruit consists of numerous fleshy berries, each with a single fleshy bract, united with an axis, forming a conical composite fruit shaped like a pinecone from which the name pineapple derives. The main diagnostic features of the fleshy component are the cells of the parenchymatous fruit flesh containing large crystals of calcium oxalate in the form of raphides, and the fibro-vascular bundles consisting chiefly of bast fibres with broad lumen and round pores and large spiral vessels.

8.4.6 Vegetables

Vegetables can be defined as the edible parts of plants, generally excluding fruits that come from the flowering part of the plant and contain seeds. So, is a tomato a fruit or a vegetable? In botanical terms, a tomato is a fruit, but in general use, it is considered to be a vegetable.

Vegetables are often grouped based on the part of the plant that is eaten, such as leaves (lettuce), stems (celery), roots (carrot), tubers (potato), bulbs (onion), flowers (broccoli), fruiting vegetables (pumpkin, cucumber, tomato, zucchini) and legumes/pulses (beans, peas, lentils). Grains, nuts, herbs and spices are generally excluded.

Tindall (1983), in his treatment of vegetables in the tropics (note, not tropical vegetables), prefers to consider crops based on botanical families rather than on groupings based on plant parts. This is the approach used below. Tindall lists 23 families and numerous species, and even this is not intended to be a complete list. Whilst many vegetables are common to a wide range of countries and societies, local knowledge is vital as many vegetables are only grown in parts of the world and often reflect on how they are used in preparing and cooking foods. Hence, only a few selected and important species will be considered in this chapter.

8.4.6.1 Family Alliaceae

This family includes chives, garlic, leeks, onions and shallots. Only onions are considered here.

Onion – *Allium cepa*: comprises swollen concentric leaf bases forming a bulb. Outer bases dry and papery, inner fleshy with membranes in between and nerves longitudinal and conspicuous. Epidermal cells elongate and are longitudinally arranged with thick walls in the dry scales.

8.4.6.2 Family Chenopodiaceae

This family includes beetroot, chard and quinoa. Only beetroot is considered here.

Beetroot – *Beta vulgaris*: comprises a fleshy, round or elongate taproot, deep red with light and dark bands when sectioned. Large parenchyma cells, up to 250 μm in diameter, with dark red sap and sucrose. Vessels average 50 μm and are pitted or reticulated, often with wide meshes.

8.4.6.3 Family Compositae

This family includes endive, Jerusalem artichoke and lettuce. Only lettuce is considered here.

Lettuce – *Lactuca sativa*: variable leaf shapes and colours. The walls of both upper and lower epidermal cells are sinuous except over veins and midribs. Stomata are numerous. Long uniseriate hairs are present on the upper epidermis, along with shorter glandular hairs with a biseriate stalk and a two-celled head.

8.4.6.4 Family Cruciferae

This family includes numerous *Brassica* species, *Brassica alba* (white mustard), *Brassica chinensis* (Chinese cabbage), *Brassica juncea* (Leaf mustard), *Brassica oleracea* (broccoli, Brussels sprouts, cabbage, cauliflower, kohlrabi, kale) and *Brassica rapa* (turnip, swede). Other members in this family include *Raphanus sativus* (radish, Chinese radish) and *Armoracia rusticana* (Horse radish).

Broccoli – *B. oleracea var. botrytis*: inflorescence of more or less perfect buds in small heads in the axils of elongate leaves. Small buds sometimes with pollen grains of maturing flowers

Cabbage – *B. oleracea var. capitata*: classed according to shape and colour, leaves rounded, inner ones blanched forming a 'head', midribs fleshy. Epidermis not beaded, stomata in groups, often in pairs

Cauliflower – *B. oleracea var. botrytis*: white dome-shaped curd of abortive flowers and inconspicuous white leaves, all encircled by large green leaves. Miniature white buds of overlapping parts

Turnip – *B. rapa*: variable shape, generally round with white flesh and purple shoulders. Very large parenchyma cells up to 500 μm in diameter. Similar to beetroot but with larger vessels with round pores

8.4.6.5 Family Cucurbitaceae

This family includes numerous genera and species with common names of cucumber, gherkin, gourd, marrow, and squash.

Cucumber – *Cucumis sativus*: green elongate fruit with a rounded triangular section, three solid locules with many flattened whitish seeds and outer skin smooth or warty. The epicarp cells are prismatic with thickened outer and radial walls. Stomata are absent. Warts emerge from the epicarp and consist of large cells with thickened, sparingly pitted walls. Attached to the apex of the wart is a broad jointed, conical hair. These hairs disappear in the early stage of growth but leave behind a hyaline spine. At the base of these warts are sclerenchyma cells. The mesocarp or fruit flesh is a mass of loose parenchyma with isolated sieve tubes, latex tubes and fibro-vascular bundles.

Cucumbers are often picked at an early stage of growth and pickled when they do not show very marked differentiation of tissues. Small pieces, however, may be recognised by the warts on their surface and the presence of thin elliptical seeds.

Pumpkin – *Cucurbita pepo*: pumpkins vary enormously in size, shape and colour. This is another example of a fruit, but treated as a vegetable, as it is picked before the seeds mature. Pumpkins may be cooked with their outer skin or pericarp layers, or these layers may be removed before cooking, along with the seeds. The main edible portion is the fleshy mesocarp. When present, the epicarp cells are prismatic, forming a palisade layer upwards of 50 μm thick. In surface view, the cells are polygonal with somewhat elongated and curved stomata. The mesocarp ground tissue is of large, rounded cells containing starch grains. Embedded in the ground tissue are strongly developed spiral vessels, often 60 μm broad, some with single strands and others with two to four strands; also, branching and anastomosing latex tubes are present.

8.4.6.6 Family Leguminosae

This is one of the most important economic plant families. There are three subfamilies, the Caesalpinoideae, Mimosoideae and the Papilionoideae.

Most of the economically important species are found in the Papilionoideae. The fruit is normally a pod, which may be round or flat, sometimes winged, straight or curved, of variable length; fibrous or fleshy and often splitting open at maturity (Tindall, 1983). From a food perspective, it is the seeds that are of most interest, although the pods are eaten in some cases, such as snap peas and green beans.

Hence, the microscopic features of the seeds are of most importance. The spermoderm typically has three layers, of which the outer two layers are each one cell thick and the inner third layer several cells thick. An inner epidermis is seldom visible. The hilum in some species is more or less elongated and pierced through its major axis by a narrow slit.

The outer epidermis or *palisade layer* is of great diagnostic significance, comprising prismatic, thick-walled cells, which are generally much higher than broad. The lumen of the inner portion of these cells is broader than the outer, which is usually seen as a thin line. The palisade layer is one cell thick except near the hilum, where it is two cells thick. In surface view, the cells are sharply polygonal and often show radiating lines.

Below the palisade layer is the next highly characteristic layer of the so-called *column* or *bearer* cells. These are shaped like an hourglass or I-shaped. In some species, these cells have no apparent contents, whilst in others, they contain crystals of calcium oxalate.

The third layer of cells comprises several layers of spongy parenchyma. The inner epidermis, when present, is made up of thin-walled cells.

An example of a cross-section of a bean is shown in Figure 8.12.

Some of the more important legume species are groundnuts, lentils, chickpeas, soya beans, lima beans, French beans, garden peas, broad beans, runner beans, black gram and mung beans. The following descriptions are a sample to illustrate some of the distinguishing features between species.

Lentil – *Lens esculenta*: seeds up to 8 mm, sharp-edged and commonly reddish brown and also green and black. Palisade cells 45 μm high by 8 μm. Bearer cells spool-shaped 12–22 × 18–35 μm with irregular brown contents

Broad bean – *Vicia faba*: seeds up to 3.5 cm, flattened, kidney-shaped and green-brown, purple or black. Distinct scar on mature seeds. Palisade cells 150–175 × 12–20 μm. Bearer cells thick-walled and hourglass-shaped. 35–50 × 35–60 μm

French bean – *Phaseolus vulgaris*: seeds ovoid, spherical or kidney-shaped, commonly flattened. French bean pods are straight or curved, with the outer green mesocarp being edible. Palisade cells up to 60 × 10 μm. Quadrilateral-shaped bearer cells contain crystals of calcium oxalate. Epicarp of pod striate with straight or hooked unicellular hairs and capitate glandular hairs with a multicellular head

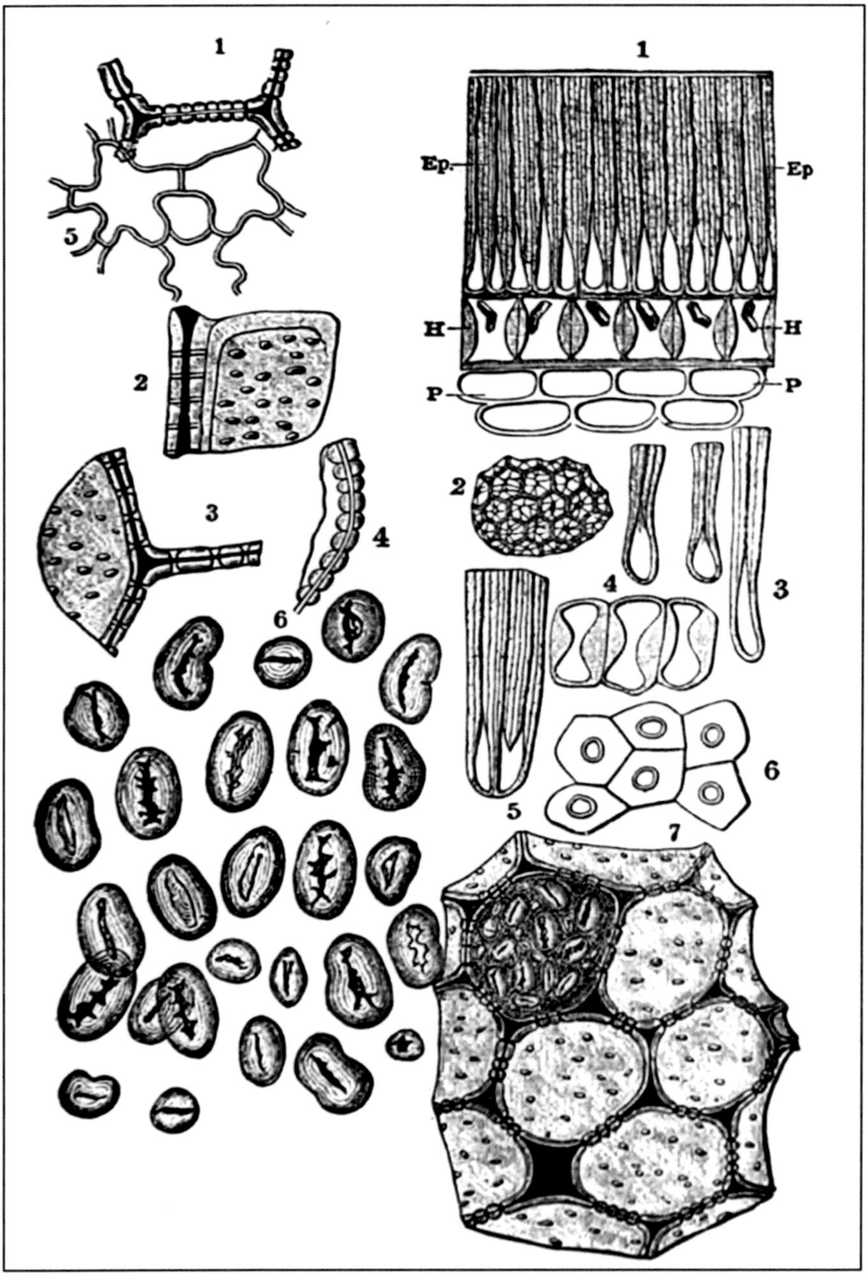

Figure 8.12 Anatomy of a bean seed showing elongate palisade and window cells.

(From *An Introduction to Pharmacognosy*. 1905. Jelliffe, S.E., page 251, figure 73. Public Domain.)

Scarlet runner bean – *Phaseolus coccineus*: flattened pod, up to 5 cm long, with flattened kidney-shaped, usually glossy pink and black seeds. Palisade cells up to 85 × 15 μm. Lumen broader at the inner end. Bearer cells, thick-walled with an hourglass-shaped lumen, 30 × 40 μm and with small crystals present.

Pea - *Pisum sativum*: seeds up to 1 cm, globular to round, smooth or wrinkled, green or buff in colour. Palisade cells 60–100 μm high, lumen narrows towards the outer end, and broader and wrinkled towards the inner end. Bearer cells hourglass-shaped, up to 20 μm high with prominent ribs (see Figure 8.13).

8.4.6.7 Family Polygonaceae

Rhubarb – *Rheum rhaponticum*: the edible part of the rhubarb plant is the pink/red-coloured petioles. These are rounded on the outer side with distinct ribs. The inner side is slightly grooved. The epidermal cells are elongate except around the stomata. Multicellular glandular hairs with a distinct fan-shaped head are present. Ground tissue with large parenchyma cells with large rosettes and much smaller crystals of calcium oxalate and spiral-shaped xylem.

8.4.6.8 Family Solanaceae

This family includes potatoes, tomatoes, a wide variety of peppers and the eggplant or aubergine.

Tomato – *Solanum lycopersicum*: tomato fruits vary widely in shape, size and colour. The epicarp has beaded polygonal-shaped cells, often thickened at the corners. The fleshy mesocarp has rounded pulp cells with lycopene. Vascular elements are spiral vessels with bundles nearer the stem being partly pitted. Bast fibres accompany these latter bundles but are not found elsewhere. Seeds are present in locules with many seeds attached to placentae. They are ovoid, flattened, 3–4 mm long and covered in short, silky hairs. The outer epidermis is highly characteristic due to pronounced ribs on the delicate radial walls, up to 500 μm long.

Potato – *Solanum tuberosum*: potatoes are tubers formed at the end of underground stolons. They vary enormously in shape, size and colour. They bear numerous eyes or buds in depressions in the surface. The outer protective coat is a cork tissue with large cells, which are polygonal in surface view. Below this layer, the outer cells of the cortex may contain pigment. The cortex is mainly composed of parenchyma cells. Most of the bulk of a potato is really wood, being the product of a cambium. However, as there is little lignification, the cells of the medulla are almost entirely thin-walled parenchyma, arranged in tangential lines in the outer layers, becoming larger and isodiametric with intercellular spaces. They are filled with starch grains, most of

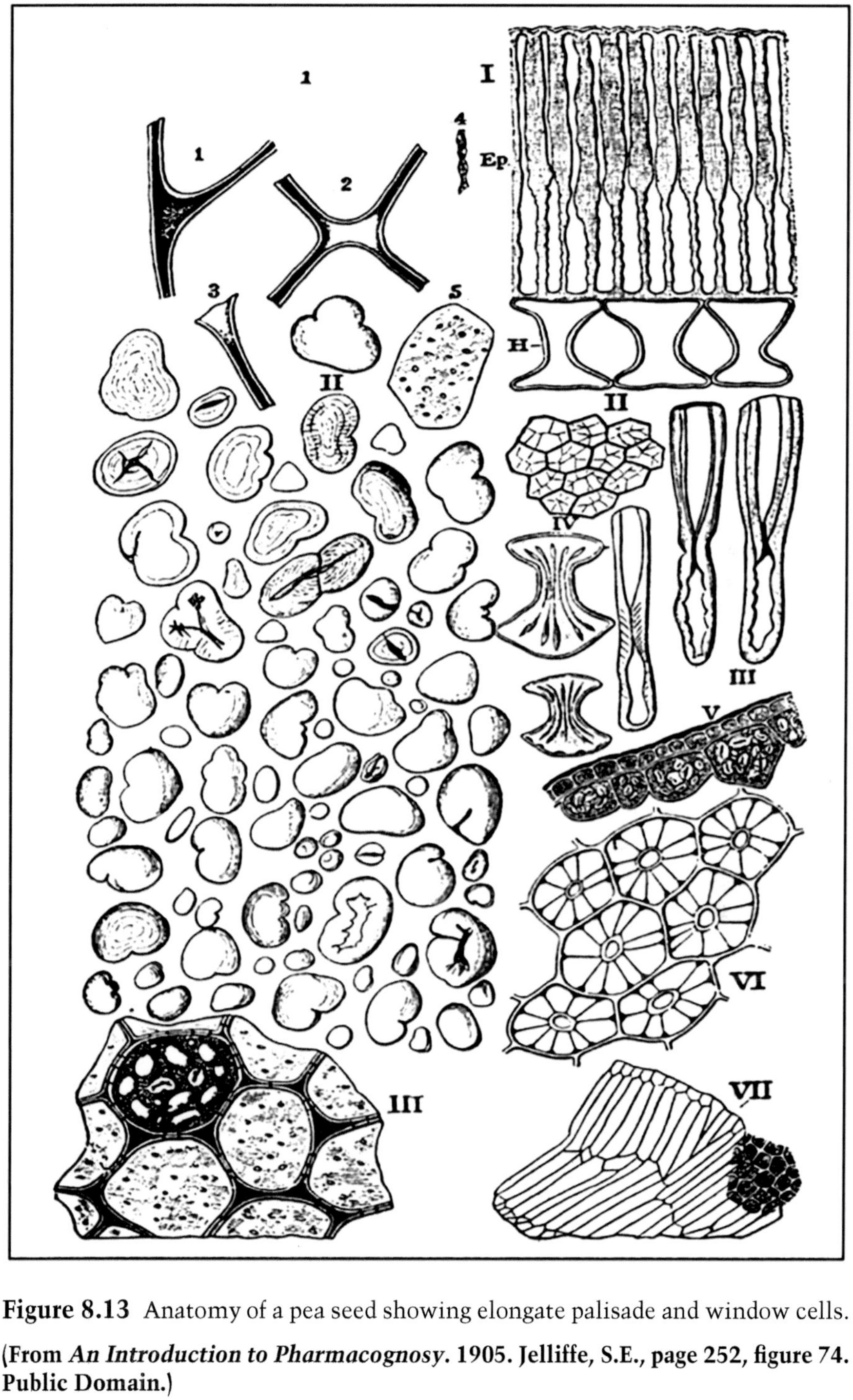

Figure 8.13 Anatomy of a pea seed showing elongate palisade and window cells.

(From *An Introduction to Pharmacognosy*. 1905. Jelliffe, S.E., page 252, figure 74. Public Domain.)

which are large, irregularly pear-shaped with distinct rings and an eccentric hilum located in the small end.

8.4.6.9 Family Umbelliferae

This family contains numerous herbs and spices in addition to several vegetables, including carrots, celery, parsley and parsnip.

Carrot – *Daucus carota*: generally orange-coloured tapering roots. Specialist varieties of other colours exist. In cross-section, zoned with radial lines. The outer cork layer consists of a zone of large cells with thick walls. The bulk of the carrot consists of parenchyma cells with some vascular tissue in the form of vessels. These are seldom over 50 μm broad and are characterised by their narrow, elongate pores, giving the vessels a reticulated pattern

Celery – *Apium graveolens*: stiff, clasping leaf stalks with an outer ridged surface and inner smooth leaves, compound with lobed leaflets. Outer epidermal cells striate, isodiametric and sometimes longitudinally arranged between ridges. Stomata numerous except over ridges and veins. Collenchyma cells under the ridges.

Parsnip – *Pastinaca sativa*: white to yellow tapering root with transverse lenticel-like marks. Outer cork layer with the bulk of the flesh being rectangular parenchyma cells and some vascular tissue. Vessels are short and reticulated and are often broader than long.

8.5 Fungi

From a strictly botanical viewpoint, fungi do not belong in a book on forensic botany as they are not plants. Fungi, however, are consumed as a foodstuff and, hence, in a forensic context, fungi may be encountered when dealing with the contents of the human digestive tract.

There are estimated to be around 1.5 million species of fungi, of which only 80,000–120,000 have been described to date (Webster and Weber, 2007). Probably, 1.5 million species is a very large underestimate, as new species are being discovered each year. The complexity of fungi is demonstrated by the fact that Webster and Weber's *Introduction to Fungi* runs to 842 pages!

Some groups of fungi produce fruiting bodies, which we recognise as mushrooms or toadstools, for example, the white button mushroom of commerce, *Agaricus bisporus* (sometimes called *Agaricus brunnescens*). There are many other species grown commercially and others that are harvested in the wild. Some cultures, such as Japan, make widespread use of mushrooms with dried shiitake (*Lentinula edodes*), the source of *umami* flavours. China is by far and away the major producer of commercial mushrooms, accounting for 75% of the world's total in 2019 FAO (Food and Agriculture Organisation) data.

In the overall life cycle of a fungus, these fruiting bodies can be very short-lived, as their sole purpose in nature is to produce spores. Spores are minute in size, measuring between 3 and 20 µm. Fruit bodies release their spores when ripe and then over a short period of time. The number of spores can be enormous, in the many hundreds of thousands, for a single fruiting body.

Figure 8.14 shows the parts of a typical mushroom. In large fungi, the size, shape and variation in fruiting bodies are diagnostic features in differentiating between species.

Unlike in plants, the flesh of mushrooms does not have well-defined cell types and consists of an integrated mass of fungal hyphae. As a result, when eaten, unless there are visible pieces of mushroom in the stomach contents, or left uneaten, it is unlikely there will be any identifiable component. In these cases, identification relies on examination of any spores present in the gastrointestinal tract.

Wiltshire et al. (2015) describe in detail a case involving the identification of fungal *Psilocybe semilanceata* spores and other material. Whilst accurate identification was an essential aspect of this case, it was the interpretation of what was present that helped reconstruct possible events leading up to a suspicious death.

Hawksworth and Wiltshire (2011) have reviewed the broader aspects of forensic mycology (the study of all kinds of fungi), including a consideration of cases of mushroom poisoning. What makes one mushroom species edible and another poisonous? Contrary to popular belief, there are no special

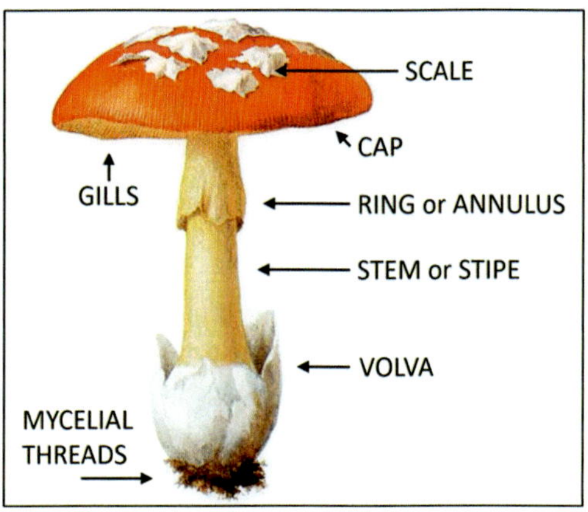

Figure 8.14 Morphology of a typical mushroom.

characteristics by which an edible fungus can be recognised, and the names mushroom and toadstool give no indication of relative toxicity and can be used almost interchangeably (Cooper et al., 2003). Identification is not easy, and great caution should be exercised when contemplating eating any fungus, as poisoning often occurs due to a misidentification.

Several varieties of mushrooms contain psychedelic compounds, the so-called magic mushrooms, whilst variously resembling non-psychoactive species. The most consumed mushrooms for recreational purposes are *Amanita muscaria* (the fly agaric) and *Psilocybe cubensis*. The most poisonous mushroom is *Amanita phalloides* (see Figure 8.15).

There are a number of excellent books with descriptions and images of mushrooms, including poisonous species (Ammirati et al., 1985; Bresinsky & Besl, 1990; Hall et al., 2003; Cooper et al., 2003). Whilst these books have their uses, they cannot be considered as comprehensive identification manuals and are not a substitute for the type of detailed forensic examination required to provide evidence that can be relied upon by the criminal justice system, including access to reference collections such as used by Wiltshire et al. (2015). As Hawksworth and Wiltshire (2011) point out, there is a critical shortage of appropriately skilled mycologists, with only six full-time fungal systematists in the United Kingdom at that time, down from 23 in 1997.

As has been discussed in almost every chapter, the future will include the use of DNA analysis. For example, Gausterer et al. (2014) reported on the rapid genetic detection of ingested *A. phalloides*, the aptly named death cap.

Figure 8.15 Amanita species. (a) *Amanita muscaria*, non-poisonous and widely consumed. (b) *Amanita phalloides*, highly poisonous species.

(a. Diego Delos, delso.photo (https://delos.photo), license CCBY_SA. https://creativcommons.org/licenses/by-sa/4.0/legalcode. b. Creative Commons Attribution – Share Alike 3.0 unported license.)

The future will almost certainly lie in the use of a forensic DNA metabarcoding approach (Schneider et al., 2021; Tedersoo et al., 2022).

The utility of DNA barcoding, however, depends on the use of genetic 'banks' such as GenBank. As Hawksworth and Wiltshire (2011) point out, only a small proportion of the known fungi are represented in existing DNA databases, with even less representation of within-species variability. Hence, a DNA approach should not be viewed as a replacement for sound microscopic analysis but rather as a complementary tool.

References

Allaby, M. 1998. *A Dictionary of Plant Sciences*, 2nd ed. Oxford: Oxford University Press.

Ammirati, J., Traquair, H.A., & Horgen, P.A. 1985. *Poisonous Mushrooms of the Northern United States and Canada*. Minneapolis: University of Minnesota Press.

Anon. 1982. *Biology Methods Manual*. Metropolitan Police Forensic Science Laboratory. In house publication.

Apling, E.C. 1980. The Optical Microscope in Food Analysis. In King, R.D. (ed.), *Developments in Food Analysis Techniques*, 2nd ed. (Chapter 4, pp. 151–184). London: Elsevier.

Beaumont, W. 1833. *Experiments and Observations on the Gastric Juice and Physiology of Digestion*. Plattsburgh: FF Allen.

Bock, J.H., & Norris, D.O. 2016. *Forensic Plant Science*. London: Academic Press.

Bowen, A.M. 2003. Starch Grain Identification Revisited. *Microscope*, 51, 21–30.

Bresinsky, A., & Besl, L. 1990. *A Colour Atlas of Poisonous Fungi: A Handbook for Pharmacists, Doctors, and Biologists*. London: Wolfe Pub.

Byard, R. 2010. A Review of the Potential Forensic Significance of Traditional Herbal Medicines. *Journal of Forensic Sciences*, 55, 89–92.

Clayton, E.G. 1909. *A Compendium of Food-Microscopy*. London: Bailliere, Tindall and Cox.

Cooper, M.R., Johnson, A.W., & Dauncey, E.A. 2003. *Poisonous Plants and Fungi: An Illustrated Guide*, 2nd ed. London: The Stationary Office.

Duckworth, R.B. 1966. *Fruit and Vegetables*. Oxford: Pergamon Press.

Evers, A.D. 1979. Cereal Starches and Proteins. In Vaughan, J.G. (ed.), *Food Microscopy* (Chapter 4, pp. 137–191). London: Academic Press.

Fashionpedia. 2021. *The Visual Dictionary of Fashion Design*. Hong Kong: Fashionary International Ltd.

Gausterer, C., Penker, M., Krisai-Greilhuber, I., Stein, C., & Stimpfl, T. 2014. Rapid Genetic Detection of Ingested *Amanita phalloides*. *Forensic Science International: Genetics*, 9, 66–71.

Greenish, H.G., & Collin, E. 1904. *An Anatomical Atlas of Vegetable Powders*. London: J & A Churchill Ltd.

Hall, I.R., Stephenson, S.L., Buchanan, P.K., Yun, W., & Cole, A.L.J. 2003. *Edible and Poisonous Mushrooms of the World*. Portland: Timber Press.

Hawksworth, D.L., & Wiltshire, P.E.J. 2011. Forensic Mycology: the use of Fungi in Criminal Investigations. *Forensic Science International*, 206, 1–11.

Jackson, B.P., & Snowdon, D.W. 1968. *Powdered Vegetable Drugs*. London: J & A Churchill Ltd.
Jackson, B.P., & Snowdon, D.W. 1992. *Atlas of Microscopy of Medicinal Plants, Culinary Herbs and Spices*. New Delhi: CBS Pub.
Jewell, G.G. 1979. Fruits and Vegetables. In Vaughan, J.G. (ed.), *Food Microscopy* (Chapter 1, pp. 1–34). London: Academic Press.
Jolliffe, G.H., & Jolliffe, G.O. 1976. Computer Aided Identification of Powdered Vegetable Drugs. *Analyst*, 101, 622–633.
Jolliffe, G.H., & Jolliffe, G.O. 1979. Computer-Aided Identification of Food Materials. In Vaughan, J.G. (ed.), *Food Microscopy* (Chapter 16, pp. 594–619). London: Academic Press.
Kent, N.L. 1975. *Technology of Cereals with Special Reference to Wheat*, 2nd ed. Oxford: Pergamon Press.
Kong, F., Oztop, M.H., Singh, R.P., & McCarthy, M.J. 2011. Physical Changes in White and Brown Rice During Simulated Gastric Digestion. *Journal of Food Science*, 76, E450–E457.
Mackie, A., Mulet-Carero, A., & Torcello-Gomez, A. 2020. Simulating Human Digestion: Developing our Knowledge to Create Healthier and More Sustainable Foods. *Food & Function*, 11, 9397–9431.
Moffat, A.C. 1980. Forensic pharmacognosy – Poisoning with Plants. *Journal of the Scientific Society*, 20, 103–109.
Panlasigui, L.N., & Thomson, L.U. 2006. Blood Glucose Lowering Effects of Brown Rice in Normal and Diabetic Subjects. *International Journal of Food Sciences and Nutrition*, 57, 151–158.
Pilling, E., & Smith, A.M. 2003. Growth Ring Formation in the Starch Granules of Potato Tubers. *Plant Physiology*, 132, 365–371.
Schneck, W.M. 2013. Stomach Contents Analysis. In: *Encyclopaedia of Forensic Sciences*, 3rd ed, (ed) Houck, M.M. (Vol. 4, pp. 671–678). London: Elsevier.
Schneider, J., Mas-Carrio, E., Jan, C., Miquel, C., Taberlet, P., Michaud, K., & Fumagalli, L. 2021. Comprehensive Coverage of Human Last Meal Components Revealed by a Forensic DNA Metabarcoding Approach. *Nature Scientific Reports*. https://doi.org/10.1038/s41598-021-88418-x
Seidemann, J. 1966. *Starch Atlas: Principles of Starch Microscopy and Descriptions of Important Varieties of Starch*. Berlin: Parey.
Sensoy, I. 2021. A Review on the Food Digestion in the Digestive Tract and the Used *in vitro* Models. *Current Research in Food Science*, 4, 308–319.
Tedersoo, L., Bahram, M., Zinger, L., Nilsson, R.H., Kennedy, P.G., Yang, T., Anslan, S., & Mikryukov, V. 2022. Bets Practices in the Metabarcoding of Fungi: From Experimental Design to Results. *Molecular Ecology*, 31, 2769–2795.
Thrasher, J.J. 1962. Chemical Identification of Bird Excrement in Egg Products. *Journal of Association of Official Analytical Chemists*, 45, 659–660.
Tindall, H.D. 1983. *Vegetables in the Tropics*. London: Macmillan Press.
Upton, R., Graff, A., Jolliffe, G., Langer, R., & Williamson, E. (eds.). 2011. *Botanical Pharmacognosy - Microscopical Characterization of Botanical Medicines*. Boca Raton: CRC Press.
Wallis, T.E. 1957. *Analytical Microscopy. Its Aims and Methods in Relation to Foods, Water, Spices and Drugs*, 2nd ed. London: J & A Churchill Ltd.
Wallis, T.E. 1960. *Textbook of Pharmacognosy*. London: J & A Churchill Ltd.

Webster, J., & Weber, R.W.S. 2007. *Introduction to Fungi*, 3rd ed. Cambridge: Cambridge University Press.

Wiltshire, P.E.J., Hawksworth, D.L., & Edwards K.J. 2015. Light Microscopy Can Reveal the Consumption of Psychotropic Plant and Fungal Material in Suspicious Death. *Journal of Forensic and Legal Medicine*, 34, 73–80.

Winton, A.L., Moeller, J., & Winton, K.B. 1916. *The Microscopy of Vegetable Foods with Special Reference to the Detection of Adulteration and the Diagnosis of Mixtures*, 2nd ed. New York: John Wiley & Sons.

Forensic Palynology and Phycology

9

9.1 Forensic Palynology

9.1.1 Introduction

Palynology is the study of pollen and spores, collectively known as *palynomorphs*. First used in 1944, the term palynology comes from the Greek *paluno*, meaning 'I strew or sprinkle' and the Greek *palynos* meaning 'dust', and close to the Latin word *pollen* meaning 'fine flour or dust' (Hesse et al., 2009; Hyde & Williams, 1944). The features of spores and pollen grains can be used to identify plant and fungal taxa, information that has been applied in a wide range of sciences by archaeologists, anthropologists, climate scientists, allergy/health studies, forensic scientists, genetic and evolutionary studies, tracing vegetation history and even honey studies, melissopalynology (Moore et al., 1991).

Despite an extensive literature dealing with palynomorphs, commentators who have written about forensic applications have observed that it is unknown and underutilised in most countries, even though in many ways pollen and spores are ideal examples of trace materials in a forensic context. Due to their small size (about 10–70 μm), they can be hard to see, and, hence, hard to remove all traces. They are produced in vast quantities, they persist in soils and other surfaces, they are highly resistant to degradation, and they have a complex and unique morphology.

The fact that they are underutilised has not stopped numerous authors from writing about the use of palynology in a forensic context. The earliest recorded forensic interest in pollen can be attributed to Locard (1930), who listed pollen as one of the types of botanical dust debris. The first recorded use of pollen analysis in case work was in 1959, where the presence of pollen in soil, taken from the shoes of a suspect, was used to show an association between the suspect and a forest where a body had been buried (Walsh & Horrocks, 2008). Other early cases during the 1960s and 1970s by Max Frei, a Swiss criminologist, used pollen to link suspects to events or crime scenes (Palenik, 1982). It is highly probable, however, at a time when forensic botany was still a significant evidence type in mainstream forensic science, that the lack of more published accounts of the use of pollen analysis in that era reflects less emphasis on publications rather than less use of pollen analysis.

Reasons put forward for why forensic palynology has been underutilised include that its crime-solving power is not well known to law enforcement,

DOI: 10.4324/9781003308508-9

there is inadequate resourcing and funding, there is a lack of control over sample collection, and it is labour-intensive and requires considerable expertise and experience (Walsh & Horrocks, 2008). Bryant and Jones (2006) add to the latter factors a lack of trained specialists, and importantly, those willing to conduct forensic pollen studies, due in part to the fact that there are no universities or other types of training facility that offers specific training in pollen forensics.

Notwithstanding, since the early 1990s, there has certainly been a renewed interest in the use of palynology for forensic applications. There are several major reviews (Bruce & Dettmann, 1996; Bryant & Jones, 2006; Mathewes, 2006; Mildenhall et al., 2006; Walsh & Horrocks, 2008), book chapters (Bock & Norris, 2016; Milne et al., 2005; Sandiford, 2012) and other academic papers dealing with forensic applications (Adams-Groom, 2018; Brown et al., 2002; Brown, 2006; Donaldson & Stephens, 2010; Horrocks et al., 1999; Horrocks & Walsh, 2001; Mildenhall, 2004; Wiltshire, 2006a, 2006b; Wiltshire et al., 2014, 2015).

Forensic palynology has been used to assess an alibi (Horrocks & Walsh, 1999; Mildenhall, 2004; Wiltshire et al., 2014), to put a suspect at the scene of an alleged crime (Horrocks & Walsh, 2001), to investigate war crimes (Brown, 2006), to determine the provenance of counterfeit cigarette products (Donaldson & Stephens, 2010; Chan et al., 2020), to determine travel history of items, including drugs (Stanley, 1992), and in numerous and diverse criminal investigations. Wiltshire (2006b) discusses many high-profile cases from the United Kingdom where forensic palynology has provided critical evidence.

Mildenhall et al. (2006) list the following uses for forensic palynology:

- Relate a suspect to the scene of a crime or discovery case
- Relate an item left at the crime scene or discovery scene to a suspect
- Relate an item at the discovery scene to the crime scene
- Prove or disprove alibis
- Narrow down a list of suspects
- Determine the travel history of items
- Provide information as to the environment that an item has come from
- Aid police in their lines of enquiry
- Help locate clandestine graves and human remains
- Help determine the peri-mortem fate of a victim
- Help determine the deposition period of human remains

Regrettably, however, and all too often, forensic pollen analysis is an afterthought when all else has failed. In the latter situation, a specialist forensic palynologist will almost never have been called to a scene in the initial stages of an investigation, and if samples have been taken at all, they will have been collected by a non-specialist crime scene examiner (Mildenhall et al., 2006).

9.1.2 Identifying Pollen and Other Spores

An essential prerequisite for any application in palynology is the accurate identification of pollen and other spores. The focus here will be on pollen, but fungal spores are another overlooked source of trace evidence that can be encountered in palynological work (Wiltshire et al., 2014; Hawksworth et al., 2016).

In Chapter 2, the basic botany and biology of the male elements of flowering plants (Angiosperms and Gymnosperms) were covered, the anthers of which contain pollen grains. Spores are also found in other plant groups, such as mosses and ferns. In this section, we shall deal with pollen, as the terminology used to describe spores differs, with elements such as the outermost wall layer (the perine or perispore), found in fern spores, being absent from pollen (Hesse et al., 2009).

In simple terms, a pollen grain comprises a tough outer wall, the *exine*, made up of a chemically stable compound, *sporopollenin*. A pollen grain may also have an interior wall, the *intine*, comprising cellulose and pectins similar to plant cells. When mature and released from the parent plant, pollen is, of course, a living structure with nuclei and other cell contents.

Pollen grains may be *binucleate* or *trinucleate*. The former contains one tube cell and nucleus, and one generative cell and nucleus. The latter contains one tube cell and a nucleus, and two sperm cells, resulting from the division of the generative cell. Trinucleate grains are relatively rare in angiosperms (Simpson, 2019).

Flowering plants and gymnosperms rely on external agents or vectors for the dispersal of pollen. These 'agents' include wind, insects and other living organisms, and water.

The aim of the following is to introduce the reader to the basics of pollen structure. The reader is referred to Erdtman (1952), Erdtman (1957), Faegri and Iversen (1989) and Moore et al. (1991), which deal with this topic in greater detail.

The identification of pollen and spores is based on structural features, including *size*, *shape*, *number of apertures* and *surface texture* and *ornamentation*. Hidden in this simple statement is the enormous variation accompanied by an extensive nomenclature used to capture and describe this information.

A glossary of pollen and spore terminology has been published under the auspices of the *International Commission for Palynology* (now the International Federation of Palynological Societies) by a working group on palynological terminology (Punt et al., 2007). Halbritter et al. (2008) have also produced a beautifully illustrated handbook of pollen terminology with light, scanning and transmission electron microscope images of mainly *angiosperm* pollen.

When released, pollen grains may not be single grains, hence the term *pollen unit*. This term refers to the number of pollen grains united at the time of release. Single pollen grains are called *nomads* and are found in most angiosperms. Pollen grains, rarely found in pairs, are called a *dyad*. When four haploid products of meiosis remain fused together, they form a *tetrad*. Tetrads can be further classified into one of five types, tetrahedral, linear, rhomboidal, tetragonal and decussate tetrads (Simpson, 2019). Pollen grains may form even larger units with eight fused units, common in the mimosoid clade of the Fabaceae (Leguminosae) family. In some instances, the entire content of one half of a stamen (a theca) can be fused. Called a *pollinium*, these are found in the families Apocynaceae and the Orchidaceae.

Individual pollen grains vary in size from less than 10 μm to over 100 μm, with most in the 10–70 μm range. Pollen size is based on the largest diameter. Pollen grains may be classified as very small (below 10 μm), small (10–20 μm), medium (26–50 μm), large (51–100 μm) and very large (over 100 μm) (Erdtman, 1952).

Smaller pollen grains are often wind-dispersed, with larger pollen grains being carried by living organisms or water.

Pollen shape and aperture location relate to *pollen polarity*. This feature is determined by the special orientation of the micropore in the meiotic tetrad and can only be examined in the tetrad stage. The polar axis of each microspore runs from the proximal end, oriented towards the tetrad centre, to the distal pole at the outer tetrad side. The equatorial plane is located at the microspore's centre, perpendicular to the polar axis and divides the pollen grain into a proximal and distal half.

Pollen grains are, of course, three-dimensional. *Shape*, however, as viewed with light microscopy, is two-dimensional. The outline shape in polar view can be *angular* (triangular, rhombic, rectangular, five-angled) or *non-angular* (circular, elliptic). Pollen shape can also be related to the ratio of the polar diameter to the equatorial diameter. Termed the *P*/*E* ratio, if greater than 1, grains are termed *prolate* and are elongate along the polar axis. If the ratio is less than 1, the grains are termed *oblate* and are compressed along the polar axis. If the ratio is approximately equal to 1, the grains are referred to as *spheroidal* (Simpson, 2019). Figure 9.1 shows pollen grain shapes in diagrammatic form.

A *pollen aperture* is a specially delimited region of the pollen grain wall, which functions as the site of the formation of a pollen tube exiting the pollen grain body on germination. The aperture may also control volume changes in the grain in response to water availability. Pollen aperture types are classified based on their shape, number, position and arrangement of the aperture. There are two basic types of apertures, pores that are circular to slightly elliptical in appearance, *porus*, and *colpi*, which appear as elongated furrows being elliptical, oblong or fusiform in outline. There are numerous other terms beyond these basic definitions used to describe aperture appearance (Halbritter et al., 2008; Simpson, 2019).

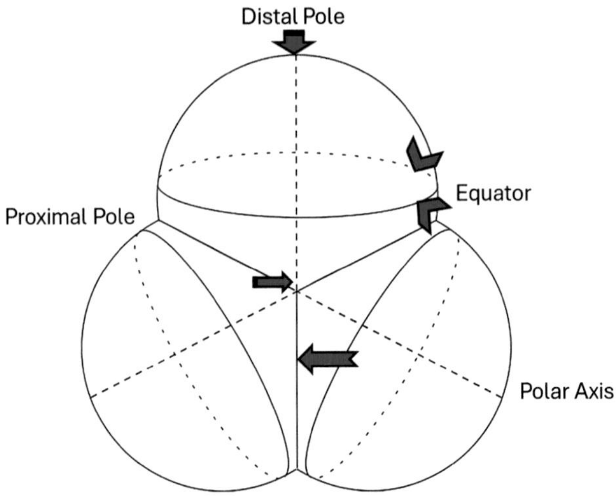

Figure 9.1 Pollen grains in diagramatic form.

The final class of feature used as a diagnostic feature for pollen grains is the *surface structure* and *sculpture*. *Structure* refers to the construction of the pollen cell wall. *Sculpture* refers to the ornamentation on the pollen surface. It is not always possible to distinguish between structure and sculpture, and hence the term *surface texture* is also used. Milne (2011) classified surface texture into five types: psilate or smooth, reticulate or net-like, satiate or roughly parallel patterns, regulate or irregular patterns and verrucate or surface lumps. However, as she also points out, the surface types are incredibly diverse, flamboyantly sculptured and vary considerably between species. Figures 9.2 and 9.3 give just a few examples of pollen surface morphology. The reader is referred to Halbritter et al. (2008) and any number of online websites for extensive images of pollen grains.

The results of a morphological analysis of a pollen grain or spore can be exhibited using a *palynogram*, a diagrammatic representation of a pollen grain or spore, its symmetry, apertures, shape, size, sporoderm stratification and sextine patterns (Erdtman, 1952).

9.1.3 Methods and Protocols for Pollen Examination

The focus of this section is on the retrieval of palynomorphs, either collected in the field or recovered from items of forensic interest. Wiltshire (2015) provides an extensive list of objects and materials that may be subjected to palynological analysis – palynomorphs can be found on almost any object, and what is examined will be informed by the individual case circumstances. Wiltshire (2015) provides guidelines for sampling vehicles, footwear and cadavers. With the latter, useful sources are beneath nails, on skin; hair

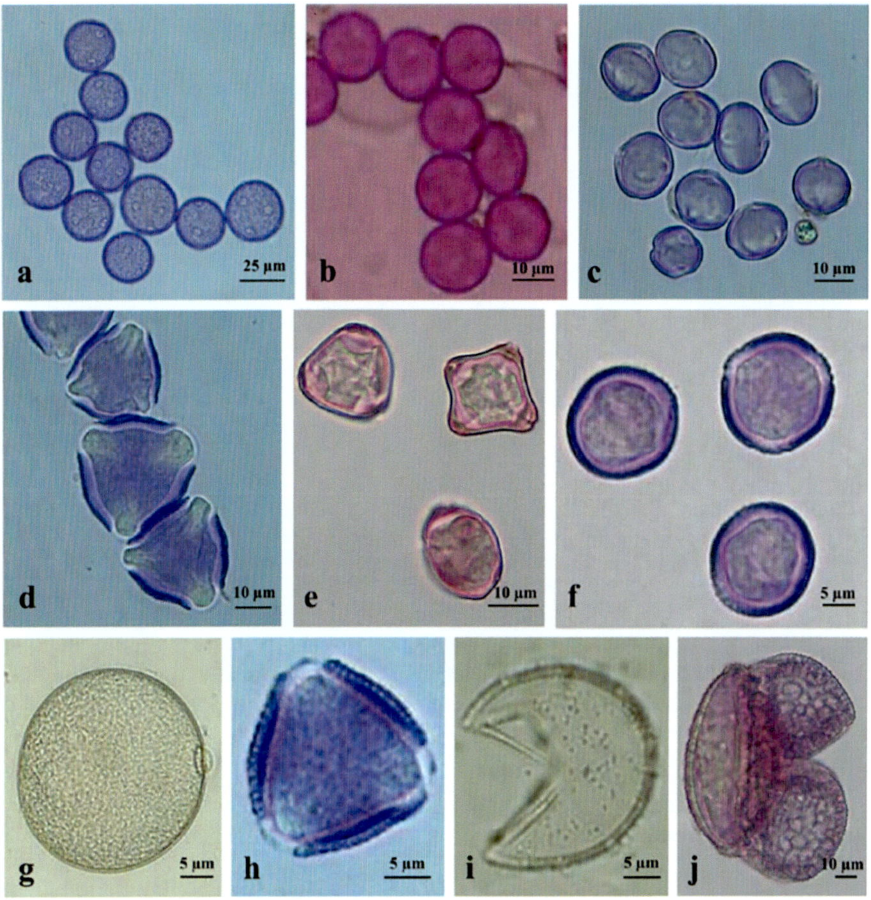

Figure 9.2 Examples of pollen grains viewed using transmitted light microscopy (TLM). (a) Plantago sp. (b) Urticaceae (Paretaria sp). (c) Castanea sp. (d) Quercus sp. (e) Myrtaceae (Callistemon sp). (f) Platanus sp. (g) Poaceae sp. (h) Oleaceae sp. (i) Cupressus sp. (j) Pinacea (Pinus sp).

(Wiltshire 2006a); stomach and lower gut contents and mucous membranes. She describes how to recover samples from the turbinate bones (thin, curved, bony plates in the nasal cavity that project into the respiratory passageway) during postmortem examinations (Wiltshire & Black, 2006).

Pollen and spores may also be recovered from the alveolar tissues of the lungs. Garcia-Mozo et al. (2021) suggest that pollen and spores are pushed down from the upper airways by force of water inhalation during drowning and that finding pollen and spores in deeper regions of the respiratory tract is an unusual occurrence.

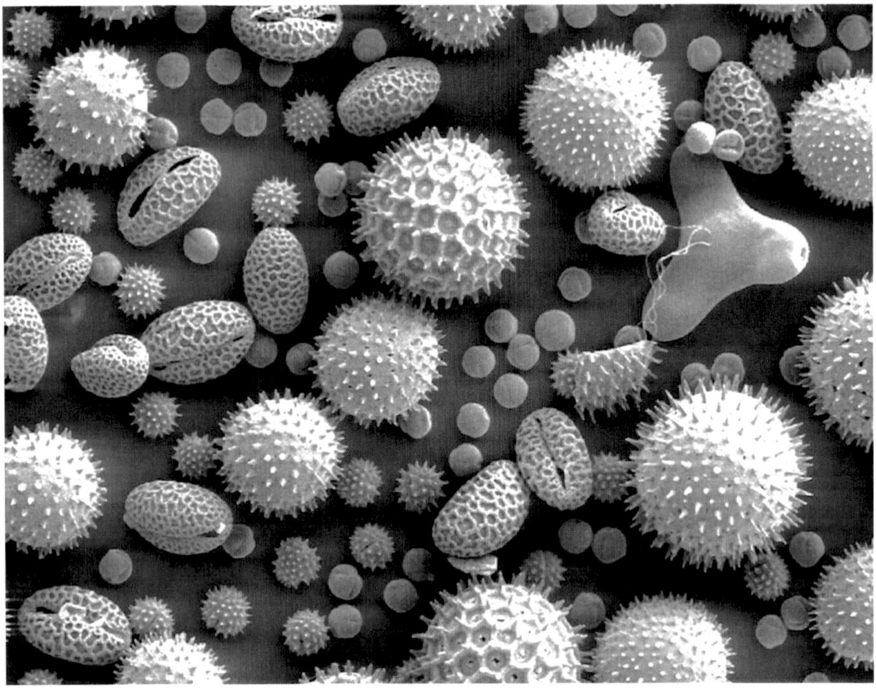

Figure 9.3 Examples of some common pollen grains as seen using scanning electron microscopy (SEM) at ×500 magnification.

(Author: Dartmouth College Electron Microscopy Facility. https://remf.dartmouth.edu/imagesindex.html)

Wiltshire (2015) recommends that appropriate items be washed and scrubbed with a hot solution of a surfactant such as Teepol. From some surfaces, pollen may be recovered using a small vacuum machine; however, the vacuum equipment must be specific for forensic use, and stringent precautions must be taken to avoid any possibility of cross-contamination.

Once initial material has been recovered, the second stage is the removal by chemical treatment, through progressive disintegration and dissolution of background material that may obscure the view of palynomorphs during microscopic examination. General methods can be found in Faegri and Iversen (1989) and Moore et al. (1991). Bryant and Wrenn (1998) outline procedures for forensic samples.

At all stages, great care must be taken to minimise the possibility of contamination by airborne palynomorphs. This includes stringent laboratory practices conducted preferably in a laboratory with positive air pressure. To be able to rely on what is recovered, it is necessary to check for contamination using jelly slide traps and include reagent and sample blanks as controls (Wiltshire, 2015).

Wiltshire (2015) outlines the following protocol for treatment of forensic samples:

- Boil in 10% sodium or potassium hydroxide to remove humic acids
- Drain the sample through a 120-μm sieve – retain sieving for possible further analysis
- Neutralise and decalcify with 10% hydrochloric acid – dissolves calcium compounds and fluorosilicates
- Dehydrate with glacial acetic acid
- Boil for not more than 4 minutes in a 9:1 mixture of acetic anhydride and concentrated sulphuric acid – removes cellulose, called *acetolysis*
- Rehydrate with glacial acetic acid
- Boil in 40% hydrofluoric acid – removes silica
- Treat with 10% hydrochloric acid – removes fluorosilicates
- Neutralise with 10% sodium or potassium hydroxide
- Stain with 0.5% safranine
- Mix in hot glycerol jelly and mix to homogenise and spread onto a hot slide on a hot plate, and add a cover slip and label

Note, the sample is washed with deionised water and centrifuged at 3,000 rpm for 4 minutes between each of the chemical treatments.

Hydrofluoric acid (HF) is *extremely* corrosive. Moore et al. (1991) emphasise the potential dangers with their comment that 'any contact with HF should be treated by thorough washing in water (20 minutes under a tap), application of burn cream, followed by *immediate hospital attention*'. As Wiltshire (2015) observes, 'only experienced workers should conduct the procedure'. The process should be conducted in a plastic fronted fume cupboard in capped polypropylene or polytetrafluoroethylene (PTFE) tubes.

Examination of recovered palynomorphs will always include the use of light microscopy, scanning each sample at ×400 using bright field light and phase contrast optics. For critical examinations, it may be necessary to use oil immersion with magnification of ×1,000. Faegri and Iversen (1989) discuss light microscopy in detail, recommending the use of phase contrast and commenting that light microscopy will always remain the main vehicle for routine pollen analysis, a view shared by Wiltshire (2015).

Many non-forensic publications, dealing with pollen and spores, make use of scanning electron microscopy (SEM) (Mander et al., 2014) and even transmission electron microscopy. Jones and Bryant (2007), in comparing pollen counts using light versus SEM, found that there was no significant difference in counts made with the two techniques, but there were significantly more taxa found with SEM. Notwithstanding, these authors concluded that it was appropriate to continue to use light microscopy with the use of SEM, depending on availability and funding.

For TLM examinations, Wiltshire (2015) recommends scanning ten or more transects for each slide examined to ensure any heterogeneity in the palynomorph is considered. An initial rapid scanning is useful to check for contamination, to check that palynomorphs are sufficiently preserved to allow identification and establish palynomorph concentration (Wiltshire et al., 2015). At least 300 pollen grains and spores should be counted, although in forensic cases, where species richness can be high and rarely encountered palynomorphs can be informative, counts of up to 1,000 or more may be necessary (Jones & Bryant, 1998).

The identification of palynomorphs makes use of standard keys, descriptions, and reference material (Erdtman, 1952; Erdtman, 1957; Faegri & Iversen, 1989; Moore et al., 1991). Identification should never be based on images or pictures of pollen in publications (Mildenhall et al., 2006) and should be based on comparisons to reference materials meeting strict standards regarding provenance. Notwithstanding, there are now online resources such PalDat (http://www.paldat.org) or the Australian Pollen and Spore Atlas (http://apsa.anu.edu.au/), amongst many others, which provide useful information.

In forensic case work, complex assemblages of pollen and other spores require a great deal of expert knowledge to understand the type of information that can be derived from often complex patterns. This phase combines identification skills and a criminalistics approach to event reconstruction.

9.1.4 Interpretation

Reports suitable for forensic use should contain raw palynological data, percentage palynological data and, where appropriate, pollen diagrams (Wiltshire, 2015). Identification must, of course, be accurate, recognising that the level of taxa may not reach the level of species in all cases.

As discussed in Chapter 11, the final interpretation of how palynomorph data translates into useful information for investigators and, ultimately, the criminal justice system (CJS) requires professional judgement based on qualifications, training and experience that underpin expertise. Expertise is not gained overnight! It has been estimated that expert levels can only be achieved after spending about 10 years in intense preparation with deliberate and structured practice of 4 hours/day (Ericsson & Lehmann, 1996).

Expertise required for forensic interpretation includes understanding how pollen is dispersed and factors such as seasonal variation (Stevenson et al., 2007), potential deterioration in soil (Havinga, 1984; Phuphumirat et al., 2009) and abundance in different environmental conditions.

As previously stated, pollen is dispersed by wind (*anemophilous* plants), by insects, by human activity, by other animals and by water.

Plants that produce wind-dispersed pollen can produce vast quantities of pollen grains, in the order of hundreds of thousands to millions from a single plant. About 95% of pollen from anemophilous plants falls within 2 km of the parent plant and, in many cases, within 100 m of the dispersal source. According to Mildenhall et al. (2006), how far pollen grains will travel, however, depends on factors including:

- height of release
- the release mechanism
- strength of wind and presence of updrafts or downdrafts
- weight, shape and aerodynamics of the grain
- atmospheric conditions (humid, wet, dry, hot, cold)
- obstructions between the plant and its surroundings

These factors will determine the *sinking speed* of each pollen grain. In a forensic context, palynological sinking models need to be applied with caution as other issues such as residuality, redeposition and local disturbance need to be considered (Mildenhall et al., 2006).

Plants that use an animal vector to disperse their pollen produce relatively few pollen grains, which are usually large, sticky, highly ornamented, thick-walled and readily preservable. They are usually found close to the dispersal source.

Some plants are self-pollinating (*autogamous* plants). The stigma and anthers develop at the same time, and there is no need for pollen to be dispersed for any great distance. These types of plants produce relatively small numbers of pollen grains.

Some plants have a closed self-pollination system (*cleistogamous* plants) in which the pollen is retained within an unopened flower. Again, few grains are produced. An example of this type of plant is cereal grasses.

Finally, pollen may be dispersed by water (*hydrogenous* plants). Pollen from this type of plant is produced in large numbers. Individual grains are typically small and unornamented. Unlike most other pollen grains, this type of pollen does not have a robust exine and sporopollenin, and they will be lost if subjected to the recovery protocol outlined above. This further emphasises the importance of a palynologist attending scenes to take their own samples and to inform the appropriate recovery protocol in the laboratory.

All the above mechanisms for pollen dispersal are focussed on outdoor environments. Pollen should also be considered where the crime scene is indoors. Pollen in this context will include pollen from outdoor plants but may also contain pollen from ornamental indoor plants or flowers (Morgan et al., 2014a; Nguyen & Weber, 2015). Morgan et al. (2014a) found that pollen from indoor flowers may be present for many days or even weeks after flowers have been removed from a room, with 25%–32% of original pollen being present in a room 20 days after the flowers had been removed.

Humans may also be the vectors to transport pollen. Jantunen and Saarinen (2011), in a study of transport of pollen by clothes, found that after walking through grasslands, 68 pollen grains/cm^2 were present on taped samples from clothes. These authors also found that shaking outdoor clothing before entering a residence removed 68% of pollen grains from trousers.

Boi (2015) studied the attachment of pollen grains to a range of common materials. Five anemophilous and one entomophilous types of pollen were transferred to cotton, denim, latex, leather, newspaper and polyester textile. Adhesion was strongest with fabric materials with an average of 275 pollen grains/cm^2.

Webb et al. (2018) also studied the retention of pollen grains on clothing and the effectiveness of retrieval methods in forensic settings, finding that there was a substantial difference in pollen retention according to wear intensity, with considerably more pollen being retained after light wear than after heavy wear. Pollen from insect-pollinated species was retained at higher concentrations than pollen from wind-pollinated species.

The efficiency of standard washing protocols for the recovery of pollen differed depending on the fabric type and plant species, and the interaction between these factors. Average efficiency was 67.7%, but ranged from a low of 21% to a high of 93%. These authors concluded that previous assumptions on the reliability of washing techniques to provide a representative sample for forensic use should be reviewed.

The primary purpose of most studies of pollen in the environment is, of course, not focussed on a forensic context but may still provide useful information in interpreting forensic scenarios. For example, a study by Hugg et al. (2007) provides information on the concentrations of pollen inside private cars during a pollen season, finding that the number of pollen grains on the inside surfaces of cars ranges from 0 to 72 grains/cm^2.

Several studies have looked at the transfer of pollen onto footwear. Horrocks et al. (1999) investigated variation in the pollen content of soil on shoes and in shoeprints in soil. Although noting the close similarity of pollen content of shoe samples with shoeprint samples, their findings also illustrated that 'perfect matches' between samples are neither found nor should be expected. A limitation of this study was that samples were taken from the shoes immediately after contact with the soil.

Wiltshire et al. (2014) point out that clothing and footwear can pick up palynomorphs from more than one place and that pollen and spores may be lost in varying amounts, so that a profile can be modified after an event and the initial transfer of palynomorphs. Hence, it is unrealistic to expect an assemblage obtained from a shoe or any other object to ever match any crime scene perfectly (Wiltshire, 2006b).

The importance of footwear being seized as soon as possible was further demonstrated in a study by Riding et al. (2007). Six rural sites were chosen with new unworn boots as well as previously worn boots to study potential mixing. Pollen adhering to footwear or in the soil samples from each site had a characteristic pollen signature. When mixing occurred, the dominant pollen content of the boots reflected that of the last site.

Nguyen and Weber (2016), in a follow-up study of indoor pollen (Nguyen & Weber, 2015), showed that pollen could be transferred to sneakers worn in a flat and correlated back to the pollen profile in that flat. After walking approximately 1.8 km, pollen could still be detected but in low numbers.

Adams-Groom (2018), in a blind study, was able to successfully determine which of 12 footwear types had been walked on a mock crime scene and to eliminate footwear that had not been exposed to the mock crime scene.

Finally, Gonzales et al. (2024) have also shown that it was possible to identify the geographic origin of pollen recovered from footwear in various non-seasonal high-altitude soils.

Wiltshire (2015) has been critical of non-palynologists using the discipline as a teaching tool and then publishing results that, in her words, 'almost invariably relate to simplistic experimental scenarios which are unlikely to occur in, or be relevant to, actual forensic situations'. One such study, cited by Wiltshire, is that of Morgan et al. (2014b), in which these authors reported on the preservation of pollen following a simulated vehicle fire. The authors of this paper, however, argue the need for peer-reviewed experiments to provide the basis for the interpretation of evidence presented to the court and claim only to this type of study, making a small contribution towards the acceptance of forensic palynology as evidence and an aid to a jury.

Most forensic science research relies on work carried out by students, and in this regard, palynology is no different. Of course, the best research is informed by real-life case work and experience. As Mildenhall et al. (2006), however, point out, an important way to gain acceptance is through public knowledge of new forensic disciplines. Hence, whilst all research should properly be the subject of critical review, and not claim more than it should, a greater interest by academics in forensic palynology surely must enhance public acceptance and contribute to meeting the demands of the CJS.

Earlier in this chapter, the importance of the involvement of a forensic palynologist at the earliest possible time was stressed. Even, if some thought has been given to the involvement of a forensic palynologist, there are simply very few forensic palynologists to call upon and who have the time to attend crime scenes. Responsibility for sample collection may rely on CSEs who are likely to have, at best, a basic knowledge of forensic palynology and, if they do collect control samples, they may not have the experience to develop an adequate sampling strategy. As Mildenhall et al. (2006) comment, 'on too many occasions subsequent investigations have determined that the scene of crime officers thought that any material from the ground would satisfy as a control'. The importance of control samples cannot be overstated, as samples collected at random are unlikely to answer the questions of importance to an investigator and, subsequently, if tendered as evidence to the CJS.

In conclusion, the detailed interpretation of palynological data is far from simple, and it is not a case of just reporting what palynomorphs are present. A perfect correspondence between control and recovered evidentiary material is a rare occurrence, and the role of the expert is to assess the quality of the correspondence. For example, how many pollen types correspond, are the percentages the same and, if not, can this be explained? How common are the palynomorphs? All of this requires professional judgement and considerable experience and proven ability on the part of the expert. As Wiltshire et al. (2014) state, 'the power of palynology is underlined by its ability to differentiate between sites, even if they are in close proximity'.

9.1.5 The Future

A recurring theme in this book has been the problem of loss of expertise in many areas of botany, with 'biologists' unable to recognise even common plants, a decline in botany students, faculty, courses, university departments and herbaria (Crisci et al., 2020; Stroud et al., 2022). We have already observed that it can take 10 years of postgraduate study to develop true expertise. Recognising the shortage of experts, the use of DNA analysis and/or the use of computational methods, machine learning or Artificial Intelligence (AI) are being proffered as possible solutions.

With respect to pollen analysis, Schield et al. (2016) studied the persistence of *Pinus echinate* on cotton fabrics and showed that pollen could persist for at least 14 days of wear. Using a portable vacuum handle for pollen collection, a high-throughput method for DNA extraction and a custom nine-locus multiplex STR system, partial or complete profiles were generated from single pollen grains.

Bell et al. (2016) reviewed the use of DNA barcoding methods in forensic palynology. According to these authors, the combination of technologies needed for successful DNA barcoding of mixed pollen samples had only just been realised, with four studies published.

A major limitation for DNA barcoding is the lack of a comprehensive database of taxonomic sequences. Whilst it can reasonably be expected that databases will continue to improve, it remains a current limitation. For pollen, there are also some specific issues. Standard plant DNA barcoding loci are on the plastid genome. It is unclear if plastid DNA is present in the pollen of all seed plant species. Furthermore, plastid DNA is typically reduced in cells as pollen develops, resulting in a low copy number that may cause issues with contamination from non-pollen plant material. Finally, as DNA analysis is destructive, it will be important to either split samples or first take microscopic images of pollen samples before lysing them for DNA analysis.

The authors concluded that whilst work was still needed on several technical issues, it was close to being feasible for forensic applications (Bell et al., 2016).

It would appear, however, that at the time of writing, despite the promise and recent advances, DNA barcoding has not yet had an impact on forensic pollen analysis. This may have less to do with the state of play with the technology and science, but rather may be the result of a lack of connectivity between forensic palynologists and DNA specialists working in other areas of palynology. For example, Banchi et al. (2020) have reviewed the use of DNA metabarcoding for aerobiology, focussing on pollen and fungal DNA. Van Haeften et al. (2024) report on the use of environmental DNA to study taxonomic diversity across seasons and climate zones for airborne grass pollen.

Pollen from the grass family (Poaceae) has a simple morphology, being generally spheroidal with a single pore surrounded by an annulus (Faegri & Iversen, 1989), and it has been suggested that pollen morphology is uniform within the family (Stromberg, 2011). For this reason, grass pollen is not normally identified below family level in routine palynological investigations. Palynologists have attempted to increase the information they can derive from grass pollen by measuring characteristics such as pollen grain length, grain width, pore diameter, annulus width and the organisation of the exine (Mander et al., 2013). Furthermore, SEM studies have shown that there is a diversity of surface ornamentation that is not visible with light microscopy.

Mander et al. (2013) have developed an approach to the identification of grass pollen based on a combination of high-resolution SEM microscopy and *computational image analysis* that provides an accuracy of 77.5%. Seven human subjects achieved accuracy levels between 68.3% and 81.75%, but consistency between subjects was low, with only 28.3% of specimens classified by all subjects. This is worrying as a major criticism of forensic feature-based disciplines has been the lack of consistency at the level of the individual analyst and between analysts (Dror, 2023).

A more in-depth study by Mander et al. (2014) measured the ability of nine analysts to identify 12 species of grass pollen using SEM images. Mean identification accuracy ranged from 46.7% to 87.5%, with consistency of each analyst ranging from 32.5% to 87.5%. Each of the analysts used different identification schemes. Unsurprisingly, but also worrying, is the author's conclusion that experts need regular exposure to pollen morphology to maintain their levels of expertise. This suggests that perhaps palynologists are not having regular exposure and, with it, a loss of expertise. If human experts will still be required to check identifications generated by computational methods, where will the experts be found?

Finally, there have been several studies to develop automatic identification methods for pollen. Li et al. (2024) list 17 references in the past 10 years, with most published in the last 5 years, dealing with the development of deep learning and computational methods for the automatic identification of pollen. None are in forensic journals! It is outside of the scope of this chapter to attempt to explain the rapidly emerging world of AI, deep learning, machine learning and all the other nomenclature used in this world. Suffice to say that forensic science will not be immune to this world and forensic applications are already emerging (Airlie et al., 2023). These authors showed that machine learning could deliver very high accuracy for simple applications, or narrow AI, but could not achieve acceptable levels of identification for more complex tasks used in the forensic examination of hair. This is likely to be the situation for the use of machine learning for pollen identification. It may be possible to achieve high levels of identification when the task is to identify only a limited range of pollen types. Forensic palynology, however, involves the identification of often complex assemblages of pollen and spores. It seems unlikely that machine learning can be trained to meet this level of complexity, and, of course, identification is only the first step in the role of a forensic palynologist, as the real value lies in the interpretation of the variables in case situations (Wiltshire, 2006b).

9.2 Forensic Phycology

9.2.1 Introduction

'Few objects are more beautiful than the minute siliceous cases of the diatomaceae: were these created that they might be examined and admired under the higher powers of the microscope?' (Darwin, 1872).

Phycology is the scientific study of algae. Described as 'the world's most important plants' (Chapman, 2013), they photosynthesise and produce 50% of all oxygen on earth. It is no exaggeration to say that life on Earth depends on algae! The major groups of algae include *big algae* or green, red and brown algae, and *small algae*, including diatoms, the primary focus of this section. It is outside the scope of this chapter to consider phycology in any detail. Van den Hoek et al. (1995) and Lee (2008) provide excellent textbooks on algae.

There are 12,000 known species of diatoms, with estimates that there may be as many as 60,000 to 600,00 species (Hasle & Syvertsen, 1997). Given that there are at least another 12,000–15,000 'big' algae and that algae are found in fresh and marine water and on land (on tree trunks, palings, leaves, roofs, walls, soils and any other damp situations), one could be forgiven for expecting that these species would be of interest to forensic scientists. Indeed in 1978, the Biology Methods Manual of the Metropolitan Police Forensic Science Laboratory (Anon, 1978) devoted eight pages to algae! Peabody (1977, 1980) has reviewed diatoms in forensic science and in drowning. The use of diatoms in drowning is the subject of its own section in this chapter.

A major use of diatoms, at least into the 1970s, was in cases of safe breaking as diatomaceous earth was used as an insulator in safe ballast during the early part of the last century. The largest deposit of diatomaceous earth, or kieselguhr, is the Lompoc area of California, dating from 25 million years ago. It is estimated that one cubic inch of diatomite contains between 25 and 50 million cells (Peabody, 1977). Peabody (1977) reported that about half of the diatomaceous safe ballast examined in his laboratory appeared to have a common origin resembling diatomaceous deposits in northern Germany. The dry powdery nature of diatomaceous earth is such that any person forcibly opening a safe containing this material could not avoid transfer of some material onto clothing, and Peabody describes a case where hundreds of diatoms were found in a smear of a suspect in a safe breaking.

Diatoms are also found in other materials such as polishes, matches and paints. The original use of diatoms was as an inert carrier for nitroglycerine in the manufacture of dynamite (Peabody, 1977).

If forensic palynology suffers from being underutilised, then this is even more the situation for any broad application of phycology in a forensic context. As with many areas of botany, interest and experience in phycology have diminished, or have entirely disappeared, from forensic laboratories as the DNA revolution emerged in the mid to late 1980s. Hence, despite it being a common occurrence for deceased persons to be found near water, or in water, unless a forensic botanist with a broad understanding of ecology attends a scene, it is highly unlikely that any thought will be given to the potential for algae to contribute to traditional forensic roles such as establishing links between people, places and event reconstruction.

Hardy and Wallace (2012), in a review of the broader forensic applications for algae, concluded that diatoms were the most used algae due to their abundance in many different habitats. Like pollen, diatoms are also very small, and at least as single cells, cannot be seen with the naked eye. They are likely to be difficult to remove as a contact trace. Diatoms are also hard and decay resistant, are highly persistent and retain species-specific diagnostic characteristics long after the actual cells have dried, died and even after passing through internal organs or digestive tracts of recovered bodies (Hardy & Wallace, 2012).

Scott et al. (2017, 2021) have studied the persistence of diatoms on clothing in simulated case scenarios. They found that diatoms adhere to clothing even after prolonged exposure to fire, indicating that it is worthwhile attempting to collect diatoms after attempts to destroy evidence through burning (Scott et al., 2017). This same group also showed that smaller diatoms (under 10 μm) were retained in significantly greater abundance with no difference between centric and pennate diatoms. A persistent species assembly was relatively stable over a period of one month of wear, but there was a significant difference between seasonal samples. The authors concluded that the complex relationships between clothing type, environmental seasonality and time since wear retention emphasised the need for diatom trace evidence to be carefully interpreted within an exclusionary framework (Scott et al., 2021).

An example of linking suspects to the scene is provided by Silver et al. (1994). In this well-documented case, sediment-encrusted sneakers from two victims and their alleged assailants were analysed for aquatic microorganisms. Numerous species of diatoms and scaled chrysophytes (planktonic algae) were recovered from the sneakers and reference samples of pond sediment from the alleged scene of the assault. Based on marked similarities between the samples, the authors concluded that the samples had originated from a common, if not the same, locality.

Hardy and Wallace (2012) also relate a case study in which a filamentous alga found on a suspect's T-shirt linked him to a brick found in a nearby pond, which the prosecution alleged was used to murder the suspect's wife, the prosecution's case being that the husband had rinsed his T-shirt in the pond where the brick had been discarded.

Casamatta and Verb (2000) studied algal colonisation of rats to explore the potential value of algal colonisation to determine the time of death. Over a 17-day period, the diversity of colonisation in both low and high rates of water flow scenarios had increased. Although diatoms were the most abundant taxa, accounting for 63 of the total of 92 taxa identified, desmids (Chlorophyta) tended to increase in diversity throughout the study, making them possible indicators of submersion time. Hardy and Wallace (2012) discuss a few other studies of a similar nature but conclude that there are too many unanswered questions for this technique to be applied in casework.

Hence, despite the potential for algae to be a useful tool for forensic investigations, the only broadly used algae are diatoms and today, almost only in cases of drowning.

9.2.2 Classification and Identification of Diatoms

Diatoms belong to the class Bacillariophyceae and are further subdivided into two orders, the Biddulphiales and the Bacillariales. This subdivision is based on the organisation or structure of the cells and several other differences.

The Biddulphiales, or Centrales, are primarily a marine planktonic group with radial or gonoid ornamentation and no raphe.

The Bacillariales, or Pennales, contain both freshwater and marine species with pennate or trellisoid ornamentation and often a raphe present.

Diatoms are unicellular cells surrounded by a rigid two-part box-like cell wall composed of silica, called a *frustule*. They can take the shape of ribbons, fans, zigzags or stars with individual cells ranging in size from 2 to 200 μm.

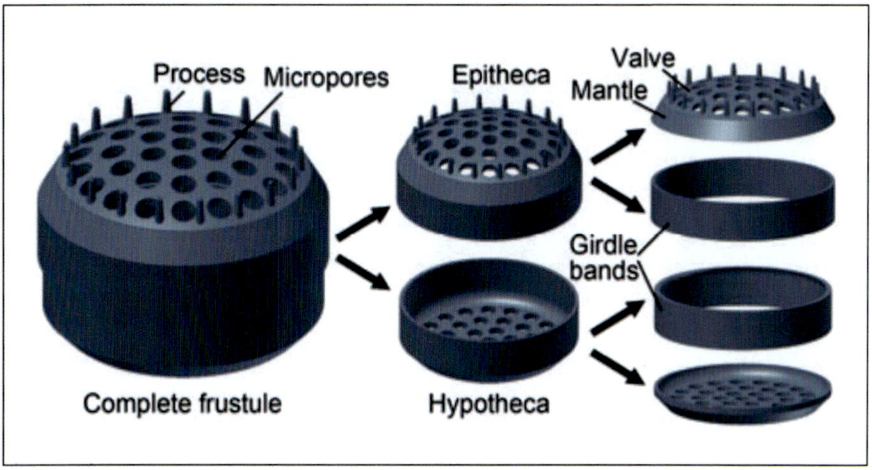

Figure 9.4 Schematic of the structure of diatom frustules.

(Creative Commons CCO 1.0 Universal Public Domain Dedication.)

Diatoms are divided into two groups based on the shape of the frustule, *centric* and *pennate* diatoms. Centric diatoms are radially symmetric, whilst pennate diatoms are bilaterally symmetric, with each one of their valves having openings that are slits along the raphes, with their shells typically being elongate parallel to these raphes. Raphes are a longitudinal slit in the theca.

Diatoms only live for a maximum of about six days, but they have regular annual fluctuations in populations due to changing environmental conditions. Living diatoms have a central nucleus suspended in a system of protoplasmic threads with chloroplasts occupying most of the cell (Lee, 2008). It is the cell wall, however, that is the characteristic feature of diatoms. It is constructed of two almost equal halves, the smaller fitting into the larger like a petri dish – see Figure 9.4.

The outer of the two half walls is called the *epitheca* and the inner the *hypotheca*. Each theca is composed of two parts: the *valve*, a flattened plate, and the *connecting* band attached at the edge of the valve. The two connecting bands, one attached to each valve, are called the *girdle band*. The girdle bands often have minute teeth, which hold the girdle bands together by their edges. The walls are porous, and they can be highly patterned with a variety of pores, ribs, minute spines, marginal ridges and elevations, which can be used to distinguish genera and species. Most diatom valves are covered in tiny pores called *punctae*, which are usually organised into lines across the valve face called *striae* (Gell et al., 1999).

Henday (1964) recognises four basic types of ornamentation:

- *Centric* or *radial*, where the structure is arranged according to a central point
- *Trellisoid*, where the structure is arranged uniformly over the surface without reference to a point or line
- *Gonoid*, where the structure is dominated by angles
- *Pennate*, where the structure is symmetrically arranged upon either side of a central line

Peabody (1977) advises that the determination of diatom species should only be attempted by experienced 'diatomatologists', whilst Hardy and Wallace (2012) recommend that police departments and other investigative agencies identify institutions that are willing and able to assist before they are needed. The challenge may be in finding an institution or individuals who are willing to become involved in criminal investigations and the CJS in general. It is preferable if the relationship is developed between partner institutions or individuals and the forensic institute, as they can assist in ensuring forensic requirements and standards are met. Partner institutions are likely to be involved in water quality analysis or a similar field of endeavour. Given the vast number of diatom species, local knowledge is vital, as are local databases and identification keys.

Standard reference works cited in Peabody (1977) include the work of Hustedt. These works are in German and are difficult to acquire. Other useful sources include Henday (1964) for diatoms of British coastal waters and Patrick and Reimer (1966) for North America. Hardy and Wallace (2012) recommend Palmer (1959), stating that the illustrations and keys are sufficient to match algal flora from a piece of evidence to a crime scene. More recent publications include Tomas (1996), Cox (1997) and Gell et al. (1999) for diatoms from Southern Australia, and Taylor et al. (2007) for diatoms from South Africa.

The UK Environment Agency also developed three interactive CD-ROMs to identify >300 diatom taxa of freshwater algae, although sadly these are no longer available (Anon, 2007). An interactive web-based resource for diatoms of North America is available at https://diatoms.org. This list of resources is not intended to be comprehensive, and there will no doubt be many more regional resources and, increasingly, online resources.

As was the case with palynology, however useful the above resources may be, they are not a substitute for reference collections with rigorous provenance and comparison against real samples.

9.2.3 Methods and Protocols

Chessman et al. (1999) have published a draft protocol for sampling and subsequent laboratory processing of diatoms from streams and surrounding land surfaces with a focus on bioassessment and biomonitoring of streams.

Hardy and Wallace (2012) discuss forensic sampling strategies in some detail. Items to be sampled will, of course, depend on the circumstances of the case and are certainly not limited to sampling streams, rivers or marine water environments. As previously discussed, algae may be present on a plethora of surfaces, transferred to suspects, victims and vehicles, to name just a few examples. As far as possible, standard forensic recovery techniques should be followed, although scrapings will probably play a larger role than for most other trace materials. Sampling in cases of drowning will be considered at Section 9.2.4. The goal of any sampling method is, of course, to ensure that the samples are representative of any variation, they are of a size that is adequate for subsequent laboratory examinations and cross-over of contamination between samples is avoided. Whilst the procedures followed by non-forensic specialists will meet many of the requirements for forensic applications, there are differences. This is again where a cooperative relationship with forensic specialists can ensure that forensic protocols are being followed.

Laboratory processing of samples involves three basic steps:

- concentration of the algae from the sample
- clearing organic matter and other debris to enable a clear view of the diatom frustules
- mounting of the cleared diatoms for microscopic examination

Peabody (1977) lists the following protocol to achieve the above aims:

1. Remove calcium carbonate and other carbonates with conc. HCl. Wash and discard supernatant
2. Add concentrated sulphuric acid to the residue. Boil until the suspension turns black
3. Allow to cool to about 50°C and add solid sodium nitrate. Reheat the suspension until it turns brown and finally clears.
4. Thoroughly wash the resultant suspension of silica diatom cells in diatom-free water and resuspend in acetone

Spot the diatom suspension in acetone onto a size 0 coverslip and dry. Place a small drop of Naphrax on a slide on a hotplate at about 50°C to melt the mountant. Place the coverslip on the mountant and heat until bubbles cease.

An alternative method is, after concentrating samples by sedimentation, to treat the sediment with hydrogen peroxide in a test tube in a water bath at 60–80°C for at least 2 hours. Following clearing, the material is allowed to settle, the supernatant hydrogen peroxide is removed, the sediment is resuspended in high-grade ethanol and the process is repeated twice. After a final rinse in ethanol, the sediment can be stored ready for mounting (Chessman et al., 1999). These authors also recommend the use of *Naphrax* as a mountant. This is based on diatoms having a refractive index (RI) of 1.44. To obtain sufficient contrast, the mountant should have a higher RI. Naphrax has an RI of 1.74 (Peabody, 1977).

Newer approaches to the extraction of diatoms have included the use of enzymatic digestion, the use of Soluene-350 and microwave digestion (Vijayan et al., 2020). The Soluene method of digestion is only suitable for freshwater diatoms, whilst the proteinase K method was found to be the most effective, with the benefit that it also preserved the microscopic features of diatom frustules for SEM analysis (Ming et al., 2007). An enzymatic approach does take longer than traditional acid digestion, but it is both safer and environmentally friendlier.

The microwave method still uses nitric acid and peroxide but has several advantages over the traditional method. It is safer, quicker and provides clean samples. Vacuum filtration replaces centrifugation and gives a maximum diatom reclaiming ratio, improving detection sensitivity. Finally, as samples are collected on a membrane filter, these can be directly placed onto adhesive carbon tape on SEM stubs (Hu et al., 2013).

Samples are then typically examined using bright field and/or phase contrast light microscopy with magnifications up to ×1,000. There are limitations even at this high magnification when examining the smaller diatoms under 10 µm, which has led to the use of SEM to reveal finer detail in the ornamentation of diatoms (Morales et al., 2001; Bortolotti et al., 2004: Scott et al., 2017). Salimon et al. (2020) have produced a mini atlas of SEM images of diatom frustules. Hagen et al. (2023) have validated and optimised a quantitative and automated SEM-based approach to examine diatoms.

9.2.4 Diatoms and Drowning

The recognition of death by drowning is a major medicolegal problem (Timperman, 1972; Piette & De Letter, 2006). According to Stephenson et al. (2019), the first problem encountered in the diagnosis of drowning is in reaching a consensus on the definition of drowning. In general, it refers to death within 24 hours of a submersion incident in a liquid medium. Often, the diagnosis is one of excluding other causes. Given these difficulties, it would be useful if there were laboratory tests that helped confirm that drowning had occurred. One such test that has been extensively studied and reviewed is the presence of diatoms.

There is no questioning that diatoms can be found in persons who have drowned, but the question is whether the presence of diatoms indicates that drowning was the cause of death. Byard (2015) concluded that laboratory testing has been generally disappointing in the evaluation of possible drowning deaths, with the use of diatoms not having been of much use in most cases.

Timperman (1972) described the fate of diatoms entering a body when a person drowns. In drowning, water is inhaled under much pressure into the lungs, leading to rupturing of the peripheral alveoli, and thus, water and other contents pass into the bloodstream. Due to their high abundance and small size, diatoms settle into the organs of drowned persons. Sea water drowning brings about haemoconcentration and pulmonary oedema, whereas fresh water drowning leads to haemodilution, hypervolemia and ventricular fibrillation.

The early literature dealing with diatoms as indicators of drowning is reviewed by Peabody (1980), with the first association between diatoms and the diagnosis of drowning being attributed to Ravenstorf (1904). A troubling finding that lies at the very credibility of the diatom method is that diatoms have been found in organs of non-drowned deceased persons. In a study of 14 deceased persons, diatoms were found in six bodies, but only with a total of nine diatoms overall (Lunetta et al., 2013). This may be explained by inhalation and/or ingestion of diatoms. A third source, however, may be through contamination of glassware and reagents used in the recovery and extraction of diatoms (Peabody, 1980).

Diatoms may be found in various organs, including the lung, liver, kidney and bone marrow. Lin et al. (2014) have suggested the use of sphenoid sinus fluid in cases of suspected drowning. Lungs and stomach contents are unsuitable in drowning diagnosis, as diatoms may occur in the lungs through means other than drowning, and stomach contents may contain diatoms from water swallowed during drowning or other sources, such as foodstuffs. For example, it has been estimated that 50 g of cleaned and cooked prawns contain between 8,000 and 29,000 diatoms! (Yen & Jayaprakash, 2007).

For the above reasons, *bone marrow* is the preferred source to examine, usually taken from the femur. As bone marrow has a high fat content, it is useful to defat first with petroleum ether (Peabody, 1977). The number of diatoms in bone marrow is typically very small. The absence, however, of diatoms in bone marrow does not mean that drowning did not occur, as drowning may have occurred in water with very few or no diatoms, such as tap water.

For example, Pollanen et al. (1996) reviewed 771 cases of drowning in Ontario, Canada. Diatoms were recovered from femoral bone marrow in 28% of freshwater drownings and in only 12% of domestic water drownings. Typically, a maximum of ten frustules were found, with less than six in most cases. A correlation was found between seasonal diatom blooms and the presence of diatoms. In a second study, Pollanen (1998) found that in 47 cases from a total of 52 case studies, the diatoms recovered were representative of the site of drowning, with small pennate frustules the most common diatoms associated with drowning.

Rapid death may also prevent diatoms from circulating through the bloodstream and reaching the bone marrow (Piette & De Letter, 2006). Ago et al. (2011), in one case of death whilst bathing due to ischemic heart disease, found no diatoms in the organs except for the lower lobe of the right lung and the left kidney.

Given the small number of diatoms that may be recovered, rigorous and standardised procedures need to be followed to demonstrate that any diatoms found are not the result of contamination (Pollanen, 1998; Lunetta et al., 2013). Even when great care is taken, contamination may still occur (Ago et al., 2011).

Nonetheless, Timperman (1972) concluded that the results of the diatom test can be considered as decisive in determining drowning when used critically and in experienced hands.

More recent reviewers have also concluded that the reliability of the diatom test has been strengthened by modern research and can be relied upon to establish a definite diagnosis of death by drowning (Caine, 2021).

The diatom test is particularly useful in cases where a body has been dismembered or has decomposed, leaving only skeletal remains (Gruspier & Pollanen, 2000). The diatom test is also the only method remaining when resuscitation has been attempted, with the proviso that there will occasionally be doubtful results that will then have to be interpreted with necessary and due caution.

Additional information above and beyond a diagnosis of drowning is to attempt to determine the location where the drowning occurred. A body or body parts may not be located where the drowning occurred, either due to movement in a body of water or because a perpetrator has moved the body. For example, Bak et al. (2018) discuss a case where an analysis of diatoms supported the proposition that a body had been moved from the original site of a water tank to where her body was discovered in a nearby river.

How accurate a determination can be made depends on several factors, including an understanding of the biodiversity of the potential water source. In a study of diatom diversity in the Douro River estuary in Portugal, Coelho et al. (2016) found that there was no significant difference in five sampling sites, making it impossible to determine the exact location of drownings. The importance of creating a diatom database of potential drowning places and the collection of water samples from drowning places is crucial.

Interpretation of diatoms may also be influenced by the previous history of the deceased person. Taylor (1994) discusses a case where the individual had swum every day for 15 years in the same body of water, making it impossible to determine if diatoms were the result of drowning in that location.

9.2.5 The Future

As with palynology and many other aspects covered in this book, the future will include the use of DNA analysis and automatic identification using deep or machine learning. With respect to DNA analysis, the forensic utility of this approach using metabarcoding technology is limited due to the lack of a reliable diatom DNA barcode reference sequence. DNA will also only be applicable for living diatoms that have not been subjected to harsh extraction conditions.

Liu et al. (2020) found that DNA identification results did not correlate well with microscopic identifications and overall performed worse than microscopic identifications. Jiang et al. (2020), however, using an 18rRNA gene array covering 168 diatom species, found that in a wide range of fresh waters in China, the diatom arrays showed better sensitivity and resulted in more consistent results than with SEM methods. They concluded that the SEM approach lacked sensitivity due to a high rate of loss of frustules in the sample preparation process.

The 100 Diatom Genome Project aims to sequence 100 diatom genomes in the next three years with 20 collaborators from different countries (Mock, 2021).

Automated diatom identification has been around for some time with software packages such as the *Automated Diatom Identification and Classification* (ADIAC) project (De Buf et al., 1997) and the *Shape Recognition Processing and Analysis*, based on the ADIAC database, using an image processing workflow focussed on the identification and measurement of diatom outlines such as shape and segmentation (Kloster et al., 2014). More recently, Puccinelli et al. (2019) have published a patented rapid method for the identification of diatom species based on digitised flow charts that guide users in the classification of diatoms.

Increasingly, papers are now emerging applying deep learning and AI to the automated identification of diatoms (Yu et al., 2021; Lambert & Green, 2020).

At the time of writing, however, the identification of diatoms for forensic purposes still relies heavily on experts and expertise.

References

Adams-Groom, B. 2018. Assessment of Pollen Assemblages on Footwear for Evidence of Pollen Deriving from a Mock Crime Scene: A Contribution to Forensic Palynology. *Grana*, 57, 223–234.

Ago, K., Hayashi, T., Ago, M., & Ogata, M. 2011. The Number of Diatoms Recovered from the Lungs and Other Organs in Drowning Deaths in Bathwater. *Legal Medicine*, 13, 186–190.

Airlie, M., Robertson, J., & Brooks, E. 2023. An Evaluation of Novice, Expert and Supervised Machine Learning Model Classifications for Forensic Hair Analysis. *Australian Journal of Forensic Sciences*. https://doi.org/10.1080/00450618.2023.2254337

Anon. 1982. *Biology Methods Manual*. Metropolitan Police Forensic Science Laboratory (MPFSL). In house publication.

Anon. 2007. *Common Freshwater Diatoms of Britain and Ireland: An Interactive Identification Key*. https://www.gov.uk/governement/publications/common-freshwater-diatoms-of-britain-and-ireland-an-intercative-key

Bak, M., Witkowski, A., Dobosz, S., Kociolek, J.P., & Pietras, L. 2018. Diatoms and Drowning: Can Diatoms Enter the Lungs After the Initial Drowning Event? *Romanian Journal of Legal Medicine*, 26, 29–36.

Banchi, E., Pallavicini, A., & Muggia, L. 2020. Relevance of Plant and Fungal DNA Metabarcoding in Aerobiology. *Aerobiology*, 36, 9–23.

Bell, K.L., Burgess, K.S., Okamoto, K.C., Aranda, R., & Brosi, B.J. 2016. Review and Future Prospects for DNA Barcoding Methods in Forensic Palynology. *Forensic Science International: Genetics*, 21, 110–116.

Bock, J.H., & Norris, D.O. 2016. *Forensic Plant Science*. London: Elsevier.

Boi, M. 2015. Pollen Attachment in Common Materials. *Aerobiology*, 31, 261–270.

Bortolotti, F., Tagliaro, F., & Manetto, G. 2004. Objective Diagnosis of Drowning by the 'Diatom Test' – A critical Review. *Forensic Science Review*, 16, 135–148.

Brown, A.G. 2006. The Use of Forensic Botany and Geology in War Crimes Investigations in NE Bosnia. *Forensic Science International*, 163, 204–210.

Brown, A.G., Smith A., & Elmhurst, O. 2002. The Combined Use of Pollen and Soil Analysis in a Search and Subsequent Murder Investigation. *Journal of Forensic Sciences*, 47, 614–618.

Bruce, R.G., & Dettmann, M.E. 1996. Palynological Analysis of Australian Surface Soils and Their Potential in Forensic Science. *Forensic Science International*, 81, 77–94.

Bryant, V.M., & Jones, G.D. 2006. Forensic Palynology: Current Status of a Rarely Used Technique in the United States of America. *Forensic Science International*, 163, 183–197.

Bryant, V.M., & Wrenn, J.H. (eds.). 1998. *New Developments in Palynomorph Sampling, Extraction, and Analysis*. Dallas: American Association of Stratigraphic Palynologists Foundation. Contributions Series, 33.

Byard, R.W. 2015. Immersion Deaths and Drowning: Issues Arising in the Investigation of Bodies Recovered from Water. *Forensic Science, Medicine, and Pathology*, 11, 323–325.

Caine, J. 2021. Diatom Analysis: Reviewing the Strengths, Weaknesses, and Impacts of Modern research. *CJUR*, 6, 11–15.

Casamatta, D.A., & Verb, R.G. 2000. Algal Colonization of Submerged Carcasses in a Mid-Order Woodland Stream. *Journal of Forensic Sciences*, 45, 1280–1285.

Chan, T., Robinson, G., Liu, J., Kurti, M., He, Y., & Von Lampe, K. 2020. Identifying Counterfeit Cigarettes Using Environmental Pollen Analysis: An Improved Procedure. *Journal of Forensic Sciences*. https://doi.org/10.1111/1556-4029.14540

Chapman, R.L. 2013. Algae: The World's Most Important 'Plants' - An Introduction. *Mitigation and Adaptation Strategies for Global Change*, 18, 5–12.

Chessman, B., Gell, P., Newall, P., & Sonneman, J. 1999. Draft Protocol; for Sampling and Laboratory Processing of Diatoms for the Monitoring and Assessment of Streams. Appendix 1, In Gell, P.A., Sonneman, J.A., Ried, M.A., Illman M.A., & Sincock, A.J. (eds.), *An Illustrated Key to the Common Diatom Genera from Southern Australia.* CRC Press. CRC for Freshwater Ecology Identification Guide No. 26.

Coelho, S., Ramos, P., Ribeiro, C., Marques, J., & Santos, A. 2016. Contribution to the Determination of the Place of death by Drowning – A Study of Diatoms' Biodiversity in the Douro River estuary. *Journal of Forensic and Legal Medicine*, 41, 58–64.

Cox, E.J. 1997. *The Identification of Freshwater Diatoms from Live Material.* London: Chapman & Hall.

Crisci, J.V., Katinas, L., Apodaca, M.J., & Hoch, P.C. 2020. The End of Botany. *Trends in Plant Science*, 25, 1173–1176.

Darwin, C. 1872. *On the Origin of the Species by Means of Natural Selection, or Preservation of Favoured Races in the Struggle for Life*, 6th ed., 502pp. London: John Murray.

De Buf, H., Bunke, H., Droop, S., Roerdink, S., Juggins, S., Cristobal, G., & Ludes. B. 1997. *Automatic Diatom Identification and Classification (ADIAC). Proceedings of the 3rd European Marine Science and Technology Conference*, Lisbon, Portugal, Vol. 3, pp. 1071–1074.

Donaldson, M.P., & Stephens, W.E. 2010. Environmental Pollen Trapped by Tobacco Leaf as Indicators of the Provenance of Counterfeit Cigarette Products; A preliminary Investigation and Test of Concept. *Journal of Forensic Sciences*, 55, 738–741.

Dror, I. 2023. The Most Consistent Finding in Forensic Science is Inconsistency. *Journal of Forensic Sciences*, 68, 1851–1855.

Erdtman, G. 1952. *Pollen Morphology and Plant Taxonomy. Angiosperms.* Stockholm: Almqvist & Wiksell.

Erdtman, G. 1957. *Pollen and Spore Morphology. Plant Taxonomy. Gymnospermae. Pteridophyta. Bryophyta.* Stockholm: Almqvist & Wiksell.

Ericsson, K.A., & Lehmann, A.C. 1996. Expert and Exceptional Performance: Evidence of Maximal Adaptation to Task Constraints. *Review of Psychology*, 47, 273–305.

Faegri, K. & Iversen, J. 1989. *Textbook of Pollen Analysis*, 4th ed. Chichester: John Wiley & Sons.

Garcia-Mozo, H., Beltran-Aroca, C.M., Badu, I.K., Jimena, I., & Girela-Lopez, E. 2021. Airborne Pollen and Spores' Deposition in Alveolar Tissues as a Tool in Drowning Forensic Diagnosis. *Aerobiology*, 37, 281–291.

Gell, P.A., Sonneman, J.A., Ried, M.A., Illman, M.A., & Sincock, A.J. 1999. *An Illustrated Key to Common Diatom Genera form Southern Australia.* CRC Press. CRC for Freshwater Ecology Identification Guide No. 26.

Gonzales, V.S., Limarino, T.O., & Duchen, P. 2024. Use of Pollen Assemblages as Forensic Evidence in Non-Seasonal High-Altitude Soils. *Science & Justice*, 64, 73–80.

Gruspier, K.L., & Pollanen, M.S. 2000. Limbs Found in Water: An Investigation Using Anthropological Analysis and the Diatom Test. *Forensic Science International*, 112, 1–9.

Hagen, D., Pittner, S., Zhao, J., et al. 2023. Validation and Optimization of the Diatom L/D Ratio as a Diagnostic Marker for Drowning. *International Journal of Legal Medicine*, 137, 939–948.

Halbritter, H., Weber, M., Zetter, R., Frosch-Radivo, A., Buchner, R., & Hesse, M. 2008. *PalDat – An Illustrated Handbook on Pollen Terminology*. https://www.paldat.org/Wein

Hardy, C.R., & Wallace, J.R. 2012. Algae in Forensic Investigations. In Hall, D.W., & Byrd, J.H. (eds.), *Forensic Botany. A Practical Guide* (Chapter 9, pp. 145–173). Chichester: Wiley-Blackwell.

Hasle, G., & Syvertsen, E.E. 1997. *Marine Diatoms*. In Thomas, C.R. (ed.), *Identifying Marine Diatoms and Dinoflagellates* (pp. 5–385). New York: Academic Press.

Havinga, A.J. 1984. A 20 Year Experimental Investigation into the Differential Corrosion Susceptibility of Pollen and Spores in Various Soil Types. *Pollen et Spores*, 26, 541–558.

Hawksworth, D.L., Wiltshire, P.E.J., & Webb, J.A. 2016. Rarely Reported Fungal Spores and Structures: An Overlooked Source of Probative Trace Evidence in Criminal Investigations. *Forensic Science International*, 264, 41–46.

Henday, N.I. 1964. *An Introductory Account of the Smaller Algae of British Coastal Waters. Part V. Bacillariophyceae (Diatoms)*. Fisheries Investigations, Series IV. London: Ministry of Agriculture, Fisheries and Food. Pub, Her Majesty's Stationary Office (H.M.S.O).

Hesse, M., Halbritter, H., Zetter, R., Weber, M., Buchner, R., Frosch-Radivo, A., & Ulrich, S. 2009. *Pollen Terminology. An Illustrated Handbook*. New York: Springer Wein.

Horrocks, M., Coulson, S.A., & Walsh, K.A.J. 1999. Forensic Palynology: Variation in the Pollen Content of Soil on Shoes and in Shoeprints in Soil. *Journal of Forensic Sciences*, 44, 119–122.

Horrocks, M., & Walsh, K.A.J. 1999. Fine Resolution of Pollen Patterns in Limited Space: Differentiating a Crime Scene and Alibi Scene Seven Meters Apart. *Journal of Forensic Sciences*, 44, 417–420.

Horrocks, M., & Walsh, K.A.J. 2001. Pollen on Grass Clippings: Putting the Suspect at the Scene of the Crime. *Journal of Forensic Sciences*, 46, 947–949.

Hu, S., Liu, C., Wen, J., Dai, W., Wang, S., Sy, S., & Zhao, J. 2013. Detection of Diatoms in Water and Tissues by Combination of Microwave Digestion, Vacuum Filtration and Scanning Electron Microscopy. *Forensic Science International*, 226, e48–e51.

Hugg, T., Valtonen, A., & Rantio-Lehtimaki, A. 2007. Pollen Concentrations Inside Private Cars During the Poaceae and *Artemisia* spp. Pollen Season – A Case Study. *Grana*, 46, 110–117.

Hyde, H.A., & Williams, D.W. 1944. The Right Word. *Pollen Analysis Circular*, 8, 6.

Jantunen, J., & Saarinen, K. 2011. Pollen Transport by Clothes. *Aerobiology*, 27, 339–343.

Jiang, L., Xiao, C., Zhao, J., Jiang, T., Lin, J., Xu, Q., & Liu, C. 2020. Development of 18SrRNA Gene Arrays for the Forensic Detection of Diatoms. *Forensic Science International*, 317, 110482.

Jones, G.D., & Bryant, V.M. 1998. Are All Counts Equal? In: Bryant, V.M., & Wrenn, J.H. (eds.), *New Developments in Palyynomorph Sampling, Extrcatio, and Analysis* (vol. 33, pp. 115–120). Dallas: American Association of Stratigraphic Playnologists Contributions Series.

Jones, G.D., & Bryant, V.M. 2007. A Comparison of Pollen Counts: Light Verus Scanning Electron Microscopy. *Grana*, 46, 20–33.

Kloster, M., Kauer, G., & Beszteri, B. 2014. An Image Segmentation and Outline Feature Extraction Tool for Diatoms and Other Objects. *BMC Bioinformatics*, 15, 218.

Lambert, D., & Green, R. 2020. *Automatic Identification of Diatom Morphology Using Deep Learning. Proceedings of the 35th International Conference on Image and Vision Computing New Zealand (IVCNZ).* Wellington, New Zealand.

Lee, R.E. (2008). *Phycology*, 4th ed. Cambridge: Cambridge University Press.

Li, J., Cheng, W., Xu, X., Zhao, L., Liu, S., Gao, Z., Ye, C., & You, H. 2024. How to Identify Pollen Like a Palynologist: A Prior Knowledge-guided Deep Feature Learning for Real-world Pollen Classification. *Expert Systems With Applications*, 237, 121392.

Lin, C., Yen, W., Hsieh, H., et al. 2014. Diatomological Investigation in Sphenoid Sinus Fluid and Lung Tissue from Cases of Suspected Drowning. *Forensic Science International*, 244, 111–115.

Liu, M., Zhao, Y., Sun, Y., Li, Y., Wu, P., & Zhou, S. 2020. Comparative Study of Diatom Morphology and Molecular Identification in Drowning Cases. *Forensic Science International*, 317, 110552.

Locard, E. 1930. The Analysis of Dust Traces. *American Journal of Political Science*, 1, 401–418.

Lunetta, P., Miettinen, A., Spilling, K., & Sajantila, A. 2013. False-Positive Diatom Teas: A Real Challenge? A Post-Mortem Study Using Standardized Protocols. *Legal Medicine*, 15, 229–234.

Mander, L., Baker, S.J., Belcher, C.M., et al. 2014. Accuracy and Consistency of Grass Pollen Identification by Human Analysis Using Electron Micrographs of Surface Ornamentation. *Applications in Plant Sciences*, 2, 1400031.

Mander, L., Li, M., Mio, W., Fowlkes, C.C., & Punyasena, S.W. 2013. Classification of Grass Pollen Through the Quantitative Analysis of Surface Ornamentation and Texture. *Proceedings of the Royal Society B*, 280, 20131905.

Mathewes, R.W. 2006. Forensic Palynology in Canada: An Overview with Emphasis on Archaeology and Anthropology. *Forensic Science International*, 163, 198–203.

Mildenhall, D.C. 2004. An Example of the Use of Forensic Palynology in Assessing an Alibi. *Journal of Forensic Sciences*, 49, 312–316.

Mildenhall, D.C., Wiltshire, P.E.J., & Bryant, V.M. 2006. Forensic Palynology: Why Do It and How It Works. *Forensic Science International*, 163, 163–172.

Milne, L.A., Bryant, V.M., & Mildenhall, D.C. 2005, Forensic Palynology. In Coyle, H.M. (ed.), *Forensic Botany. Principles and Applications to Criminal Case Work* (Chapter 14, pp. 217–252). Boca Raton: CRC Press.

Ming, M., Meng, X., & Wang, E. 2007. Evaluation of Four Digestive Methods for Extracting Diatoms. *Forensic Science International*, 170, 29–34.

Mock, T. 2021. *Embarking on the '100Diatom Genomes Project'.* https://jgi.doe.gov/csp-2021-100-diatoms-genomes. Accessed 26 March 2024.

Moore, P.D., Webb, J.A., & Collinson, M.E. 1991. *Pollen Analysis*, 2nd ed. London: Blackwell Scientific.

Morales, E.A., Siver, P.A., & Trainor, F.R. 2001. Identification of Diatoms (Bacillariophyceae) During Ecological Assessments: Comparison Between Light Microscopy and Scanning Electron Microscopy Techniques. *Proceedings of the Academy of Natural Sciences of Philadelphia*, 151, 95–103.

Morgan, R.M., Allen, E., King, T., & Bull, P.A. 2014a. The Spatial and Temporal Distribution of Pollen in a Room: Forensic Implications. *Science & Justice*, 54, 49–56.

Morgan, R.M., Flynn, J., Sena, V., & Bull, P.A. 2014b. Experimental Forensic Studies of the Preservation of Polle in Vehicle Fires. *Science & Justice*, 54, 141–145.

Nguyen, P., & Weber, M. 2015. Forensic Value of Pollen from Ornamental Indoor Plants. *Grana*, 54, 236–246.

Nguyen, P., & Weber, M. 2016. Can Pollen Match Shoes to a Previously Visited Indoor Location? *Grana*, 55, 164–172.

Palenik, S. 1982. Microscopic Trace Evidence – the Overlooked Clue: Part II, Max Frei – Sherlock Holmes with a Microscope. *Microscope*, 30, 163–168.

Palmer, C.M. 1959. *Algae in Water Supplies: An Illustrated Manual on the Identification, Significance, and Control of Algae in Water Supplies*. Washington, DC: US Dept. Health, Educ. & Welfare Pub., Health. Ser. Pub., no. 657.

Patrick, R, & Reimer, C. 1966. *The Diatoms of the United States*. Philadelphia: Livingston Pub. Co.

Peabody, A.J. 1977. Diatoms in Forensic Science. *Journal of the Forensic Science Society*, 17, 81–87.

Peabody, A.J. 1980. Diatoms and Drowning - A review. *Medicine, Science, and the Law*, 20, 254–261.

Phuphumirat, W., Mildenhall, D.C., & Purintavaragul, C. 2009. Pollen Deterioration in a Tropical Surface Soil and its Impact on Forensic Palynology. *The Open Forensic Science Journal*, 2, 34–40.

Piette, M.H.A., & De Letter, E.A. 2006. Drowning: Still a Difficult Autopsy Diagnosis. *Forensic Science International*, 163, 1–9.

Pollanen, M.S. 1996. The Diagnostic Value of the Diatom Test for Drowning, II. Validity: Analysis of Diatoms in Bone Marrow and Drowning Medium. *Journal of Forensic Sciences*, 42, 286–290.

Pollanen, M.S. 1998. Diatoms and Homicide. *Forensic Science International*, 91, 29–34.

Pollanen, M.S., Cheung, C., & Chiasson, D.A. 1996. The Diagnostic Value of the Diatom Test for Drowning, I. Utility: A Retrospective Analysis of 771 Cases of Drowning in Ontario, Canada. *Journal of Forensic Sciences*, 42, 281–285.

Puccinelli, C., Marcheggiani, S., & Mancini, L. 2019. A Patented Rapid Method for Identification of Italian Diatom Species. *International Journal of Environmental Research and Public Health*, 16, 3933.

Punt, W.P., Hoen, P., Blackmore, S., Nilsson, S., & Le Thomas, A. 2007. Glossary of Pollen and Spore terminology. *Review of Palaeobotany and Palynology*, 143, 1–81.

Ravenstorf, V. 1904. Der Nachweis der Aspirierten Ertankungs Flussigkeit als Kriterium des Todes Durch Ertrinken. *Vierteljahresschrift für Gerichtliche Medizin und Öffentliches Sanitätswesen*, 28, 274–279.

Riding, J.B., Rawlins, B.G., & Coley, K.H. 2007. Changes in Soil Pollen Assemblages on Footwear Worn at Different Sites. *Palynology*, 31, 135–151.

Salimon, A.I., Sapozhnikov, P.V., Everaerts, J., et al. 2020. A Mini-Atlas of Diatom Frustule Electron Microscopy Images at Different Magnifications. *Materials Today Proceedings*, 33, 1924–1933.

Sandiford, A. 2012. Palynology, Pollen, and Spores, Partners in Crime: What, Why, and How. In Hall, D.W., & Byrd, J.H. (eds.), *Forensic Botany. A Practical Guide* (Chapter 8, pp. 127–144). Oxford: Wiley-Blackwell.

Schield, C., Campelli, C., Sycalik, J., Randle, C., Hughes-Stamm, S., & Gangitano, D. 2016. Identification and Persistence of Pinus Pollen DNA on Cotton Fabrics: A Forensic Application. *Science & Justice*, 56, 29–34.

Scott, K.R., Jones, V.J., Cameron, N.G., Young, J.M., & Morgan, R.M. 2021. Freshwater Diatom Persistence on Clothing II: Further Analysis of Species Assemblage Dynamics over Investigative Timescales. *Forensic Science International*, 326, 110897.

Scott, K.R., Morgan, R.M., Jones, V.J., Dudley, A., Cameron, N., & Bull, P.A. 2017. The Value of an Empirical Approach for the Assessment of Diatoms as Environmental Trace Evidence in Forensic Limnology. *Archaeological and Environmental Forensic Science*, 1, 49–78.

Silver, P.A., Lord, W.D., & McCarthy, D.J. 1994. Forensic Limnology: The Use of Freshwater Algal Community Ecology to Link Suspects to an Aquatic Crime Scene in Southern New England. *Journal of Forensic Sciences*, 39, 847–853.

Simpson, M.G. 2019. *Plant Systematics*, 3rd ed. Amsterdam: Elsevier.

Stanley, E.A. 1992. Application of Palynology to Establish the Provenance and travel History of Illicit Drugs. *Microscope*, 40, 140–152.

Stephenson, L., Van Den Heuvel, C., & Byard, R.W. 2019. The Persistent Problem of Drowning- A Difficult Diagnosis with Inconclusive Tests. *Journal of Forensic and Legal Medicine*, 66, 79–85.

Stevenson, J., Haberle, S.G., Johnston, F.H., & Bowman, D.M.J.S. 2007. Seasonal Distribution of Pollen in the Atmosphere of Darwin, Tropical Australia: Preliminary Results. *Grana*, 46, 34–42.

Stromberg, C.A.E. 2011. Evolution of Grasses and Grassland Ecosystems. *Annual Review of Earth and Planetary Sciences*, 39, 517–544.

Stroud, S., Fennell, M., Mitchley, J., Lyndon, S., Peacock, J., & Bacon, K.L. 2022. The Botanical Education Extinction and the Fall of Plant Awareness. *Ecology and Evolution* https://doi.org/10.1002/ece3.9019

Taylor, J. 1994. Diatoms and Drowning- A Cautionary Note. *Medicine, Science, and the Law*, 34, 78–79.

Taylor, J.C., Harding, W.R., & Archibald, C.G.M. 2007. *An Illustrated Guide to Some Common Diatom Species from South Africa*. Water Research Commission, Report TT 282/07.

Timperman, J. 1972. The Diagnosis of Drowning. A Review. *Forensic Science*, 1, 397–409.

Tomas, C.R. 1996. *Marine Diatoms and Dinoflagellates*. San Diego: Academic Press.

Van Den Hoek, C., Mann, D.G., & Johns, H.N. 1995. *Algae. An Introduction to Phycology*. Cambridge: Cambridge University Press.

Van Haeften, S., Campbell, B.C., Milic, A., Addison-Smith, E., Al Kouba, J., Huete, A., Beggs, P.J., & Davies, J.M. 2024. Environmental DNA Analysis of Airborne Poaceae (grass) Pollen Reveals Taxonomic Diversity Across Seasons and Climate Zones. *Environmental Research*, 247. https://doi.org/10.1016/j.envres.2023.117983

Vijayan, A., Kallumpurat, A., & Christal, L.G. 2020. Diatoms: A Review on Its Forensic Significance. *Journal of Forensic Dental Sciences*, 12, 197–200.
Walsh, K.A.J., & Horrocks, M. 2008. Palynology: Its Position in the Field of Forensic Science. *Journal of Forensic Sciences*, 53, 1053–1060.
Webb, J.C., Brown, H.A., Toms, H., & Goodenough, A.E. 2018. Differential Retention of Pollen Grains on Clothing and the Effectiveness of Laboratory Retrieval Methods in Forensic Settings. *Forensic Science International*, 288, 36–45.
Wiltshire, P.E.J. 2006a. Hair as a Source of Forensic Evidence in Murder Investigations. *Forensic Science International*, 163, 241–248.
Wiltshire, P.E.J. 2006b. Consideration of Some Taphonomic Variables of Relevance to Forensic Palynological Investigation in the United Kingdom. *Forensic Science International*, 163, 173–182.
Wiltshire, P.E.J. 2015. Protocols for Forensic Palynology. *Palynology* https://doi.org/10.1080/01916122.2015.1091138
Wiltshire, P.E.J., & Black, S. 2006. The Cribriform Approach to the Retrieval of Palynological Evidence from the Turbinates of Murder Victims. *Forensic Science International*, 163, 224–230.
Wiltshire, P.E.J., Hawksworth, D.L., Webb, J.A., & Edwards, K.J. 2014. Palynology and Mycology provide Separate Classes of Probative Evidence from the Same Forensic Samples: A Rape Case from Southern England. *Forensic Science International*, 244, 186–195.
Wiltshire, P.J., Hawksworth, D.L., & Edwards, K.J. 2015. A Rapid and Efficient Method for Evaluation of Suspect Testimony: Palynological Scanning. *Journal of Forensic Sciences*, 60, 1441–1450.
Yen, L.Y., & Jayaprakash, P.T. 2007. Prevalence of Diatom Frustules in Non-vegetarian Foodstuffs and its Implications in Interpreting Identification of Diatom Frustules in Drowning Cases. *Forensic Science International*, 170, 1–7.
Yu, W., Zue, Y., Knoops, R., et al. 2021. Automated Diatom Searching in the Digital Scanning Electron Microscopy Images of Drowning Cases Using the Deep Neural Networks. *International Journal of Legal Medicine*, 135, 497–508.

Grass Family 10

10.1 Introduction

Grasses are monocots (one cotyledon) within the Angiosperms, belonging to the Order Poales and the family Poaceae, also known as the Gramineae. With an estimated 780 genera and over 12,000 species, the Poaceae is the fourth or fifth largest Angiosperm family (Christenhusz & Bying, 2016).

Grasses are incredibly important, playing a central role in human and animal nutrition, accounting for about 50% of all human dietary energy, with maize, wheat, rice, barley, millet and sorghum the main cereal grains. Included in the grass family are also sugar cane and bamboo. The contribution of cereal grains has already been considered in Chapter 8, dealing with starch present in the endosperm of grains.

Grasses may be native or introduced and may become invasive weeds. Grasses are vital in agriculture as animal fodder, either grazed *in situ* or harvested. Grasslands such as savannah, steppes and prairie are estimated to constitute around 40% of the land area of the Earth (Reynolds, 2016). Grasses are also found in sports fields, golf courses, gardens, roadsides, salt marshes, indeed almost anywhere one could imagine, including Antarctica, where a grass species, *Deschampsia antarctica*, is one of only two plant species native to the western Antarctic peninsula.

Given that grasses are so widespread, almost ubiquitous, and, as we shall see, produce seeds that are almost designed to attach themselves to humans and persist on clothing or inside vehicles, etc., the dearth of reported forensic cases where grasses have been a key factor is somewhat surprising. As covered in Chapter 9, an exception is cases including grass pollen.

Could this be because they are seen as ubiquitous, of less value than some other 'rarer' species, or is this yet another example of the underappreciated and underutilised resource that is forensic botany?

The decision to include a chapter on grasses in this book stems to some extent from my early studies in agricultural botany. I recall spending time as a student collecting faeces for studies into the grazing habits of sheep and being introduced to the analysis of epidermal fragments to identify plant fragments (see Martin, 1964). Later, as a lecturer in forensic science, I included grasses in my forensic botany lectures. Before the emergence of DNA technologies, I published a study on the use of isoelectrophoresis of storage proteins to identify grass seeds of interest (Carroll et al., 1983) that later led to the

DOI: 10.4324/9781003308508-10

development of a molecular identification system for grasses as DNA technologies developed (Ward et al., 2005, 2009). Hence, I have had a long-standing interest in grasses as a potential forensic trace material.

10.2 Classification of Grasses

The evolutionary relationships within and between taxonomic levels of grasses, or the phylogeny of grasses, is an ever-evolving space. As a student in the late 1960s, only six subfamilies were recognised. As late as 1982, the Royal Botanic Garden, Kew, still listed five subfamilies (Anon, 1980). Today, largely because of advances in the use of DNA analysis, 12 subfamilies are recognised. In alphabetic order, these are the Anomochlooideae, Arisidoideae, Arundinoideae, Bambusoideae, Chloridoideae, Danthonioideae, Ehrhartoideae (=Oryzoideae), Micrairoideae, Panicoideae, Pharoideae, Pooideae and the Puelioideae (GPWG, 2001). A total of 52 tribes and 768 genera are recognised within these families (Soreng et al., 2017). For a detailed treatment of the taxonomy and phylogeny of the grass family, the reader is referred to Jacobs and Everett (2000) and Kellogg (2002).

10.3 Morphology

The mature 'grain' in grasses has a fused pericarp and testa and is an elongated *sheroid* or *caryopsis*. Botanically, a grain is a fruit, but in commerce is referred to as a seed. The fruit may also have protective structures called *pales* (*lemma* and *palea*), either loosely enclosing the fruit, as in oats, or fused to the fruit, as in barley.

The first formed root system, the *seminal* root system, arises from the embryo and does not persist for long before new roots are produced. The main *coronal* root system is adventitious and typically fibrous. Once the first leaves reach the soil surface, they pierce the plumule and emerge to give a young seedling.

Except for the bamboos, grasses are herbaceous plants and most grow in a tufted form with each plant consisting of several leafy shoots or *tillers* clustered together at the base. Shoots may be fully erect, ascending or semi-prostrate. A robust, densely tufted plant is commonly referred to as a *tussock*. Tussock-forming plants are common in the arid inland of Australia, alpine areas and temperate grasslands (McCusker, 2002). Figure 10.1 shows a typical grass plant in diagrammatic format.

Many species have the tillers arising, usually singly, at the nodes of horizontal stems. These *stolons* or *rhizomes* form dense mats or long runners and

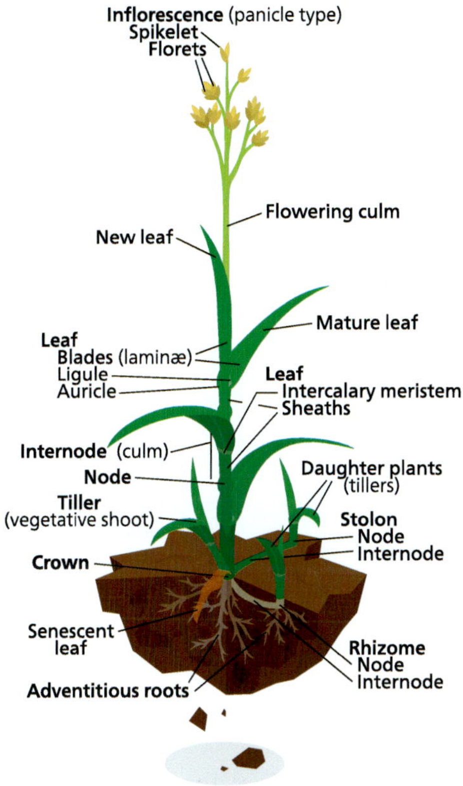

Figure 10.1 Schematic of the morphology of a typical grass plant.

(Creative Commons Attribution 3.0 unported license.)

form roots at some of the nodes. Many grasses also form upright stems (*culms*) as part of the vegetative plant. Culms usually have solid nodes and hollow internodes. Exceptions to the latter are found in sugar cane, where the nodes are pithy, and in rice and other aquatic grass species, where the internodes are *aerenchymatous*.

Grass leaves are borne singly, one at each node. Leaves in grasses are almost always alternate and *distichous* (in one lane). They are quite different to the leaves of dicots. They do not have petioles, comprising a basal cylindrical *sheath* and an upper *lamina* or blade, which is typically elongate or lanceolate with parallel veination. Successive leaves arise from the sheathing of previous leaves. The sheath may be entire (unsplit) or split.

At the area where the blade and sheath join on the adaxial surface, there is often an outgrowth of tissue called the *ligule* (see Figure 10.2). This may consist of a row of hairs or teeth, a fringed member or a distinct tongue-like flap of tissue. In split sheath grasses, there are often outgrowths from the edge

Figure 10.2 *Avena fatua*, wild oat, showing an example of a ligule.

(Creative Commons – Share Alike 2.0 France license. Author: Portas, M.)

of the sheath/blade area called *auricles*, the shape and size of which are characteristic of the species.

Environmental factors such as photoperiod determine when a grass plant will flower. The apice develops floral primordia, and at the same time, there is a rapid elongation of the main axis to form the stem or *culm*. The last leaf before the inflorescence is called the *flag leaf*.

Grass inflorescences are terminal *racemes*, that is, having indefinite growth with no terminal spikelet. The basic form of grass inflorescence is a *panicle* with secondary and often higher orders of branching of the primary axis. In some grasses, however, the inflorescence has a single axis, and this is called a *spike*. In a spike, the spikelets are attached to the stem by short stalks or pedicels. Where branches are short and crowded together, the panicle is narrow and dense and is called a *spike-like* panicle.

The degree of branching in panicles, however, is highly variable and is a useful taxonomic character. McCusker (2002) describes the variations seen in open panicles. The ultimate branches bear several sessile or pedicellate spikelets, which contain small and inconspicuous flowers.

The *spikelet* is a unit found only in grasses, in the family Cyperaceae and some Restionaceae. Understanding the internal components of this unit is central to the identification of grasses. A typical spikelet consists of two basal *glumes* and one or more *florets* borne in an alternative arrangement in two

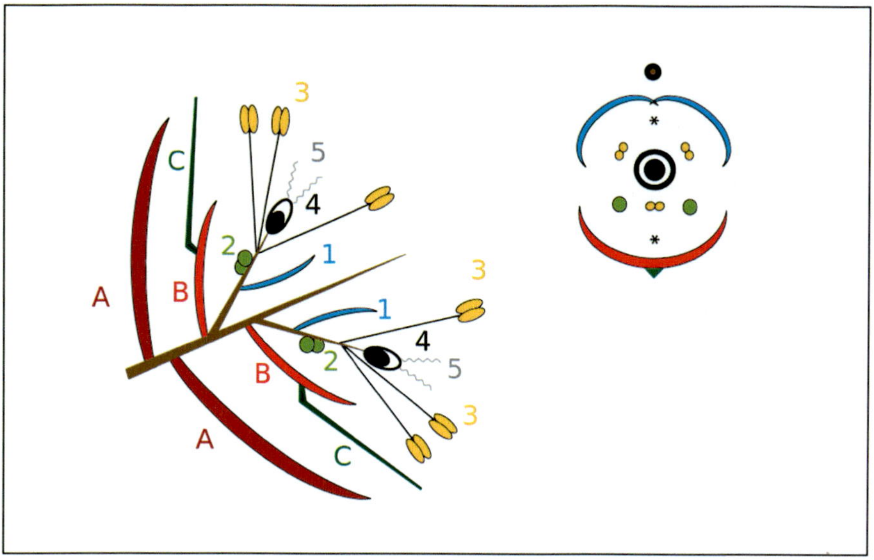

Figure 10.3 Schematic of a typical grass inflorescence.

(Public domain. Author: Grasbluete.png.)

rows on the axis of the spikelet, known as the *rachilla* (see Figure 10.3). The rachilla is bent alternately in opposite directions (flexuose or zig-zag).

Glumes are reduced leaves or bracts. The outer bract is called the *lemma*, and the inner bract the *palea*. The lemma usually encloses the other parts and typically has a single keel running down the middle of the back surface. When immature, lemmas are often green and membranous, becoming harder and less green as they mature. There is considerable variation in size, shape, colour, texture, compression, hairiness and presence of appendages such as *awns*. Awns are long, slender appendages attached to the tip, back or base of a glume (see Figure 10.4).

At the base of many florets is a hardened structure, often sharply pointed, that develops at maturity and can be clearly seen when the floret separates from the rachilla. This structure, the *callus*, is useful in seed dispersal and to penetrate soil. The callus can have less useful characteristics, causing blindness in lambs and damage to animal hides.

Paleas are usually two-keeled and two-nerved and are softer and less variable in structure than lemmas. In the floret, a flat side lies towards the rachilla whilst the keels touch the sides of the lemma, thus enclosing the flower. Paleas of rice and their relatives only partly enclose their lemmas.

The flowers of grasses lack sepals and petals with two or three small structures at the base of the ovary, called *lodicules*, representing a reduced perianth. Lodicules play a role in the opening of the flower at anthesis when they rapidly become turgid and force the separation of lemma and palea for the

Figure 10.4 Avena fatua, wild oat, showing elongated awns.

exertion of stigmas and anthers, supporting wind pollination. Stebbins (1956) recognised four types of lodicules.

Attached to the base of the flower are one to six stamens, with three stamens being common. These are attached by long, delicate filaments, allowing the anther to swing with the wind, enabling pollen to be shed in all directions.

The single ovary contains a single loculus or chamber with one ovule attached near the base on the palea side. At the apex of the ovary are two lateral styles bearing hairy stigmas.

Most grasses are bisexual, but some grass species are *dioecious* (male and female parts are on separate plants) and others are monoecious (male and female parts on different parts of the same plant – maize is an example where the female components are axillary inflorescences and the male flowers are terminal). Due to the variable number of female fertile florets in the spikelet, there is considerable variation in propagule structure.

Figure 10.5 shows two examples of important cereal grasses, barley and wheat.

Members of two other families are often confused with grasses, these being the sedges (Cyperaceae) and the rushes (Juncaceae) (Anon, 1980). The major differences in these three families are as follows:

Grasses: stems round and usually hollow; leaves 2-ranked; flowers, subtended by two bracts, the lemma and palea; perianth, reduced to two small lodicules and fruit, a one-celled caryopsis

Sedges: stems, triangular and solid; leaves, three-ranked; flowers, subtended by one bract; perianth, reduced to scales, bristles or absent and fruit, one-celled nut

Rushes: stems, round, hollow or solid; leaves, in a basal tuft; flowers, no bracts; perianth, six segments in two whorls and fruit, one- to three-celled capsule

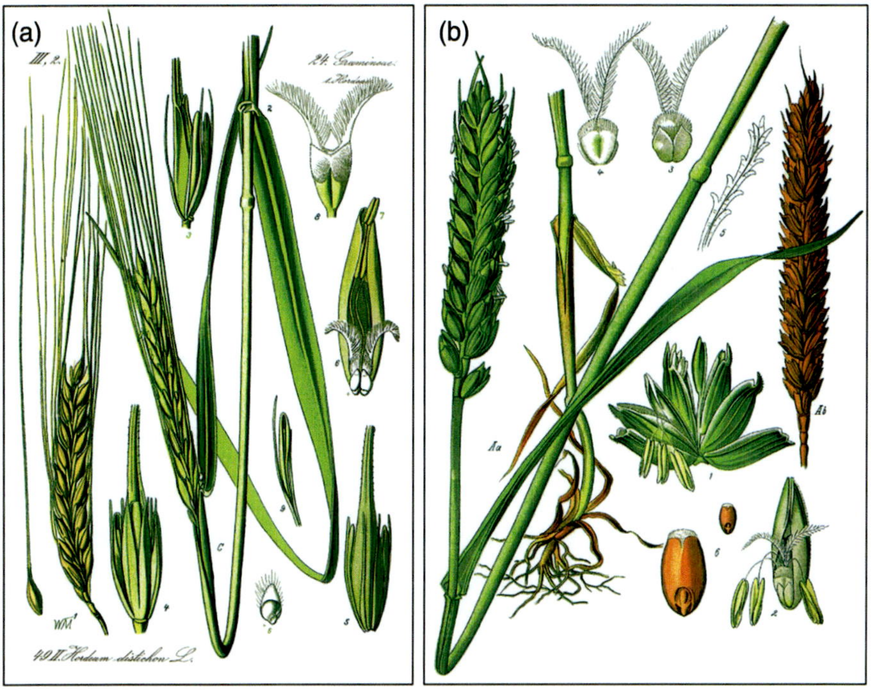

Figure 10.5 Illustration of two common cereal grasses. (a) *Hordeum distichon*, barley. (b) *Triticum aestivum*, wheat.

(Public domain.)

10.4 Leaf Anatomy

The following discussion of grass leaf anatomy draws on two major publications, those of Metcalfe (1960) and Renvoize (2002).

According to Renvoize (2002), most anatomical studies have focussed on the leaf blade due to the amount of variation with specific characteristics identified and corresponding to major taxonomic groups. This short introduction will, hence, focus on the leaf epidermis.

The blade portion of grass leaves is usually elongate. Descriptions of leaf anatomy are based on the long axis of the leaf lying horizontally in the field of view when examined under a microscope. Thus visualised, for descriptive purposes, the zones and files of cells that lie parallel to the long axis of the leaf can be referred to as 'horizontal' and those at right angles to the long axis as 'vertical' (Metcalfe, 1960).

Two zones are usually visible, the *costal* zone, where epidermal cells are over the underlying veins, and the *intercostal* zone, where the cells between the veins overlie the mesophyll. The two main types of cells in the epidermis are called *long cells* and *short cells*.

In *long cells*, the long axis is elongated horizontally parallel with the long axis of the blade. Variation in long cells includes length, having straight or wavy walls/sinuous walls. Straight cells usually have thin walls, with wavy walls having comparatively thicker walls.

A further cell type, called a *bulliform cell*, is sometimes found in the centre intercostal zone. These cells are found in files interrupting the long cells. Bulliform cells always have straight walls and are usually equilateral or broadly oblong and shorter than long cells.

The epidermis over the veins usually consists of files of long cells interspersed with files of short cells, which may or may not contain *silica bodies*.

Short cells usually appear equilateral. In the intercostal zone, short cells occur singly, occasionally paired or in short rows and are generally unspecialised cells. The exception to this generalisation is when short cells are filled with silica bodies, although silica bodies are not exclusively restricted to short cells (Metcalfe, 1960).

Silica bodies assume very characteristic shapes and are of considerable value for diagnostic and taxonomic purposes (Wynn Parry & Smithson, 1964; Mulholland & Rapp, 1992). According to Mulholland and Rapp (1992), grass silica bodies conform to three general geometric shapes: a truncated or bevelled pyramid, a short cylinder or cone shape and a saddle-like shape. From these basic shapes, these authors developed a classification scheme and a key to the major types of grass silica bodies based on eight silica body shape types.

When grasses decompose in soils, these highly resistant silica bodies are deposited in soils and are referred to as *phytoliths* (Smithson, 1958).

Stomata occur in the intercostal zone of both the adaxial and abaxial surfaces. They are arranged in files, usually alternating with oblong interstomatal cells, which may or may not overlap with the guard cells. Each stomatal band may have one or more rows of stomata, with the number of rows varying from one grass to another and sometimes in different parts of a single blade or in leaves taken from different levels of the plant (Metcalfe, 1960). Stomata vary in size from 15 to 50 μm long and have a characteristic appearance depending on the shape of the two guard cells, which may be domed, triangular or parallel-sided. Grass stomata, when fully mature, can be classified according to the shape of their subsidiary cells and are of value as a diagnostic feature and as a taxonomic character.

Finally, surface appendages are often present. Unlike with dicots, glandular hairs are rare and are only found in the subfamily Chloridoideae, tribe Pappophoreaea (Renvoize, 2002). Dermal appendages can be classified as *macro-hairs*, *micro-hairs*, *prickle-hairs* and *papillae*.

Macro-hairs are normally unicellular and may occur in the intercostal zones, but are more common over the veins or at the leaf margins. Often these hairs form an indumentum on the grass leaf, which is visible to the naked eye. When the base of the hair is sunken and surrounded by epidermal cells that are larger and more inflated, these hairs are referred to as *cushion hairs*. Intercostal macro-hairs are often cushion hairs or at least have a sunken base (Metcalfe, 1960). Metcalfe (1960) recognises some 13 types of macro-hairs based on length, frequency, thickness of cell wall and the extent to which they are rigid and straight or flexible and bent.

Micro-hairs differ from macro-hairs in being very much smaller and mostly two-celled. Metcalfe (1960), however, points out that it may not be obvious that there are two cells because the thin-walled distal cell is easily damaged or may be somewhat obscured in outline. The wall of the basal cell is nearly always much thicker and more durable than that of the distal cell. One-celled micro-hairs occur in species of the genus *Sporobolus*. Micro-hairs occur most commonly either in the stomatal bands or in those parts of the intercostal zones that lie between the stomatal bands and the veins. Metcalfe (1960) describes nine subcategories of micro-hair, with Bambusiod micro-hairs being worthy of a separate grouping.

Prickle-hairs, as the name implies, are robust, sharply and shortly pointed structures with swollen bases. Their swollen bases form an integral part of the epidermis. They generally point upwards in the direction of the apex of the leaf and have thickened lignified cell walls. Prickle-hairs vary in size, shape, the appearance of the base and length of the barb. Like prickle-hairs, but recognised as a separate entity, are called *hooks*. The base of hooks is generally more rounded than the bases of prickle-hairs. It should be noted that there are also intermediates between prickles and hooks. Prickle-hairs are more common above the veins with hooks between them. Angular, strongly pointed prickle-hairs are common on the leaf margins.

Finally, *papillae* are variously shaped protrusions from the outer walls of the epidermal cells. Papillae occur mostly on the long cells, particularly in the intercostal zones, with one or more in a single cell. For descriptive purposes, Metcalfe (1960) recognises six subcategories of papillae.

The identification of grasses based on the above epidermal features requires attention to very fine detail within the various categories. For example, Ellis (1979) recognises about 340 characters of the epidermis with 'ample room for expansion'.

10.5 Grass Identification

There are numerous books and online resources providing descriptions and keys that can be used to describe and identify grasses. Most of these resources are aimed at identifying grasses in the vegetative and flowering growth stages.

A number of useful references, but by no means the only resources, include Hubbard (1968), dealing with grasses in the British Isles; Wheeler et al. (1990), grasses of New South Wales, Australia, a major three-part series in the Flora of Australia (Mallett, & Orchard, 2002; Mallett, 2005; Wilson, 2009) and Hitchcock (1971) on grasses of the United States. Irish members of the Poaceae are described in a series of six monographs (Farragher, 1972a, 1972b, 1973a, 1973b, 1974, 1975).

Metcalfe (1960) is the most comprehensive resource with respect to the surface features of grass epidermises. In general, epidermal features are assessed with standard transmitted light microscopy, although this can be challenging as grass leaf epidermises are not easily peeled off and samples may have to be treated using leaching, staining and scraping techniques (Metcalfe, 1960; Barkworth, 1981). Hilu and Randall (1984) describe the use of clear nail polish to create an impression of the epidermis, although this is just a variation on an older paper by North (1956), where cellulose acetate films were used. Making impressions has advantages for forensic samples in that it is essentially non-destructive.

An alternative approach is to use scanning electron microscopy (SEM) (Palmer & Tucker, 1981). SEM produces visually very attractive images (see Figure 10.6), but it is unlikely to be the method of choice for forensic samples for the reasons discussed in previous chapters.

Originally available as a CD (Sharp & Simon, 2002), AusGrass is now a free online tool covering some 1,323 Australian grasses, which uses either

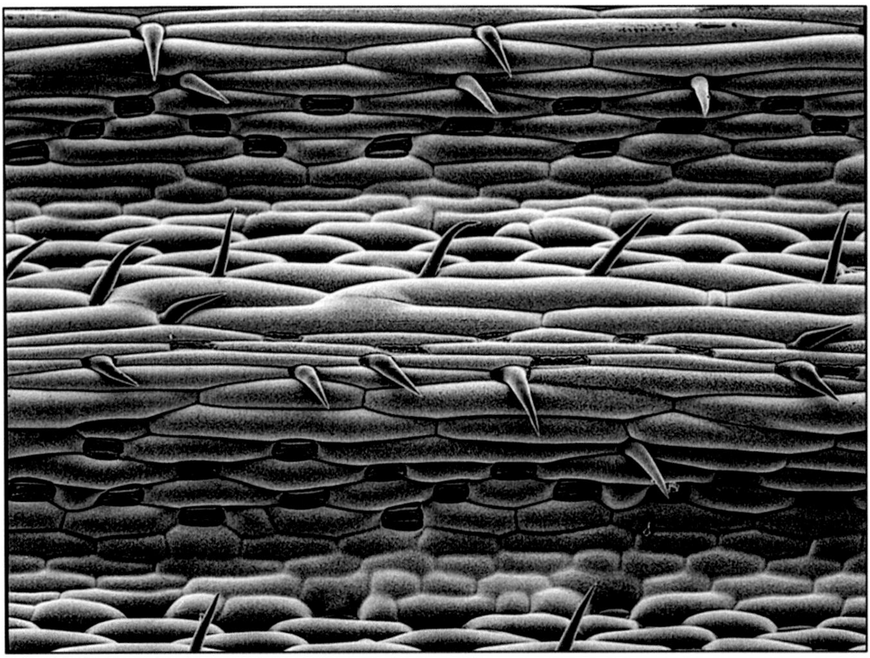

Figure 10.6 SEM image of epidermis of *Festuca arundinacea*, tall fescue.

interactive or dichotomous keys to identify an unknown specimen of grass irrespective of whether the specimen is in flower or fruit. Keys are based on 226 characters. The user is first guided through an initial 24 'best and simple' characters, which are considered generally easy to score and which have strong discriminating power. Identification solely based on vegetative features should be used with care due to their plastic nature, but within this group, ligule features are reasonably reliable. Whilst this resource is undoubtedly useful, the authors caution users not to expect that they will end up with a single taxon.

GrassBase is described as the online world grass database (Clayton et al., 2006), and this also provides the 'basic kernel' for Grassworld, which incorporates further information (see https://grassworld.myspecies.info, accessed 29 April 2024).

In a forensic context, the use of this tool or indeed any other tool does not substitute for direct comparison of any questioned material with voucher specimens from a herbarium. Typically, forensic samples will either be highly compromised or limited in size and features.

Many national or major botanical gardens and research institutes offer identification services, including the world-renowned Kew Gardens.

10.6 Future Potential for the Use of DNA Analysis

As has been discussed, DNA studies in the last two decades have revolutionised the classification of grasses. Without making any claim to be a comprehensive list of studies of DNA for grasses, the following studies give some insights into the approaches being developed and some applications.

Peterson et al. (2014) reported on steps to produce a standard DNA barcode library for grasses in North America to facilitate easy identification of small, vegetative leaf or culm samples of grass. Four barcodes were included in this study: *matK*, *rbcL*, *rpl32-trnL* and internal transcriber space (ITS). Over 50 taxa, the discriminatory power of ITS was found to be 96%, three times greater than *rpl32-trnL* and 33 times greater than *matK*. These authors concluded that the ITS region is indispensable, especially when used in tandem with *rpl32-trnL*.

In a study of six candidate DNA barcode loci for the identification of five invasive grass species, Wang et al. (2017) found that two plastid loci (*alp*F and *matK*) used as a dual locus could distinguish most of the taxa examined. In their study, low polymerase chain reaction success rates limited the value of two nuclear loci (ETS and ITS).

Birch et al. (2017) studied 408 individuals representing 94% of the grasses in the tribe Poeae found in Australia. Using the internationally accepted plant barcoding markers of *rbcl*, *matK* and ITS, the success rate for specimen identification of genera was between 87.5% and 99.5%, but only 25.6% and 44.6% at the species level. The ITS marker achieved the highest success rate for specimen identification at both the generic and species levels. The authors concluded that the standard DNA barcode markers are not variable enough for accurate identification at the species rank.

Finally, Wang et al. (2022) screened four candidate plant DNA barcoding regions (*rbcl*, *matK*, *trnl*-F and ITS) for seven genera and 14 accessions. *MatK* and ITS outperformed the other two loci and, in this limited study, were able to identify the seven genera and showed the highest species discriminatory power.

At the current time, identification of grasses using DNA analysis can only be reliably achieved at the level of genus, with *mat*K and the ITS region being the loci of choice.

I am unable to find any recent study of grass DNA with a specific forensic focus. To date, the only forensic studies of grass DNA remain Ward et al. (2005, 2009). These authors developed a molecular identification system based on indel and single-nucleotide polymorphism variation at multiple taxonomic levels and evaluated its strengths and limitations. A minimum of four markers were required to identify a sample to the subfamily level and nine markers for species identification. The accuracy of the system was confirmed with blind testing. Regrettably, this proof-of-concept study was never operationalised.

The most realistic scenario for the future use of DNA to identify grass species undoubtedly lies in the application of a barcoding approach, but the current reality falls well short of meeting the requirements for forensic application, one of which would be rigorous validation for forensic samples. Academic studies would not typically deal with the often compromised or small samples encountered in forensic examinations.

DNA analysis should not be viewed as a replacement for microscopic examination of morphological and anatomical features, but may become an additional tool in appropriate circumstances.

It is likely that the use of machine learning/artificial intelligence will also find some role in the future examination and identification of grasses, but, again, this is someway off for forensic application and should not be viewed as a replacement for microscopic examinations.

References

Anon. 1980. *Grasses. A Guide to Their Biology and Classification*. London: Her Majesty's Stationery Office.

Barkworth, M.E. 1981. Foliar Epidermis and Taxonomy of the North American Stipeae. *Systematic Botany*, 6, 136–152.

Birch, J.L., Walsh, N.G., Cantrill, D.J., Holmes, G.D., & Murphy, D.J. 2017. Testing Efficacy of Distance and Tree-based Methods for DNA Barcoding of Grasses

(Poaceae tribe *Poeae*) in Australia. *PLoS ONE*, 12. https://doi.org/10.1371/journal.pone.0186259

Carroll, D.J., Cambell, M., Thorpe, J.W., & Robertson, J. 1983. The Potential Use of Isoelectric Focusing for the Identification and Discrimination of Grass Species on the Basis of Their Seed Proteins. *Journal of the Forensic Science Society*, 23, 297–305.

Christenhusz, M.J.M., & Bying, J.W. 2016. The Number of Known Plant Species in the World and its Annual Increase. *Phytotaxa*, 261, 201–217.

Clayton, W.D., Vorontsova, M.S., Hartman, K.T., & Williamson, H. 2006. *GrassBase – The Online World Grass Flora*. London: The Board of Trustees, Royal Botanic Gardens, Kew. https://www.kew.org/data/grasses-db.htlm. Accessed 29 April 2026.

Ellis, R.P. 1979. A Procedure for the Standardizing Comparative Leaf Anatomy in the Poaceae. 11. The Epidermis as Seen in Surface View. *Bothalia*, 12, 641–671.

Farragher, M.A. 1972a. Poaceae – Irish Members. Part 1. Artificial Key for the Identification of the Inflorescence Phase of Irish Grasses. *Scientific Proceedings of the Royal Dublin Society: Series A*, 4, 275–293.

Farragher, M.A. 1972b. Poaceae – Irish Members. Part 2. Artificial Key for the Identification of the Vegetative Phase of Irish Grasses. *Scientific Proceedings of the Royal Dublin Society: Series A*, 4, 295–314.

Farragher, M.A. 1973a. Poaceae – Irish members. Part 3. Artificial Key for the Identification of the Fruits of Irish Grasses. *Scientific Proceedings of the Royal Dublin Society: Series A*, 5, 37–54.

Farragher, MA. 1973b. Poaceae – Irish Members. Part 4. Classification. *Scientific Proceedings of the Royal Dublin Society: Series A*, 5, 55–76.

Farragher, M.A. 1974. Poaceae – Irish Members. Part 5. Anatomy. *Scientific Proceedings of the Royal Dublin Society: Series A*, 5, 159–198.

Farragher, M.A. 1975. Poaceae – Irish Members. Part 6. Morphology and Ecological Notes. *Scientific Proceedings of the Royal Dublin Society: Series A*, *5*, 331–383.

Grass Phylogeny Working Group (GPWG). 2001. Phylogeny and Subfamilial Classification of the Grasses (Poaceae). *Annals of the Missouri Botanical Garden*, 88, 373–457.

Hilu, K.W., & Randall, J.L. 1984. Convenient Method for Studying Grass leaf Epidermis. *Taxon*, 33, 413–415.

Hitchcock, A.S. 1971. *Manual of the Grasses of the United States, Volumes 1 and 2, Revised by Chase, A.* New York: Dover Pub.

Hubbard, C.E. 1968. *Grasses*, 2nd ed. Harmondsworth, UK: Penguin Books Inc.

Jacobs, S.W.L., & Everett, J. (eds.). 2000. *Grasses: Systematics and Evolution*. Collingwood, Australia: CSIRO Pub.

Kellogg, E.A. 2002. Classification of the Grass Family. In *Flora of Australia, Vol. 43. Poaceae 1. Introduction and Atlas* (pp. 19–36). Melbourne: ABRS/CSIRO.

Mallett, K. (ed.). 2005. *Flora of Australia*. Vol. 44B. Melbourne: ABRS/CSIRO.

Mallett, K., & Orchard, A.E. (eds.). 2002. *Flora of Australia. Vol. 43. Poaceae 1. Introduction and Atlas*. Melbourne: ABRS/CSIRO.

Martin, D.J. 1964. Analysis of Sheep Diet Utilizing Plant Epidermal Fragments in Faeces Samples. In *Grazing in Terrestrial and Marine Environments* (pp. 173–188). Dorking: Blackwell Sci. Pub.

McCusker, A. 2002. Structure and Variation in the Grass Plant. In *Flora of Australia, Vol. 43. Poaceae 1. Introduction and Atlas* (pp. 3–36). Melbourne: ABRS/CSIRO.

Metcalfe, C.R. 1960. *Anatomy of the Monocotyledons. 1. Gramineae*. Oxford: Clarendon Press.
Mulholland, S.C., & Rapp, G. 1992. *A Morphological Classification of Grass Silica-Bodies*. In Rapp, G. et al. (eds.), *Phytolith Systematics* (Chapter 4, pp. 65–89). New York: Springer Sciences.
North, C. 1956. A Technique for Measuring Structural features of Plant Epidermis Using Cellulose Acetate Films. *Nature*, 178, 1186.
Palmer, P.G., & Tucker, A.E. 1981. A Scanning Electron Microscope Survey of the Epidermis of East African Grasses, 1. *Smithsonian Contributions to Botany*, 49, 1–84.
Peterson, P.M., Romaschenko, K., & Soreng, R.J. 2014. A Laboratory Guide for Generating DNA Barcodes in Grasses: A Case Study of *Leptochloa* s.I (Poacaea: Chloridoideae). *Journal of Plant Taxonomy and Geography*, 69, 1–12.
Renvoize, S. 2002. Grass Anatomy. In Mallett, K., & Orchard, A.E. (eds.), *Flora of Australia, Vol. 43, Poaceae 1. Introduction and Atlas* (pp. 71–132). Melbourne: ABRS/CSIRO Australia.
Sharp, D., & Simon, B.K. 2002. *AusGrass: Grasses of Australia*. Melbourne: CSIRO Pub. Now an online tool at https://www.dcceew.gov.au/science-research/abrs/publications/cds/ausgrass
Smithson, F. 1958. Grass Opal in British Soils. *Journal of Soil Science*, 9, 148–154.
Soreng, R.J., Peterson, P.M., Romaschenko, K., et al. 2017. A Worldwide Phylogenetic Classification of the Poaceae (Gramineae) 11: An Update and a Comparison of Two 2015 Classifications. *Journal of Systematics & Evolution*, 55, 259–290.
Stebbins, G.L. 1956. Cytogenetics and Evolution of the Grass Family. *American Journal of Botany*, 43, 890–905.
Suttie, J.M., Reynolds, S.G., & Batello, C. (eds.). 2005. *Grasslands of the World* (p. 514). Rome: Food and Agriculture Organisation of the United States.
Wang, A., Gopurenko, D., Wu, H., & Lepschi, B. 2017. Evaluation of Six Candidate DNA Barcode Loci for Identification of Five Important Invasive Grasses in Eastern Australia. *PLoS ONE*, 12. https://doi.org/10.1371/journal.pone.0175338
Wang, J., Yan, Z., Zhong, P., Shen, G., Yang, G., & Ma, L. 2022. Screening of Universal DNA Barcodes for Identifying Grass Species of Gramineae. *Frontiers in Plant Science*, 13. https://doi.org/10.3389/fpls.2022.998863
Ward, J., Gilmore, S.R., Robertson, J., & Peakall, R. 2009. A Grass Molecular Identification System for Forensic Botany: A Critical Evaluation of the Strengths and Limitations. *Journal of Forensic Science*, 54, 1254–1260.
Ward, J., Peakall, R., Gilmore, S.R., & Robertson, J. 2005. A Molecular Identification System for Grasses: A Novel Technology for Forensic Botany. *Forensic Science International*, 152, 121–131.
Wheeler, D.J.B., Jacobs, S.W.L., & Norton, B.E. 1990. *Grasses of New South Wales*, 2nd ed. Armidale, NSW: Univ. New England Pub.
Wilson, A. (ed.). 2009. *Flora of Australia. Vol. 44A. Poaceae 2*. Melbourne: ABRS/CSIRO.
Wynn Parry, D., & Smithson, F. 1964. Types of Opaline Silica Depositions in the Leaves of British grasses. *Annals of Botany*, 28, 169–185.

Evidence Evaluation, Interpretation and Reporting

11

11.1 Introduction

Drawing on Australian Standard AS 5388-3 interpretation and AS 5388-4 reporting, to guide our discussion, the purpose of this chapter is to provide a general framework for the evaluation, interpretation and reporting of botanical evidence, aimed at specialists conducting forensic botanical examinations and scientists who may be unfamiliar with the requirements of the CJS. For the purposes of this chapter, I will refer to those individuals as 'occasional experts'.

In addition, some aspects, such as DNA reporting, have very specific guidelines and standards. For example, the Scientific Working Group on DNA Analysis Methods has produced validation guidelines for forensic DNA analysis methods and a number of interpretation guidelines. The primary focus of these Scientific Working Group guidelines is the analysis and interpretation of human DNA; however, the validation guideline has been used to validate a DNA system for cannabis (Howard et al., 2008). Section 4 of the *Wildlife Forensic Science* standards and guidelines for forensic botany also includes general guidance for DNA analysis. The interpretation of plant DNA data is very different from that of human DNA data.

11.2 Interpretation

11.2.1 Transforming Data Into Information

Forensic botanical examinations almost inevitably will make use of macroscopic and microscopic observations and rely heavily on individual expertise and professional judgement.

Figure 11.1 shows a visual representation of the pathway that transforms data into 'useful' information.

In this figure, the vertical pathway starts by posing the relevant question. This is a key element as, put simply, if you do not ask the right question, then you cannot get the right answer.

In AS 5388-2 analysis, it is clear that the starting point in formulating a request is the application of the scientific method. Put simply, this consists of the collection and assessment of relevant data through objective observation, followed by the formulation of a hypothesis and testing of that hypothesis.

DOI: 10.4324/9781003308508-11

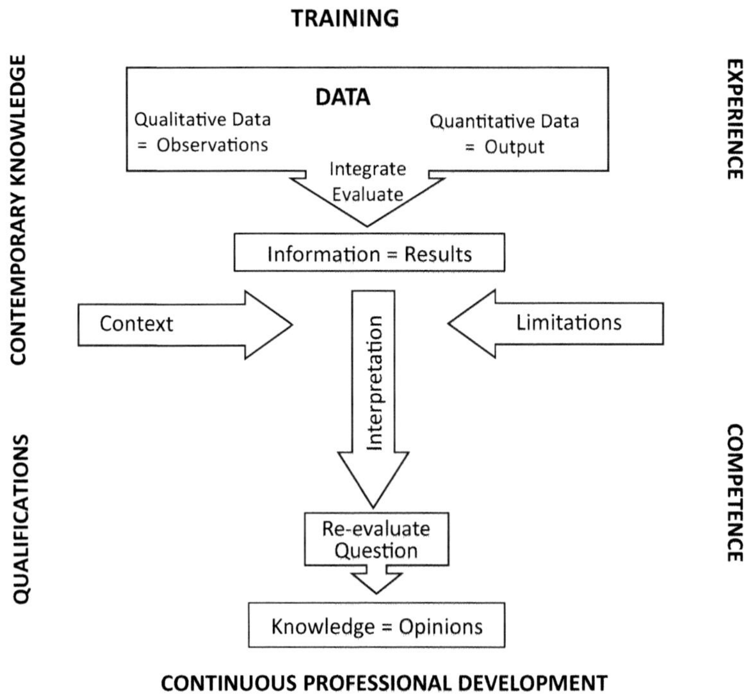

Figure 11.1 Transforming data into information.

The formulation of the hypothesis should be informed by the request for analysis. In cases where the request does not easily translate into a hypothesis, the parties need to maintain communication to ensure no relevant aspect of the case is overlooked and to avoid conducting meaningless or unnecessary work. Resources are finite, and it is important that they are focussed on what is realistic given the circumstances of each case. There has to be a balance between the temptation to turn each case into a mini research project and more pragmatic considerations, such as timeliness and cost.

The hypothesis is normally stated as the *null hypothesis* and its *alternative*. Some forensic authors frame these as competing propositions.

NULL HYPOTHESIS

- When comparing two objects or sets of data, the hypothesis states that an observed difference is due to chance alone
- The general or default position. For example, the null hypothesis may be that there is no relationship between the two items

In some case types, such as the identification of cannabis, the formulation of a hypothesis is simple: substance X is or is not cannabis. Often, the hypothesis is, do two or more objects have a common source or do they not have a common source?

Following the scientific method, challenging the null hypothesis requires data that may include both *qualitative* and *quantitative* observations and analytical input. Forensic botany will almost always include observational data. As previously emphasised, the key challenge is to capture observations through a well-structured, systematic and thorough process that maximises the accuracy and repeatability of these observations. Once data have been considered and finalised, this becomes a result.

The rest of the vertical axis in Figure 11.1 involves the process of turning a *result* into *meaningful information* or *knowledge*. Transforming results into knowledge requires the interpretation of data in the context of the case and any limitations. This may result in re-evaluating the question, but ultimately, the expert is required to formulate an opinion. As professional judgement informs that opinion, the *weight* that can be placed on an opinion will be based on a number of factors. These include the qualifications, training, maintenance of contemporary knowledge involving continuous professional development and proven competence of the individual, all of which determine what value can be placed on the experience of the expert.

11.2.2 Mitigating Against Cognitive Bias

Cases start with a request from a client, which will contain information thought to be relevant by the client, but which may contain information that is not relevant to the task, the so-called *task-irrelevant* information. The latter has the potential to cause bias. It is now widely accepted in mainstream forensic science that cognitive bias is unavoidable and is a factor that requires positive steps to mitigate or minimise its potential impact (Mattijssen et al., 2016).

Dror (2020) points out that cognitive bias does not usually arise from dishonesty, intentional discrimination or a deliberate act arising from unethical behaviour, but stresses that experts are not immune to bias and may not even be aware of their own biases. Indeed, experts can fall into the trap of believing they can overcome bias by 'mere willpower'. This can, in fact, have the opposite effect and actually increase its effect through the so-called 'ironic rebound'.

Many forensic organisations have introduced case management strategies aimed at mitigating or minimising the potential for cognitive bias in forensic examinations. The aim of these strategies is to ensure the integrity of the data collection by protecting the examiner from receiving task-irrelevant information. Information that is irrelevant for performing the task may, however, be necessary for the interpretation of the findings.

The terms *context information management* and *context information manager* have been proposed by Mattijssen et al. (2016). The release of information needs to be efficient, at the same time not leading to depriving the examiner of relevant information. This process involves *sequential unmasking*.

Dror (2020) has discussed eight factors that can lead to bias and has proposed six ways in which bias can be minimised as follows:

- using blinding and masking techniques that prevent exposure to task-irrelevant material
- using methods, such as *linear sequential masking* (LSU), to control the sequence, timing and linearity of exposure to information, so as to minimise 'going backwards' and being biased by the reference materials
- using case managers that screen and control what information is given to whom and when
- using blind, double blind and proper verifications where possible
- having a number of hypotheses with competing and alternative conclusions
- adopting a differential diagnosis approach, where all different conclusions and their probability are presented, rather than one conclusion

The occasional expert should be aware of context bias when dealing with an instructing client, avoid as far as practical receiving task-irrelevant information and, where practical, examine questioned/recovered materials before examining known and/or reference materials.

Where a client expressly requests an examiner to evaluate their results in a particular context, then the context and circumstances of the request should be documented and clearly stated in any subsequent report. The possibility of alternative explanations should also be acknowledged.

11.2.3 Formulating an Opinion

AS 5388.3 Interpretation states that interpretation often involves the application of *professional judgement* to assign a degree of probability to an opinion. Although the opinion must be owned by the individual, that does not preclude the individual from consulting with peers at any stage of the process and considering their input as outlined in the standard. The process of formulating an opinion should include:

- consideration of relevant questions and their related explanations
- evaluating the possible explanations according to their ability to explain the information

- ranking the conclusions according to the explanations and support provided by the information
- giving an opinion as to the best supported conclusion

The standard also states that any identifiable and reasonable alternatives to the final opinion should be documented, as should reasons for their rejection. Where there is more than one explanation that could explain the data, the explanations should be ranked and the reasons for each ranking included in case notes and any final report.

The underlying purpose is to ensure that the basis for all opinions is clear, not only to peers but also such that it can be understood by a non-expert.

11.2.4 The Role of Statistics and Estimating Probabilities

In the forensic legal context, we talk about the *weight* of evidence. The weighting of an opinion may be expressed numerically as a probability or ratio of probabilities, or qualitatively, in a way that demonstrates the practitioner's certainty of that opinion.

AS 5388.4 reporting states that where statistical calculations or weightings are used to support an opinion, the basis of the statistical calculations (including assumptions) should be fully explained in the report and supporting documentation.

The use of *evaluative reporting* is becoming a widely accepted approach as a way of providing the strength of support for competing propositions or hypotheses. Although not strictly synonymous, evaluative reporting may also be referred to as the likelihood ratio approach, logical thinking or Bayesian inference.

Guidelines and position statements on evaluative reporting have been issued by the National Institute of Forensic Science, Australia and New Zealand (NIFS), the European Network of Forensic Science Institutes and the Royal Statistical Society (Jackson et al., 2014).

NIFS defines evaluative reporting as

> a formalised thought process that enables evaluation of scientific findings given two opposing or competing propositions, which, if properly used, has the additional benefit of minimizing cognitive bias and allows for opinions to be updated in a logical way on receipt of new information.

The NIFS guideline lists the body of knowledge and experience as relevant to informing the evaluation process. The body of knowledge can include published literature, databases of characteristic frequencies and understanding of analysis limitations. Experience can include professional experience or personal knowledge, beliefs and assumptions.

An expert, however, who draws heavily on personal knowledge, beliefs and assumptions has an obligation to demonstrate that any opinion offered has a sound science-based foundation.

AS5388.3 states that where a suitably representative and statistically valid database exists from which reproducible numerical values can be obtained, a numerical probability value may be calculated, and that this may be expressed as part of a likelihood ratio or as a frequency. This standard goes on to state that where no formal database exists, practitioners may estimate the frequency of occurrence of an evidence type 'based on their experience within their area of expertise and expressed as a part of a likelihood ratio and used in a qualitative way'. The NIFS guideline also says that practitioners can use their experience and knowledge of the relevant population to subjectively estimate likelihood, recognising that it is critical that consistency is evaluated and checked.

It should be clear and transparent in case records and reports the basis for any conclusion and opinion reached based on an evaluative or any other approach, whether or not an actual likelihood ratio has been calculated.

AS 5388.3 states that the terminology used should provide a clear indication of the level of support for a conclusion. This may include the use of a numerical and verbal scale as follows:

Verbal Conclusion	Likelihood Ratio (LR)
Extremely strong support against	<0.000001
Very strong support against	0.000001–0.001
Strong support against	0.001–0.01
Moderate support against	0.01–0.1
Slight support against	0.1–1
Neutral	1
Slight support for	1–10
Moderate support for	10–100
Strong support for	100–1,000
Very strong support for	1,000–1,000,000
Extremely strong support for	>1,000,000

The use of verbal scales is not, however, without its own challenges and limitations. Mullen et al. (2014), in a survey of 400 volunteers, concluded that the terms used in verbal scales are unlikely to be understood properly by lay people, with participants attributing a greater weight of evidence to lower weights on the verbal scale and lesser weight to the higher weights on the verbal scale. Martire et al. (2014) also found a *weak evidence effect* and concluded that, given the high likelihood of miscommunication associated with low-strength verbal expressions of uncertainty, in instances where numerical values for low-strength evidence cannot be provided, it would seem appropriate to question whether expert opinions in the form of a verbal scale should be offered at all. Martire and Watkins (2015) concluded that "if the intention of verbal conclusion scales is to facilitate effective and accurate communication of opinions regarding evidential weight, then that aim has not been achieved' and that 'the high potential for miscommunication given the verbal conclusion scale is sobering'. Van Straalen et al. (2020) found that criminal justice professionals in general overestimated their actual understanding of all conclusion types.

Marquis et al. (2016), whilst presenting arguments in favour of the use of verbal qualifiers, emphasise that they do not replace the use of numbers. They recognise, however, that verbal qualifiers are prone to misunderstandings and cannot be coherently combined with other evidence. Hence, they recommend that verbal qualifiers are not used alone in reports and suggest that scientists should use examples to let the recipient of information understand how the scientific evidence affects the probabilities of propositions whilst also pointing out what the results do not mean. They express the opinion that if experts were able to coherently articulate numbers and recipients could properly handle such numbers, then verbal qualifiers could be abandoned all completely – in this authors view an unlikely outcome anytime in the near future.

The NIFS guideline to evaluative reporting recognises that for some disciplines, it may not be possible to consistently and accurately separate between levels on the verbal scale and proposes that a reduced scale may assist in achieving greater consistency between practitioners.

The relevance of the above to forensic botanical applications will depend on the nature of the question being asked. Where the primary question is of identity, for example, is the sample of questioned material cannabis or is a questioned sample of wood a particular species, then the use of verbal scales is really irrelevant. Where the question for cannabis is, can two or more samples be compared for a common origin, this is more complex, as there may not be a relevant database. For example, Howard et al. (2009) developed a short tandem repeat DNA database for Australian cannabis seizures. As with all databases, unless databases are kept up to date, they gradually lose whatever value they may have had when created, and, of course, the interpretation of plant DNA databases is not comparable to that for human DNA.

For complex assemblages of plant material, such as seen in forensic ecology applications, interpretation will depend heavily on individual expertise and professional judgement. An individual may be highly respected as an expert in their own field of study – it does not follow that this will easily translate across into forensic applications, especially when there is a criminalistic aspect to the questions being asked.

Finally, opinions do not always have to be conclusive. Where results do not provide sufficient information regarding the questions being answered, it may not be possible to reach an exclusionary or inclusionary finding, and an inconclusive opinion is justified. An inconclusive opinion may result for many reasons, including poor sampling, poor storage of materials resulting in sample decay and lack of suitable comparative material. The occasional experts should resist the desire to assist or be helpful before having all of the data necessary to form an opinion. As Spencer (2019) puts it, 'I have learned not to offer final conclusions until I am confident that I have done all I need to do to substantiate them'.

11.3 Reporting

11.3.1 Report Formats

The format of a report will be determined by factors including jurisdictional legal requirements, the intended purpose of the report and the type of information the report contains.

Oral reports may include any communication given to an investigator, lawyer or other relevant party over the phone, by video or any other electronic means. In general, such communications are intended to be for information and can be at any stage in an investigation or in the proceedings that follow. Where possible, oral communications should be recorded, but where this is not practical, a written record of the communications should form a part of the case file. A follow-up email to confirm that both parties agree on the outcomes of the communication is *strongly* recommended. As this type of communication can have the potential to influence the direction of an investigation or in the preparation for a trial, it is vital that there are no misunderstandings between the parties involved.

Oral communications are important in ensuring relevant information is shared in a timely manner, but by their very nature, information may be given before analysis is complete. This needs to be understood by all parties to such communications. Oral communications will not generally substitute or replace a final written report. In circumstances where examinations have not resulted in any meaningful information, an oral report may be sufficient and fit for purpose.

Written reports include email and text messages, interim reports, supplementary reports, laboratory reports and a full report intended for court purposes.

The use of *text messages* and *email* to communicate information is an everyday reality. It is, however, important to remember that all such communications should not be considered 'informal' as they are open to disclosure and discovery. Hence, it is important to ensure that messaging meets the same standards as any other form of reporting.

In some circumstances, an *interim report* may be issued prior to the finalisation of a full report and is usually intended to deliver preliminary results or a summary of final results. Such reports are most often issued to investigators to assist with the investigative process, but in some circumstances, lawyers may ask for an interim report in preparing for a court hearing. Great care needs to be taken in formulating conclusions and opinions to include in an interim report, as these may later need to be modified or changed in light of further examinations and analysis. All reports should include consideration of limitations, and this is even more important with interim reports.

A *laboratory report* is usually a written report provided to investigators, where a full report for court purposes is not required. This may be in a circumstance where the primary purpose of such a report is for intelligence purposes or where a decision has been made not to proceed with an investigation. Regardless of the 'intended purpose', this does not preclude such reports being used for court purposes, and, hence, although they may not fully conform with the format or other content of a report intended for court purposes, they should meet all other quality requirements.

A *full report* intended for court purposes should comply with all of the quality requirements for such a report. These requirements are discussed in Section 11.3.2.

A *supplementary report* may be issued to any other form of report where further testing or information becomes available.

For some forensic botany applications, *certificates* may be issued. These usually deal with the identification of an alleged illicit drug. The format and wording of certificates are usually mandated and defined by relevant jurisdictions.

Finally, jurisdictions may require that the findings be written in the format of an *affidavit* or equivalent statement. An affidavit is a written statement that contains a declaration of the truth of a statement under penalty of perjury. It includes an oath or affirmation taken by a notary public or other authorised person.

In some jurisdictions, a statement written under penalty, but without oath-taking, may be considered equivalent to an affidavit, such as a statutory declaration. Affidavits and signed statements containing declarations of truth are binding and may serve in place of oral testimony.

11.3.2 Report Contents

The precise format and contents of reports are often subject to jurisdictional requirements with formal procedures and practice guidelines that must be complied with. The following is a general guide as to what would reasonably be expected in a full report for court purposes and is guided by AS5388.4 reporting requirements.

A written report can be expected to include the following:

- the date of issue
- the name of the forensic facility or organisation
- a unique case identifier
- the name of the person responsible for the report
- a means of ensuring that each page is a part of the report
- a means of signifying the end of the report

A report intended for legal proceedings should, in addition, include the following:

- information about the collection and continuity of forensic material
- analysis and comparisons conducted
- results
- limitations
- conclusions and opinions
- relevant protocols
- definitions or explanations for technical terms used in a report

A report whose primary purpose is not intended for court may not include all of the above. The order of the above may vary according to jurisdictional requirements. It is often helpful to have a summary of the main findings, conclusions and opinions at the front of the report.

The intended recipients for forensic reports are investigators, lawyers and, ultimately, members of juries. It follows that conclusions and opinions should be written in language that will be understood by these recipients. The available evidence would suggest that this is not always the case (Howes et al., 2013). These authors make some suggestions to assist scientists in writing conclusions with increased readability for non-scientists.

Wiltshire (2015) provides reporting guidelines for palynology, which are discussed in more detail in Chapter 9.

In some jurisdictions, reports for legal purposes will be attached to a formal statement that may contain further information such as qualifications and relevant experience.

Some information can also be addressed by the use of appendices. As previously discussed, experts may be required to produce a report in the form of an affidavit.

Forensic organisations also need to have documented procedures for the issuing and control of reports. Reports are subject to both *technical* and *administrative* reviews before being issued. A technical review aims to ensure that all relevant experimental procedures have been followed, including the evaluation of data or observations and the interpretation of results. A technical review does not shift responsibility for the scientific findings from the examiner to the reviewer. The reviewer, however, does share responsibility for any scientific findings they endorse. An administrative review ensures the report complies with the administrative requirements of the organisation under whose authority the report is issued.

Where possible, the sole practitioner should try to access a suitably qualified peer to perform the role of technical reviewer.

11.3.3 Testimony and Giving Evidence

AS5388.4 reporting classifies testimony as a form of oral reporting.

The responsibilities of the scientist as an expert witness are often captured in professional or legal codes of practice for expert witnesses. Although the details regarding responsibilities will vary across jurisdictions, a common central element in such codes is an *overriding duty of the expert to assist the court* impartially on matters relevant to their expertise and not as an advocate for any party. An expert has a responsibility to ensure they comply with specific jurisdictional requirements.

What constitutes an 'expert' witness as opposed to a 'normal' witness? A 'normal' witness is only allowed to speak to factual information and is not allowed to offer an opinion. Experts are allowed to offer an opinion, but only in those areas where the court accepts that they have relevant expertise. Most codes of practice or rules of evidence will spell out that such evidence must be based wholly or in part on a body of specialist knowledge. The expert also needs to be able to demonstrate that they have training, qualifications and experience relevant to that body of specialist knowledge. In leading evidence from an expert, the party calling the expert may ask questions to establish the latter, although in practice, unless the area of expertise is novel, the level of specialist knowledge of experts is rarely rigorously examined or challenged.

It should go without saying, but it is worth a reminder, that an expert should not offer an opinion that is clearly outside their area of expertise and knowledge. On occasion, an expert may be invited to stray beyond their area of expertise to 'assist the court' – this is invariably not helpful!

In the continental or inquisitorial system of justice, an expert is usually appointed by the court and not the parties (prosecution and defence). In the adversarial system, experts may be instructed by either the prosecution or the defence. None of this reduces the absolute requirement that the duty of the expert is to assist the court – there is no ownership in witnesses.

From my own experience over five decades of giving evidence for both prosecution and defence, although not common, there can be direct or more subtle pressure made to make one feel you are part of the prosecution or defence team! The occasional expert should be conscious of this potential and not confuse being helpful with remaining neutral as to the eventual outcome of the judicial process.

In some cases, opposing experts will be instructed by both parties. In many jurisdictions, an increasing trend has been to introduce formal requirements that require experts to work cooperatively and, if so directed, to meet pretrial, discuss their findings and attempt to reach a joint position or, at least, be clear as to areas where agreement cannot be reached. The aim of this approach is to ensure that juries are not placed in the position of having to choose which expert to believe where the science is complex. Where agreement is reached, experts may not be called as witnesses and the jury receives a written agreed statement of facts.

It is beyond the scope of this chapter to consider all aspects of being an expert witness. The following is offered as a practical guide to assist the scientist facing their first or early appearance as an expert witness.

If employed by a forensic organisation, the organisation will be required to provide training on being an expert witness. In some countries, privately run commercial programmes exist providing training on being an expert. For example, in Australia, Forensic Foundations International delivers a two-day workshop for expert evidence (www.ffint.com.au/for-experts/workshop).

Prospective witnesses should be familiar with the legal system under which they are appearing, including court protocol and procedure. This includes being familiar with the physical layout of the court room, including where lawyers, judicial officers and, where appropriate, jurors sit and, most importantly, where the witness box is located. All of this is just common sense, but in a busy world, the responsibility to be prepared lies with the individual – do not expect to be looked after by others! It is invaluable to attend courts and just observe proceedings. The court is a very unfamiliar place for the uninitiated. Court rooms are also not all the same. Hence, every effort should be made to visit the 'actual' courtroom before being called as a witness. Over time, it is also helpful to watch experienced experts giving evidence.

Obvious as it may seem, it is also important to dress in a presentable and professional way so as to be seen as meeting appropriate professional standards.

If you are called as an expert witness, how your evidence is taken will depend on the legal system and jurisdiction. In the inquisitorial system, experts may not be called to present evidence in person and are unlikely to experience cross-examination. Indeed, even in jurisdictions with an adversarial system, experts appear on a less frequent basis than was previously the case. In the past, experts would 'cut their teeth' giving evidence in lower courts during committal proceedings. This was a very different experience from giving evidence to a higher court, but very useful as experts were often subject to quite rigorous cross-examination.

In the adversarial system, you will be asked to either take an oath or to affirm that the evidence you give will be the truth, the whole truth and nothing but the truth, or similar words. You should be clear whether you will take an oath or affirm.

The party calling you as a witness will lead your evidence in chief. Initially, you can expect that you will be asked your name, your occupation and employer and some information about your qualifications. Perhaps surprisingly, it is rare that you will be asked any detailed questions about your qualifications and experience.

At this point, you may be shown your report and be asked to confirm that it is your report, and sometimes asked if it is a true and honest account of your work. In some instances, the party calling you may then tender your report and ask no further questions. The 'other' party may then accept your report or they may cross-examine you. There are quite strict rules regarding cross-examination, but this is out of scope for this chapter.

When asked a question from either party, you should pay close attention to the question you are being asked and take time to think about your answer before speaking. You do not need to give an instantaneous response. If you are unsure or confused about the question, then you should seek clarification. Questions asked by lawyers are not always clear or unambiguous. In general, the judicial officer will ask the lawyer to rephrase their question.

Your answer to questions should be as concise as possible and unambiguous. You should resist offering unnecessary information. It is important to speak clearly, at a reasonable volume and at a speed slightly slower than normal conversation. In an ideal world, the person asking you questions will lead you through your evidence in a series of closed questions. You should resist trying to second-guess the next question in formulating your answer. However, the situation will not always be ideal, and there will be occasions where you will need to decide how much background information is necessary for your evidence to be understood. Whilst your role is not to deliver an academic lecture, you may need to lay the ground for case-specific information.

You will almost always wish to refer to your original case file, but should not do so until asking formal permission. It would be a very rare occasion that you were not given permission to refer to your case file. Your case file should be your friend. You should not rely on memory.

Whilst being polite and professional to the lawyers, in a jury trial, the jurors are the fact finders, and you should address your answers to them. Usually, courtrooms are set up so that you will be facing the jury, although at some distance. This allows you to make eye contact and often will give you a sense of whether or not a jury is following what you are saying.

In an inquisitorial setting, there may be one or more judicial officers (examining magistrates or judges) to whom you should address your answers.

If you are asked a question to which you do not know the answer, it is better to say so and explain why, if necessary, rather than attempt to answer the question and later realise that your answer is not accurate or correct. If you do give an answer that you realise is wrong or potentially misleading, you should correct this answer as soon as possible.

On completion of your evidence in chief, you may be surprised that you have not been asked about information in your report. There may be a sound reason why you have not been asked a question. You should not attempt to correct what you think may be an error of 'omission'. If, however, you feel that a question may lead you to providing an answer that could be misleading, then it is your duty to correct an error of 'commission'. This may be that the lawyer asking the question does not have a sound grasp of the subject matter.

During giving evidence, you may find yourself in a *voir dire* situation. This is where the opposing party challenges the question being asked. Often referred to as 'a trial within a trial', the jury will normally be excused whilst the two parties argue the point before the judge. In expert evidence, this usually arises when one party wishes to stop the other party from asking a particular question or attempting to have evidence excluded and may revolve around a question of law. Sometimes, the challenge may be an attempt to exclude your evidence on the basis that the prejudicial value of your evidence outweighs its probative value. Once decided by the judge, the jury will be recalled, and you will then be asked or not asked the question.

After your evidence has been led in chief, you may be cross-examined by the opposing party and then re-examined, although this is rare. There are quite strict legal rules about what you can be asked in cross-examination, but that is a matter for the lawyers.

Regrettably, on some occasions, parties will try to portray you as not being impartial. This may include suggesting directly, or by implication, that you owe some greater loyalty to your employer, especially if you work for law enforcement. If given the opportunity, you should stress that what is important is *impartiality*, not *independence* from the party calling you. If you are a member of a relevant professional body with a code of professional practice, you should point out that your professional obligation is to adhere to that code. If you work for an agency that holds relevant accreditation, you should make this clear and point to the various checks and balances, such as administrative and technical reviews and audit processes, that ensure that your approach as a witness is impartial.

Impartiality can also be reinforced as a witness if your demeanour gives an appropriate impression that you are being thoughtful in your use of language and not being defensive in your answers. The latter, however, does not mean that you should accept whatever is put to you, as this is not helpful to the fact finder. You should continue to support your conclusions and opinion whilst considering alternative explanations for your findings.

Finally, you may leave the witness box not feeling especially comfortable with how your report/information was led, or after cross-examination. This does not necessarily mean that you have not performed in an adequate manner as a witness. The adversarial system is called adversarial for a reason! However, you should be receptive to any feedback on your performance as a witness. Indeed, AS5388.4 reporting requires that an organisation have a documented system of testimony monitoring. This may include feedback forms from the parties, along with an assessment by a peer who has observed the evidence.

References

Dror, I.E. 2020. Cognitive and Human Factors in Expert Decision Making: Six Fallacies and Eight Sources of Bias. *Analytical Chemistry*, 92, 7998–8004.

Howard, C., Gilmore, S., Robertson, J., & Peakall, R. 2008. Developmental Validation of a *Cannabis sativa* STR Multiplex System for Forensic Analysis. *Journal of Forensic Sciences*, 53, 1061–1067.

Howard, C., Gilmore, S., Robertson, J., & Peakall, R. 2009. A *Cannabis sativa* STR Genotype Database for Australian Seizures. Forensic Applications and Limitations. *Journal of Forensic Sciences*, 54, 556–563.

Howes, L.M., Kirkbride, K.P., Kelty, S.F., Julian, R., & Kemp, N. 2013. Forensic Scientist's Conclusions: How Readable Are They for Non-Scientist Report-Users? *Forensic Science International*, 231, 102–112.

Jackson, G., Aitken, C., & Roberts, P. 2014. Case Assessment and Interpretation of Expert Evidence. Practitioner Guide no. 4, Working Group on Statistics and Law of the Royal Society.

Marquis, R., Biedermann, A., Cadola, L., Champod, C., Guueissaz, L., Massonnet, G., Mazella, W.D., Taroni, F., & Hicks, T. 2016. Discussion on How to Implement a Verbal Scale in a Forensic Laboratory: Benefits, Pitfalls and Suggestions to Avoid Misunderstandings. *Science & Justice*, 56, 364–370.

Martire, K.A., Kemp, R.I., Sayle, M., & Newall, B.R. 2014. On the Interpretation of Likelihood Ratios in Forensic Science Evidence: Presentation Formats and the Weak Evidence Effect. *Forensic Science International*, 240, 61–68.

Martire, K.A., & Watkins, I. 2015. Perception Problems of the Verbal Scale; A Reanalysis and Application of a Membership Function Approach. *Science & Justice*, 55, 264–273.

Mattijssen, E.J.A.T., Kerkhoff, W., Berger, C.E.H., Dror, I.E., & Stole, R.D. 2016. Implementing Context Information Management in Forensic Casework: Minimizing Contextual Bias in Firearms Examination. *Science & Justice*, 56, 113–122.

Mullen, C., Spence, D., Moxey, L., & Jamieson, A. 2014. Perception Problems of the Verbal Scale. *Science & Justice*, 54, 154–158.

Spencer, M.A. 2019. *Murder Most Florid. Inside the Mind of a Forensic Botanist.* London: Quadrille.

Van Straalen, E.K., De Poot, C.J., Malsch, M., & Elffers, H. 2020. The Interpretation of Forensic Conclusions by Criminal Justice Professionals: The Same Evidence Interpreted Differently. *Forensic Science International*, 313. https://doi.org/10.1016/j.forsciint.2020.110331

Wiltshire, P. 2015. Protocols for Forensic Palynology. *Palynology*. https://doi.org/10.1080/01916122.2015.1091138

Standards and Guidelines

Australian Standard AS 5388.2 – 2012. Forensic Analysis. Part 2: Analysis and Examination of Material.
Australian standard AS 5388.3 – 2013. Forensic Analysis. Part 3: Interpretation.
Australian Standard AS 5388.4 – 2013. Forensic Analysis. Part 4: Reporting.
European Network of Forensic Science Institutes (ENFSI). ENFSI Guideline for Evaluative Reporting in Forensic Science. https://enfsi.eu/up-content/uploads/2016/09/m1_guideline.pdf. Accessed 29 April 2026.
National Institute of Forensic Sciences Australia New Zealand (NIFS). Evaluative Reporting Fact Sheet. June 2017. https://www.anzpaa.org.au/ArticleDocuments. Accessed 14 August 2024.
Wildlife Forensic Science. Standards and Guidelines for Forensic Botany Identification. https://www.wildlifeforensicscience.org. Accessed July 2016.

12 Forensic Botany – Is There a Future?

12.1 Introduction

It was Winston Churchill who famously stated that 'the longer you look back, the farther you can look forward' (Anon, 2011).

Even though recent authors, writing about forensic botany, think it is a new application of botany, this is far from being historically accurate. As discussed in Chapter 1, in the mid-1970s, forensic botany, and botanists, were employed in core forensic organisation, such as the Metropolitan Police Forensic Laboratory (MPFL) and many others. The forensic programme at the University of Strathclyde had developed in response to the needs of the local police to be able to identify wood fragments from safe blowings. Forensic botany occupied about one-third of the MPFL method manual in the early 1980s and about one-third of the content of the master's degree in forensic science at Strathclyde.

As discussed in Chapter 1, in Australia, forensic botany first came to the notice of the public in the kidnapping for ransom of an eight-year-old Sydney schoolboy, Graeme Thorne (Tedeschi, 2015).

Whilst there have been advances in some aspects of forensic botany, such as applications involving palynology and ecology, and of course the application of DNA analysis, it is safe to say that today there would be very few examples of forensic botany in academic programmes or being practised in 'core' forensic organisations. How has this situation come about, and should it, or could it, be reversed? This is where we need to look back over the past 50 years, not only at forensic science but science in general and what has shaped where we are today.

12.2 How Has Science Evolved Over the Past 50 Years, and What Have Been the Implications for the Discipline of Botany?

In discussing this previously in the context of hair examinations, I identified four **eras** in the past 100 years that, in my view, have defined how the world of science, and flowing on to forensic science, has evolved (Robertson & Brooks, 2022).

DOI: 10.4324/9781003308508-12

Era 1 started well before the early 1900s and lasted until the 1950s, perhaps a little later. From a technology perspective, it was an era of *observation* with microscopy at its centre, with very few of the analytical technologies that are taken for granted today. As seen in previous chapters, many of the classic textbooks of this era have never been replaced. Some are now available as print-on-order facsimiles, making them more readily available for the interested scientist of today.

Era 2 started in the 1950s but really came into its own in the 1960s and 1970s. This was the era of the rapid expansion in *analytical chemistry* that impacted both biology and chemistry. Forensic biology developed from basic blood grouping to being able to analyse a variety of informative proteins due to the technology of electrophoresis and specifically, isoelectric focussing (IEF). Murch and Budowle (1986) reviewed the applications for IEF in forensic serology and demonstrated the central role for this technique, even as we were on the cusp of the next seismic change in forensic biology, the ability to analyse DNA from forensically relevant samples. Of course, electrophoresis remains the core technology to separate DNA until today.

Chemistry and forensic chemistry were, of course, the major beneficiaries of this explosion in technology. The emergence of plant physiology in the 1970s to a dominant position in botany owed much to the emergence of gas chromatography and mass spectrometry, with, some would argue, a detrimental flow-on effect to other traditional aspects of academic botany. As an example, at Glasgow University in 1970, the first thing that the newly appointed Professor of Botany did was to get rid of the herbarium and use the space for a plant physiology research laboratory. Personally, I benefited from this when I was a PhD student at Glasgow from 1972 to 1975, working on plant hormones and seed dormancy. I believe Glasgow is the first botany department to have its own mass spectrometer, an AEI MS 30 single-beam, double-focussing instrument.

And this brings us to **Era 3** and the emergence of *molecular biology*. Only a Luddite would argue that developments in molecular biology have not brought immeasurable benefit to mankind across health and many other areas, including its application in forensic science through DNA analysis. But has it all been positive? From a purely forensic perspective, I have argued that it has not all been positive. Certainly, in the early days of DNA for forensic science, the focus was squarely on its value in identifying human suspects with significant investments in creating large DNA databases. Only slowly has a more nuanced approach emerged with an increasing recognition that an uncritical application of DNA analysis does not address the broader criminalistics question of how trace biological material is deposited (not DNA as such!), and when and how this might relate to what is alleged to have taken place (Gill, 2014). There is now an extensive literature dealing with criminalistics questions of transfer and persistence of biological trace material.

The historical problem is that, in a world of limited funding, other areas of forensic science, which were viewed as being less useful, lost funding (Robertson & Roux, 2010). I do not subscribe to the view that the rise of molecular biology alone can be causally related to the reduction, indeed in many cases, disappearance of forensic botany from core forensic organisations, but I do believe it has been a contributing factor and not just from a funding perspective.

When I started university at Glasgow University in 1968, this was the year that separate botany and zoology first year subjects were replaced by a biology unit with a heavy focus on emerging molecular biology. I recall Dr Alex Berrie (to whom this book is dedicated) attending some of our lectures. Alex ran the agricultural botany subjects, and he was worried, correctly, that much of what had previously been in the first year would fall to him to remediate in our second-year subject! Fast forward to the current era at Glasgow, there is no identifiable entity that is a botany department with plant science being within the School of Molecular Biosciences. The current BSc/MSc is titled 'Molecular and Cellular Biology (with Plant Science)'. And this is by no way unique to my alma mater. Quite recently, Oxford University combined plant sciences with zoology to form a department of biological sciences.

Stroud et al. (2022) found that named courses, such as botany, are now practically non-existent in the United Kingdom and that it has been a decade since a student was enrolled in a solely botanical degree. This is by no means unique to the United Kingdom and appears to be a global trend (Crisci et al., 2020). Of even greater concern is the decline in taxonomy, where a review in the United Kingdom found that 'taxonomy and systematics in the UK is unsatisfactory …in some areas to the point of crisis' with university training failing to produce appropriately competent graduates or postgraduates (Boxshall & Self, 2011). Crisci et al. (2020) concluded that biologists were unable to recognise even common plants with a decline in botany students, faculty, courses, university departments and herbaria, highlighting the current erosion of botany. To be fair to these authors, they did declare that they had chosen the title of their paper, 'the end of botany', as an ambiguous prophecy whilst advocating for the critical importance of botany, now and in the future.

Allied to this decline, however, is that those left are an ageing population, described as the 'extinction of experience' (Miller, 2005). Three of the most prominent names in forensic palynology, Vaughn Bryant, Pat Wiltshire and Dallas Mildenhall, were born respectively in 1940, 1942 and 1944. We have already lost Vaughn Bryant, who passed away in 2021.

Hawksworth and Wiltshire (2011), with respect to forensic mycology, also identify the shortage of appropriately skilled mycologists able to reliably identify fungi across different systematic groups. In the United Kingdom, there has been a reduction in systematic mycology positions in the last 10–15 years, with the number of full-time fungal systematists falling to just eight from 23 in 1997. These authors point out that there is no quick fix, as it takes at least a PhD plus ten years' experience to produce a competent professional systematic mycologist.

In the absence of appropriate courses and experienced individuals, the challenge is, even if there were to be an unlikely resurgence of botany courses, who would teach plant identification, ecology and systematics? Not one systematic mycologist is employed by a university in the United Kingdom (Hawksworth & Wiltshire, 2011).

The impact of molecular biology on other areas of biology can be debated, but what is clear is that in the last 50 years, we have seen a catastrophic decline in traditional botany, leaving a huge problem, of which focussing on forensic botany would be far from the top of the list of priorities.

The flow from where we are today becomes all too apparent when we turn our attention to **Era 4**, the emergence of *artificial intelligence* (AI). This overarching term conceals a plethora of terms such as machine learning, narrow AI, artificial general intelligence and artificial super intelligence. Narrow AI, powered by machine learning, artificial neural networks and deep learning, has no knowledge, intelligence, self-awareness or consciousness (Saghiri et al., 2022). Proponents of AI, of course, always stress that there will always be a need for experts, but from where will these come?

12.3 Where to Now and What Might Be the Future

12.3.1 Where Are We Now?

'Science has not yet mastered prophecy. We predict too much for next year and yet far too little for the next ten' (Armstrong, 2024).

In summary, where we are now is that botany as an academic subject as it existed 50 years ago is unrecognisable today. Botany as a separate entity in academia has all but disappeared to be replaced by plant science, often as a minor component in broader undergraduate biology subjects, which are usually dominated by animal science. The foundations upon which undergraduate botany relied, anatomy, morphology, taxonomy and systematics, are now so weakened that any attempt to rebuild expertise must lie at the postgraduate level.

It would be unsafe to assume that any graduate today with an undergraduate degree will come with a solid foundation in traditional botany. There is no realistic prospect that there will be a resurgence in undergraduate academic botany, and the university sector will not invest at the postgraduate level unless there is a strong industry demand and support for postgraduate courses. I think all of this can be predicted with regrettable confidence.

So, where might botany be in the next 10–20 years, and, specifically, is there a brighter future for forensic botany ahead?

12.3.2 Artificial Intelligence

The last section finished by talking about era 4, the emergence and potential impact of AI. AI is defined as

> a field of computer science and engineering concerned with information processing technologies that integrate models and algorithms which produce a capacity to learn and to perform cognitive tasks leading to outcomes such as predictions and decision making in material and virtual environments.
>
> (Shapiro, 1992)

Frankly, by the time this book is published, any prediction about AI is likely to look very dated, such is the speed at which AI is evolving, truly almost like a virus!

Can we be more confident as to how AI might play out in forensic botany? There are already many examples in the literature where AI methods have been developed for plant identification. The following is not intended to be comprehensive, but should give the reader a feel for where AI currently sits in its evolution.

For example, AI has been used for the recognition of medicinal plants based on leaf shapes (Begue et al., 2017: Gavhale & Thakare, 2020) and more broadly for plant identification using leaf shapes (Banzi & Abayo, 2021; Beikmohammadi et al., 2022). Boston and Van Dijk (2019) reported on the use of AI for Australian plants using convoluted neural networks (CNNs). This is one of the most comprehensive studies where model performance was tested using 9,612 images of whole plants from 392 species and 3,384 images of flowers from 271 species. Of the three models, the best performing model achieved 80.6% overall accuracy and 88.4% for flowers. In reviewing the literature between 2015 and 2020, Sk and Wadhawan (2021) discussed the use of CNNs for image acquisition, image pre-processing, feature extraction, classification and testing. Accuracy in reported studies ranged from almost 100% to as low as 60%. These authors concluded that CNN is the best approach for plants and leaves.

Barash et al. (2023) have reviewed the use of machine learning for human DNA profiling.

Finally, Galante et al. (2023) have reviewed AI in forensic science in general, assessing the potential benefits, limitations and perspectives. A generic challenge for forensic applications is how do we validate the use of AI when the algorithms internalise data in ways that are not completely understood by humans or are capable of being audited, making AI a black box. These models also have to be trained with quality large data sets to achieve an effective performance.

Through the previous chapters, the potential use of AI in areas such as pollen, diatom and wood identification has been discussed. For forensic applications, it is always stressed that AI should not be relied upon on its own but viewed as a tool to assist experts; in other words, the final decision should still lie with human expertise applying professional judgement. Herein lies the most significant challenge that forensic science will face: where will that expertise come from in a world where, as we have seen, botanical expertise is being lost?

In the immediate future, the most likely scenario will be that applications will be limited to narrow AI, that is, relatively simple tasks where expertise is not critical, for example, field applications for non-specialists, such as timber identification, where the number of options is limited, and large databases do not need to be constructed.

From the literature cited above, it should be obvious to the reader that forensic scientists will have to broaden the scope of the journals they read if they wish to keep informed about developments in AI!

12.3.3 DNA Futures

The other science that is touted as the future of forensic botany is, of course, DNA analysis. Throughout this book, the application and status of DNA analysis have been discussed in each chapter.

Robertson et al. (2021), in reviewing DNA-based analysis of plant material in forensic investigations, dealt with sample types, collection and DNA isolation and forensic applications such as taxonomic identification, geographic origin and individualisation. Reports of studies using techniques such as random amplified polymorphic DNA and AFLP (amplified fragment length polymorphism) largely stopped around the early 2010s, with microsatellites and single-nucleotide polymorphisms (SNPs) being the dominant technologies between 2010 and 2020. Throughout the same time frame, the use of metagenomics (barcodes and metabarcoding) and environmental or eDNA started to emerge. Metabarcoding is a spin-off of barcoding and is performed exclusively on massively parallel sequencing (MPS) instruments.

The short- and medium-term future for plant DNA taxonomy would appear to be increasingly in the use of barcode approaches (Letsiou et al., 2024).

The term DNA barcode was first introduced in 2003 and refers to standardised sequences of DNA, ideally unique, from the genome either of the organism, or from its organelles, with a length between 400 and 800 base pairs, that is used to then identify or classify a group of organisms after reference to a database containing the relevant sequences from different species (Hebert et al., 2003). This approach has been successfully used for species identification in animals where a 648-base pair (bp) fragment in the mitochondrial gene cytochrome c oxidase subunit 1 has been selected as the standard barcode (Coissac et al., 2016). So far, there is no single ideal barcode for plants.

The Consortium for the Barcode of Life (CBOL) Plant Working Group have evaluated seven candidate genes and has proposed the combination of two chloroplast genes, *matK* (maturase K) and *rbcl* (ribulose bisphosphate carboxylase/oxygenase large subunit) as the best plant barcode (CBOL, 2009). This combination, however, does not offer the best discrimination at the level of species, achieving only 70%–75% species level discrimination (Hollingworth et al., 2009). The intergenic space, trnH-psbA, has been proposed as a supplementary locus. Several other combinations of potential markers have been suggested by various research groups, but none have yet reached universal consensus (Letsiou et al., 2024).

In earlier chapters, we saw the use of the Internal Transcriber Spacer (*ITS*) found in nuclear ribosomal DNA (rDNA), where they separate several coding regions within rDNA. *ITS* consists of two subregions, *ITS1* and *ITS2. ITS* has a widespread distribution and a high rate of evolution, which enables the comparison of relatively divergent taxa. Again, the CBOL Plant Working Group recommend *ITS* as a supplementary locus.

There are, however, remaining limitations in applying a barcode approach in the forensic context. A prerequisite for effective DNA barcoding is the establishment of a database of the sequences of DNA barcodes. The *Barcode of Life Data Systems* and GenBank are the two main public databases of DNA barcode data.

Ideally, sequence data should consist of sequences generated for high-quality vouchered specimens accessed from museum collections and herbaria and where their quality has been checked before inclusion. Regrettably, the latter is not always the reality with incorrectly labelled taxonomic sequences reported because of poor-quality DNA sequences, contamination of DNA samples, polymerase chain reaction (PCR)–based errors and obsolete or dated nomenclature (Robertson et al., 2021). This can result in searching algorithms such as GenBank's basic alignment search tool arriving at misleading or ambiguous identifications.

A second challenge for forensic samples is that the DNA extracted may be highly degraded and fragments recovered lower than the 400–800 bp region typically targeted. This can make the selection of primer pairs for PCR difficult and impact data recovery (Robertson et al., 2021).

When it comes to individualising and comparing samples, there are a range of challenges. A knowledge of the reproductive cycle of plants is essential, as this will determine whether individualisation is technically possible.

For example, plants that have self-breeding strategies or are clonally propagated, as is the case for a significant proportion of contemporary cannabis cultivations, the resulting progeny will be genetically very similar or identical. On the other hand, plants that reproduce through cross-breeding will generally produce variable offspring. Collecting representative samples in the field to identify a source with certainty may be difficult.

Then there is the challenge that there are estimated to be over 420,000 plant species globally (IUCN, 2022). Markers typically used for plant individualisation have included AFLPs, short tandem repeats, SNPs and INDELS (an insertion or deletion of bases in the genome of an organism) from the nuclear genome. For forensic samples, the method selected to extract the DNA is critical and should only be selected after thorough consideration of the sample type and potential inhibitors (Robertson et al., 2021).

Future developments in barcoding, including the so-called *Meta Barcoding* and *Super Barcoding*, have the potential to impact forensic applications of DNA analysis. Both involve the use of next-generation sequencing (NGS) or MPS.

The concept of meta barcoding involves identifying numerous species in a single sample using universal PCR primers to simultaneously amplify DNA barcodes. NGS allows for the parallel reading of DNA sequences. In contrast to traditional barcoding, which attempts to identify intact specimens up to the species level, metabarcoding aims to identify various samples or degraded samples, like environmental samples (eDNA), up to the family level or higher (Letsiou et al., 2024). Young and Linacre (2021) discuss the potential for eDNA to be applied in the forensic context, including botanical applications. Although the methodology for obtaining complex biological signals from environmental samples is now generally accepted by the scientific community, specific validation measures will be required before it can be applied in forensic science. Some progress is being made in this direction with Elwick et al. (2024) and Damaso et al. (2024) having recently published protocols for species identification and guidelines for the analysis of DNA barcoding/metabarcoding sequencing data and interpretation of publicly available databases.

Young and Linacre (2021) make a highly pertinent point when they raise the broader issue of applying this technology in the forensic context, that is, interpreting the data from a criminalistics viewpoint. They point out that there will be a microbial interaction between the sample surface and different environments over time if that surface is moved from its original location, presenting a layer of complexity when applying eDNA as a forensic tool.

Finally, a super barcode refers to the concept of using the entire chloroplast (cp) genome. A complete cp genome would generate the same amount of information as the COI (cytochrome c oxidase subunit) gene used in animals. The first cp genome was sequenced in 1986, with the number of cp genomes sequenced since then expanding significantly due to NGS technology (Chen et al., 2020). The creation of a complete cp genome database, reducing the cost of sequencing and obtaining better DNA quality and quantity, is seen as the key to the future potential of super barcoding (Letsiou et al., 2024).

12.4 A Working Model to Promote and Enhance the Use of Forensic Botany

Many authors writing about forensic botany have predicted an exciting future, but is this based on a realistic assessment of the current status?

Current reality is that, as an academic discipline, there is a crisis in botany with probably no undergraduate programmes teaching traditional subjects such as anatomy and morphology to any serious depth.

I am not aware of any degree programmes with any significant forensic botany content, and certainly not programmes aimed at producing forensic botanists. I am confident in also stating that crime scene examiners will have been given minimal, if any, exposure to forensic botany in their training.

Forensic botany is not recognised as a discipline within any of the major forensic science practitioner organisations and societies.

Excluding groups limited to the identification of cannabis, I know of only one major forensic organisation that has a forensic botany section and that is the Victoria Police Forensic Laboratory in Australia. I am happy to learn if there are other dedicated forensic groups embedded in core forensic organisations, but I am sure it will be a very small number. Hence, forensic botany, which, if not embedded, was more common 50 years ago, is no longer part of mainstream forensic organisations.

As a result of the above, to a very large extent, forensic botany relies on the use of experts coming from botanical institutions or academics who have developed an interest having been engaged for their specific expertise at some point in their careers. For some, this 'engagement 'may be a one-off experience, but for others, they have gone on to develop a career in forensic botany. Many of the latter are an ageing population, and far be it from me to be ageist, as I am also in my mid-1970s, but many of these individuals are at the end of their professional lives. There are no systematic efforts to develop the next generation of specialists in forensic botany.

In my view, the status quo will remain the future unless there is some way to re-embed botany into mainstream forensic science and organisations. As Bock and Noris (2016) put it, 'forensic botany will remain an ad hoc forensic activity'. This will not be an easy pathway, but it is a pathway that has been travelled by other forensic disciplines with similar challenges, such as soil examination. This involves a four-stage model for effective triage and management of trace evidence and has been applied for soils and hair examination disciplines (Woods et al., 2016; Robertson & Brooks, 2022).

In the context of forensic botany, the four stages would look something like the following:

- Stage 1. Enhanced awareness for crime scene examiners of the potential for forensic botany and training on scene aspects of forensic botany.
- Stage 2. Basic forensic botany in the laboratory, such as cannabis identification, natural fibre identification and triage of other potential botanical materials, ensuring integrity for more detailed examinations.
- Stage 3. For those organisations that move to establish a forensic botany group or at least have a suitably qualified botanist(s) on staff, develop expertise in one or more aspects of forensic botany.
- Stage 4. Specialist botanical expertise. Whilst it may be possible to reach this level of expertise in one or more aspects of botany, there will always be a need to involve deep subject experts.

To establish this model would require significant investment in educating and training crime scene examiners to recognise the potential for forensic botany, liaise and advise investigators; ensure appropriate recording, recovery and forensic handling of samples in the event a specialist is either not available or has not been identified and to work with specialist botanists to ensure compliance with the organisation's forensic standards.

If the organisation employs one or more (preferably more!) botanists, over time in-house expertise is developed in identified areas of strength. A key role would be to identify organisations and individuals in academia and/or core botanical organisations who may be willing to assist in forensic investigations, and over time, develop a cooperative model to ensure expertise is used efficiently and external experts are provided with specialist support to make best use of their time and expertise.

Where stages 2 and 3 are missing, then the role of crime scene examiners becomes even more critical. Generally, crime scene examiners remain within police organisations, and it would be incumbent on those organisations to establish the direct connections with appropriate experts and organisations and develop those relationships.

The worst outcome would be crime scene examiners doing a great job to promote the use of forensic botany, only for it to fail due to there being no experts willing to become involved.

An alternative approach for some countries or jurisdictions might be to establish a *centre of specialisation* in forensic botany along the lines of the US Fish and Wildlife Laboratory, which specialises in animals but not plants.

This could even be a virtual centre by bringing together a network of specialists working with a mainstream forensic provider.

These pathways to enhancing the potential for forensic botany will only happen if there is interest and leadership within mainstream forensic organisations and public policy engagement and commitment from relevant authorities. A university-based model would only succeed if a robust business plan could be established. History would suggest that ongoing financial commitment from police or the forensic industry for such an approach is unlikely to be forthcoming.

Robertson et al. (2021) observed that DNA analysis of plant evidence is not widely implemented by forensic laboratories. This remains the situation today, but the application of DNA analysis, and AI, will undoubtedly increase in the future. They are not, nor should they be viewed, however, as a replacement for traditional botany and the central role of microscopic observation. Unless there is a re-emergence of forensic botany, then the reality will become a reliance on these two technologies, with the focus on identification with little or no thought to context viewed from a criminalistics perspective.

References

Anon. 2011. *Churchill in 'Quotes'*. Lewes, UK: Ammonite Press.

Armstrong, N. 2024. *Brainyquote.com*. https://www.brainyquote.com/quotes/neil_armstrong_363176. Accessed 20 August 2024.

Banzi, J., & Abayo, T. 2021. Plant Species Identification from Leaf Images Using Deep Learning Models (CNN-STM Architecture). *Tanzania Journal of Forestry and Nature Conservation*, 90, 93–103.

Barash, M., Mcnevin, D., Fedorenko, V., & Giverts, P. 2023. Machine Learning in Forensic DNA Profiling: A Critical Review. *Forensic Science International: Genetics* 69. https://doi.org/10.1016/j.fsigen.2023.102944

Begue, A., Kowlessur, V., Mahomoodally, F., Singh, U., & Pudaruth, S. 2017. Automatic Recognition of Medicinal Plants Using Machine Learning Techniques. *International Journal of Advanced Computer Science and Applications*, 8, 166–175.

Beikmohammadi, A., Faez, K., & Motallebi, A. 2022. SWP-LeafNET: A Novel Multistage Approach for Plant Leaf Identification Based on Deep CNN. *Expert Systems with Applications*, 202, 117470.

Bock, J.H., & Noris, D. 2016. *Forensic Plant Science*. London: Academic Press.

Boston, T., & Van Dijk, A. 2019. *Some Experiments in Automated Identification of Australian Plants Using Convoluted Neural Networks. 23rd International Congress on Modelling and Simulation (MODSIM2019)*. https://mssanz.org.au/modsim2019

Boxshall, G., & Self, D. 2011. *UK Taxonomy & Systematics Review-2010. Results of Survey Undertaken by the Review Team at the Natural History Museum Serving as Contractors to the Natural Environment research Council (NERC)*. Swindon, UK: Natural Environmental Research Council.

CBOL Plant Working Group. 2009. A DNA Barcode for Land Plants. *Proceedings of the National Academy of Sciences of the United States of America*, 106, 12794–12797.

Chen, Q., Wu, X., & Zhang, D. 2020. Comparison of the Abilities of Universal, Super, and Specific DNA Barcodes to Discriminate Among the Original Species of Fritillariae cirrhosae Bulbus and Its Adulterants. *PLoS One*, 15, e0229181.

Coissac, E., Hollingsworth, P.M., Lavergne, S., & Taberlet, P. 2016. From Barcodes to Genomes: Extending the Concept of DNA Barcoding. *Molecular Ecology*, 25, 1423–1428.

Crisci, J.V., Katinas, L., Apodaca, M.J., & Hoch, P.C. 2020. The End of Botany. *Trends in Plant Science*, 25, 1173–1176.

Damaso, N., Elwick, K.E., & Robertson, J.M. 2024. Guidelines for the Analysis of DNA/Metabarcoding Sequencing Data and Interpretation of Publicly Available Databases. In DeSalle, R. (ed.), *DNA Barcoding Methods and Protocols* (pp. 391–402). New York: Humana.

Elwick, K.E., Damaso, N., & Robertson, J.M. 2024. DNA Barcoding and Metabarcoding Protocols for Species Identification. In DeSalle, R. (ed.), *DNA Barcoding Methods and Protocols* (pp. 155–169). New York: Humana.

Galante, N., Cotroneo, R., Furci, D., Lodetti, G., & Casali, M.B. 2023. Applications of Artificial Intelligence in Forensic Sciences: Current Potential benefits, Limitations and Perspectives. *International Journal of Legal Medicine*, 137, 445–458.

Gavhale, N., & Thakare, A.P. 2020. Identification of Medicinal Plants Using Machine Learning Approach. *International Research Journal of Engineering and Technology*, 7, 1116–1119.

Gill, P. 2014. *Misleading DNA Evidence. Reasons for Miscarriages of Justice*. London: Elsevier.

Hawksworth, D.L., & Wiltshire, P.E.J. 2011. Forensic Mycology: The Use of Fungi in Criminal Investigations. *Forensic Science International*, 206, 1–11.

Hebert, P.D., Cywinskas, A., Ball, S.L., & DeWaard, J.R. 2003. Biological Identifications Through DNA Barcodes. *Proceedings of the Royal Society of London. Series B, Biological Sciences*, 270, 313–321.

Hollingworth, P.M., Forrest, L.L., Spouge, J.L., et al. 2009.A DNA Barcode for Land Plants. *Proceedings of the National Academy of Sciences of the United States of America*, 106, 12794. https://doi.org/10.1073/pnas.0905845106

IUCN. 2022. *The IUCN Red List of Threatened Species. Version 2022-1*. https://nc.iucnredlist.org

Letsiou, S., Madesis, P., Vasdekis, E., et al. 2024. DNA Barcoding as a Plant Identification Method. *Applied Sciences*, 14, 1415. https://doi.org/10.3390/app14041415

Miller, J.R. 2005. Biodiversity Conservation and the Extinction of Experience. *Trends in Ecology & Evolution*, 20, 430–434.

Murch, R.S., & Budowle, B. 1986. Applications of Isoelectric Focusing in Forensic Serology. *Journal of Forensic Sciences*, 31, 869–880.

Robertson, J., & Brooks, E. 2022. *A Practical Guide to the Forensic Examination of Hair. From Crime Scene to Court*. Baton Rouge: CRC Press.

Robertson, J., & Roux, C. 2010. Trace Evidence: Her Today, Gone Tomorrow? *Science & Justice*, 50, 18–22.

Robertson, J.M., Damaso, N., & Meiklejohn, K.A. 2021. DNA-Based Analysis of Plant material in Forensic Investigations. In Dash, H.R., Shrivastava, P., & Lorente, J.A. (eds.), *Handbook of DNA Profiling*. Singapore: Springer. https://doi.org/10.1007/978-981-15-9364-2_59-1

Saghiri, A., Vahidipour, S., Jabbarpour, M., Sookhak, M., & Forestiera, A. 2022. A Survey of Artificial Intelligence Challenges: Analysing the Definitions, Relationships, and Evolutions. *Applied Sciences*, 12. https://doi.org/10.3390/app12084054

Shapiro, S.C. 1992. Artificial Intelligence. In Shapiro, S.C. (ed.), *Encyclopaedia of Artificial Intelligence*, 2nd ed. (Vol. 1, pp. 89–93). New York: Wiley.

Sk, R., & Wadhawan, A. 2021. *Identification of Plants Using Deep learning: A Review. International Seminar and International Conference (ISIC'21).* https://ceur-ws.org/Vol-2786/Paper51.pdf

Stroud, S., Fennell, M., Mitchley, J., Lydon, S., Peacock, J., & Bacon, K.L. 2022. The Botanical Education Extinction and Fall of Plant Awareness. *Ecology and Evolution*, 12, e9019. https://doi.org/10.1002/ece3.9019

Tedeschi, M. 2015. *Kidnapped. The Crime That Shocked the Nation*. Sydney: Simon & Schuster.

Woods, B., Lennard, C., Kirkbride, K.P., & Robertson, J. 2016. Soil Examination for a Forensic Trace Evidence Laboratory – Part3: A Proposed Protocol for the Effective Triage and Management of Soil Examinations. *Forensic Science International*, 262, 46–55.

Young, J.M., & Linacre, A. 2021. Massively Parallel Sequencing Is Unlocking the Potential of Environmental Trace Evidence. *Forensic Science International: Genetics*, 50, 102393. https://doi.org/10.1016/j.fsigen.2020.102393

Index

Pages in *italics* refer to figures and pages in **bold** refer to tables

D

E

I

J

K

P

T

X

Y

Z

For Product Safety Concerns and Information please contact our EU representative GPSR@taylorandfrancis.com Taylor & Francis Verlag GmbH, Kaufingerstraße 24, 80331 München, Germany

Batch number: 10756178

Printed by Printforce, the Netherlands